The World Food Economy

The World Food Economy

**Douglas Southgate,
Douglas H. Graham, and
Luther Tweeten**

Department of Agricultural, Environmental,
and Development Economics
The Ohio State University, OH, USA

© 2007 by Douglas Southgate, Douglas H. Graham, and Luther Tweeten

BLACKWELL PUBLISHING
350 Main Street, Malden, MA 02148-5020, USA
9600 Garsington Road, Oxford OX4 2DQ, UK
550 Swanston Street, Carlton, Victoria 3053, Australia

The right of Douglas Southgate, Douglas H. Graham, and Luther Tweeten to be identified as the Authors of this Work has been asserted in accordance with the UK Copyright, Designs, and Patents Act 1988.

All rights reserved. No part of this publication may be reproduced, stored in a retrieval system, or transmitted, in any form or by any means, electronic, mechanical, photocopying, recording or otherwise, except as permitted by the UK Copyright, Designs, and Patents Act 1988, without the prior permission of the publisher.

First published 2007 by Blackwell Publishing Ltd

1 2007

Library of Congress Cataloging-in-Publication Data is available for this title.

ISBN-13: 978-1-4051-0596-5 (hardback)
ISBN-10: 1-4051-0596-8 (hardback)
ISBN-13: 978-1-4051-0597-2 (paperback)
ISBN-10: 1-4051-0597-6 (paperback)

A catalogue record for this title is available from the British Library.

Set in 10/12.5 pt Palatino
by Charon Tec Ltd (A Macmillan Company), Chennai, India
www.charontec.com
Printed and bound in the United Kingdom
by TJ International Ltd, Padstow, Cornwall

The publisher's policy is to use permanent paper from mills that operate a sustainable forestry policy, and which has been manufactured from pulp processed using acid-free and elementary chlorine-free practices. Furthermore, the publisher ensures that the text paper and cover board used have met acceptable environmental accreditation standards.

For further information on
Blackwell Publishing, visit our website:
www.blackwellpublishing.com

Contents

Preface ix
Acknowledgments xii

1 Introduction 1
   1.1 Our Focus 4
   1.2 Chapter Outline 7
   Study Questions 9

2 The Demand Side: How Population Growth and Higher Incomes Affect Food Consumption 10
   2.1 Classic Malthusianism, Its Modern Variants, and Its Critiques 11
   2.2 Demographic Transition 16
   2.3 Trends in Human Numbers, Past and Present 25
   2.4 Food Consumption and Income 28
   2.5 Demand Trends and Projections 31
   2.6 Summary and Conclusions 33
   Study Questions 34
   Appendix: The Fundamental Economics of Demand 34

3 The Supply Side: Agricultural Production and Its Determinants 41
   3.1 The Nature of Agriculture 42
   3.2 Increases in Agricultural Supply 49
   3.3 Has Intensification Run Its Course? 62
   3.4 Trends in Per-Capita Production 66
   Study Questions 67
   Appendix: The Fundamental Economics of Supply 68

4 Aligning the Consumption and Production of Food Over Time 74
   4.1 The Desirability of Competitive Equilibrium 76
   4.2 The Market Impacts of Commodity Programs 78
   4.3 Historical Trends in the Scarcity of Agricultural Products 83
   4.4 Outlook for the Twenty-First Century 86

Study Questions 89
Appendix: The Coordination of Decentralized Decision-Making 89

5 Agriculture and the Environment 98
5.1 Environmental Trade-Offs 99
5.2 Market Failure 102
5.3 Environmental Deterioration in the Absence of Agricultural Intensification 109
5.4 Agricultural Development and the Environment 119
Study Questions 123

6 Globalization and Agriculture 124
6.1 The Theory of Comparative Advantage 125
6.2 The Net Costs of Trade Distortions 128
6.3 The Debate Over Globalization 132
6.4 Agricultural Trade: Recent Trends and the Current Debate 135
6.5 Why Not More Trade? 142
Study Questions 143
Appendix: A Two-Country Illustration of Comparative Advantage 143

7 Agriculture and Economic Development 147
7.1 Growth and Economic Structure 148
7.2 Agriculture's Role in Economic Development 155
7.3 Trying to Develop at Agriculture's Expense 158
7.4 Agricultural Development for the Sake of Economic Growth and Diversification 161
7.5 Summary and Conclusions 164
Study Questions 165

8 Striving for Food Security 166
8.1 What is Food Security? 166
8.2 Who and Where Are the Food Insecure? 167
8.3 Achieving Food Security 170
8.4 The Food Security Synthesis and Economic Development 176
8.5 The Standard Model, Communitarian Values, and Economic Equity 183
Study Questions 185

9 A Synopsis of Regional Trends in the Global Food Economy 187
9.1 Economic Growth and Income Distribution 187
9.2 Population Dynamics 191
9.3 Agriculture's Response to Demand Growth 198
9.4 Summary 202
Study Questions 205

10 Affluent Nations — 206
- 10.1 Standards of Living — 207
- 10.2 Population Dynamics — 210
- 10.3 The Food Economy — 212
- 10.4 Dietary Change and Consumption Trends — 220
- 10.5 Summary — 225
- Study Questions — 227

11 Asia — 228
- 11.1 Trends in GDP Per Capita — 228
- 11.2 Population Dynamics — 233
- 11.3 Agricultural Development — 237
- 11.4 Dietary Change, Consumption Trends, and Food Security — 242
- 11.5 Summary — 248
- Study Questions — 248

12 Latin America and the Caribbean — 250
- 12.1 Trends in GDP Per Capita — 251
- 12.2 Population Dynamics — 254
- 12.3 Agricultural Development — 258
- 12.4 Dietary Change, Consumption Trends, and Food Security — 265
- 12.5 Summary — 272
- Study Questions — 272

13 The Middle East and North Africa — 273
- 13.1 Trends in GDP Per Capita — 274
- 13.2 Population Dynamics — 278
- 13.3 Agricultural Development — 281
- 13.4 Dietary Change, Consumption Trends, and Food Security — 288
- 13.5 Summary — 293
- Study Questions — 294

14 Eastern Europe and the Former Soviet Union — 295
- 14.1 Patterns of Economic Growth since the Fall of Communism — 296
- 14.2 Demographic Trends — 302
- 14.3 The Agricultural Sector — 306
- 14.4 Dietary Change, Consumption Trends, and Food Security — 313
- 14.5 Summary — 318
- Study Questions — 319

15 Sub-Saharan Africa — 321
- 15.1 Trends in GDP Per Capita — 324
- 15.2 Demographic Trends — 329
- 15.3 Agricultural Development — 337

	15.4	Consumption Trends and Food Security	345
	15.5	Summary	354
	Study Questions		355

16 The Global Food Economy in the Twenty-First Century — **357**

    16.1 Victims of Our Own Success? — 360
    16.2 The New Food Economy — 364
    16.3 The Changing Role of Government — 366
    Study Questions — 369

Abbreviations and Acronyms — *370*

Map Annex — *372*

References — *378*

Index — *390*

Preface

This book is a product of the three authors' experience with a course about global trends in population and food demand, the agricultural development required to satisfy demand growth, and the environmental challenges that this development creates. Like comparable offerings at other universities and colleges, this course consistently attracts sizable enrollments.

Our students are a mixed group. Many of them are majors in the Department of Agricultural, Environmental, and Development Economics. However, the majority are undergraduates from non-agricultural departments at Ohio State University. The largest cohort comprises people with an interest in international studies and who have little or no prior training in economics. Professors teaching similar classes at other institutions report that the backgrounds of their students are just as diverse.

There is a sizable literature, of course, on the demand side of the food economy, with considerable attention given to nutritional issues and the economics of food insecurity. Likewise, a number of books focus on agricultural development, particularly in less affluent parts of the world. Much has been written about agricultural markets and policy as well. What has been lacking, however, is an up-to-date and integrated text on the global food economy as a whole.

In light of this purpose, this book does not really substitute for more specialized readings. A single chapter on general trends in food demand, for example, draws on what economists have to say about consumer behavior, without dwelling on finer points of theory. Likewise, fundamental economic ideas about production are harnessed in another chapter to interpret the broad course of agricultural development – again, without going into theoretical detail. The same approach is followed in other chapters, on market trends and public policy, agriculture and the environment, agriculture and globalization, the sector's role in economic development, and food security.

An advantage of opting for a non-specialized approach is that the book is accessible to a general audience. All three authors are economists, each with several decades' experience. Nevertheless, we avoid using professional jargon and, where technical terms are employed, these are defined in ways that a person with no background in economics can understand. No doubt, readers who have taken a course or two in the subject will grasp some material a little quicker than those with no prior exposure to economics. However, this book has been written to be read and appreciated by a general audience. For those who want to learn a little economics or refresh their knowledge of the "dismal science" as they read this book, four of the chapters have appendices in which basic concepts such as demand curves and the theory of comparative advantage are reviewed and explained.

Like practically all economists, we think that goods and services are best allocated in free markets. Although we are not shy about pointing out the advantages of markets in this book, the value of a functional government is not neglected. For one thing, the state is in a unique position to satisfy the institutional prerequisites of a market economy, such as reliable enforcement of contracts and property rights. Moreover, governments play a central role in providing public goods, such as transportation infrastructure and research and development. The authors highlight the importance of governmental contributions such as these to a successful, market-based economy.

The sequence of this book's chapters reflects the order in which we cover various topics in the course at Ohio State. To begin, trends in food demand are examined in the second Chapter 2, with attention paid to population growth as well as the impacts on food consumption of improved living standards. Chapter 3 is about agricultural development, with an emphasis on improvements in the technology for commodity production. After these separate treatments of the demand and supply sides of the food economy, Chapter 4 focuses on markets in which consumption and production are brought together and which are influenced by public policy.

With the basic elements of the food economy surveyed in these three chapters, the authors proceed to examine environmental issues connected with crop and livestock production (Chapter 5), the central role that agriculture is playing in the debate over globalization (Chapter 6), and the equally pivotal position of agriculture in economic development (Chapter 7). This is followed by Chapter 8, which is about food insecurity and its alleviation.

Chapters 9–15 have a regional, as opposed to topical, focus. Differences among Latin America, Sub-Saharan Africa, and other parts of the world are examined in Chapter 9. The next six chapters are about individual regions and are organized identically. Economic trends in specific countries are compared and contrasted with one another. Next, we assess demographic realities, agricultural development, and food consumption, always in this order.

The main discrepancy between the book's chapter-order and the sequence of topics in the syllabus for our course has to do with Chapter 9, which is often the first reading assignment. The advantage of beginning with this material is that students are introduced to ways that various parts of the world – affluent, poverty stricken, and in-between – differ from one another as well as common features or characteristics shared by many or all regions. In addition, Chapter 9 can be used to introduce general class themes, thereby facilitating students' understanding of the earlier chapters.

This book closes in Chapter 16 with a survey of new and emerging issues in the global food economy. One of these is the problem of overeating. This is obviously an issue in affluent nations. However, it is also a growing worry in the developing world, in which the traditional concern has been inadequate food consumption.

An unpublished version of this book has been used successfully at Ohio State for a number of years. During the Spring 2006 semester, this book was used as a core text in a course taught by Professor Robert Thompson at the University of Illinois. Chapters of this book also have been assigned at the Universidad de Belgrano, in Buenos Aires, Argentina, as well as the Universities of Natal, Pretoria, and Stellenbosch, in South Africa.

The book's value as a core text is enhanced by review questions provided at the end of each chapter. A detailed subject index and extensive list of bibliographic citations are also included. An instructor intending to adopt this book for a class can obtain PowerPoint slides containing tables and figures by contacting Douglas Southgate, at southgate.1@osu.edu.

Acknowledgments

As we wrote this book, we benefited enormously from advice and insights generously shared by numerous colleagues and students. Various faculty members at The Ohio State University enlightened us on key facets of the global food economy, suggested useful readings, or both. Others reviewed draft chapters or served as a sympathetic sounding board. We are grateful to all of them: Wen Chern, William Flinn, D. Lynn Forster, Claudio González-Vega, Frederick Hitzhusen, David Kraybill, Rattan Lal, Donald Larson, Richard Meyer, Alan Randall, Norman Rask, Ian Sheldon, Brent Sohngen, Donald Thomas, and Stanley Thompson. Similar assistance was provided by colleagues at other institutions: Dennis Avery, who directs the Center for Global Food Issues; Robert Gitter, of Ohio Wesleyan University; Donald Hertzmark, an economic consultant based in Washington; and Sarah Scherr, of the University of Maryland. We also appreciate very much Robert Thompson's willingness to give this book a "test-drive" in an undergraduate course on the global food economy he taught at the University of Illinois in Spring 2006.

Special thanks are due to Barry Goodwin, formerly the Anderson Professor of Marketing, Trade, and Policy at Ohio State and now at North Carolina State University, who provided funds used to hire a pair of students – Elizabeth Myre and Myriam Elizabeth Southgate – who assembled much of the data reported in Chapters 9–15. These two individuals as well as the authors received valuable assistance from Susan Logan, a reference librarian at Ohio State. We are particularly grateful to four former graduate teaching students: Sarah Lowder (now working for the United Nations Economic and Social Commission for Asia), Sharon May (a faculty member at Maryville College), Maria Pagura (on the staff at the United Nations Food and Agriculture Organization), and Mehnaz Safavian (currently at the World Bank). Other students who dug up information, caught stylistic lapses, or contributed in other ways were Mohammad Ashfaq, John Gossom, and Hammad Qureshi. Also greatly appreciated has been the feedback received from students in courses at Ohio State and Ohio Wesleyan Universities in which draft chapters of this book were assigned.

We particularly appreciate the assistance provided by two members of the support staff of Ohio State's Department of Agricultural, Environmental, and Development Economics. With only a couple of exceptions, all the book's figures were drawn by

Janice DiCarolis. Likewise, most of the book's tables were prepared by Judy Luke, who also typed much of the text. Advice on map preparation was provided by Stephen Rogers, a map librarian at Ohio State.

This book never would have been written had Seth Ditchik, our editor at Blackwell Publishing, not shown an interest. His counsel has been deeply appreciated, as has the assistance provided by Laura Stearns, Seth's successor at Blackwell, and Elizabeth Wald, Seth's former colleague. Comments on draft chapters provided by Dale Colyer (Professor Emeritus, West Virginia University) and Valerie Askren (Professor Emeritus, University of Kentucky) also have been of great value.

Finally, we have received steadfast support from our wives: Myriam, Jane, and Eloyce. Sincerely grateful for their patience, we dedicate *The World Food Economy* to them.

1

Introduction

Severe hunger, which most human beings rarely if ever experience, is sometimes the subject of headlines. The most heart-wrenching reports are about mass starvation in countries suffering civil war, ethnic cleansing, or both. Just as alarming as scenes from refugee camps in places like the Horn of Africa, however, are estimates issued periodically by the UN Food and Agriculture Organization (FAO) about the large numbers of people who do not eat well enough for a healthy and productive life.

The headlines used to be much worse, warnings of imminent famine being especially rife during the late 1960s and early 1970s. Environmentalist Paul Ehrlich, for one, insisted that many parts of the world were beyond hope. Giving food to countries with no reasonable chance of feeding their populations from domestic production and imports was futile; to the contrary, he advocated an immediate cessation of all donations to these countries (Ehrlich, 1968, pp. 141–149). By no means was Ehrlich the only advocate of drastic action. The Club of Rome, for example, argued that, barring immediate curbs on human reproduction and economic growth, the world economy would collapse soon after the turn of the twenty-first century (Meadows *et al.*, 1972).

Alarm of this sort was not justified by later events. People have not bred themselves to oblivion and there is no prospect of a global shortfall in agricultural production. Nevertheless, popular thinking about the race between food demand and food supply continues to be swayed by purveyors of gloom and doom from a former generation. For those still influenced by the likes of Ehrlich and the Club of Rome, a magnificent human accomplishment remains obscured. In the face of demographic expansion on an historically unprecedented scale, food supplies have not just kept pace. In fact, supply growth since the end of the Second World War consistently has outstripped increases in demand.

Production has not risen steadily because farmers have been responding to higher prices. To the contrary, food has become cheaper, with inflation-adjusted market values going down a lot over time. As reported in Chapter 4 of this book, real prices of cereals such as corn, rice, and wheat – which account for two-thirds of the human diet if the feed grains eaten by livestock are accounted for – declined by 75 percent during the second half of the twentieth century.

Can increased supplies of the products that nourish us be attributed to an increase in the agricultural workforce? No. In many countries, that workforce actually has shrunk. This holds in the United States, where crop and livestock output is enormous and yet where the farm population dropped from 30 million in 1940 to under 5 million in 1990 (Gardner, 2002, p. 94). In many other parts of the world, the farm population has not declined, although percentage growth in output has greatly exceeded that in agricultural labor.

The same holds for geographic expansion of the agricultural economy. Due to major increases in yields (i.e., output from a given area of farmland), increases in agriculture's geographic domain have been negligible relative to those in production. Indeed, farmed area around the world has held steady since the middle 1990s (Chapter 3). This has created an environmental benefit of great importance since there has been less encroachment on tropical forests and other natural habitats, which harbor much of the world's flora and fauna.

Why has agricultural production gone up, thereby making food less scarce in the face of rising demand, without corresponding increases in the use of labor and land? The answer to this critical question has to do with technological change that allows for substitution away from the traditional inputs of farming and, of greater importance, gains in productivity. The development and adoption of agricultural machinery, which began in the 1800s, has reduced the demand for farm labor, not to mention draft animals. Likewise, yields from farmland have risen as farmers have switched to improved seeds and agricultural chemicals, such as fertilizers and pesticides. Coinciding with input substitution has been an increase in what economists call total factor productivity (TFP). Today, much more output is obtained from a given amount of inputs (or, equivalently, the same output is produced with far fewer inputs) than was formerly the case.

The causes and consequences of diminished food scarcity are examined in this book. At the core of our findings are lessons about the numerous advantages of specialization and trade. Where commercial exchange is minimal and self-reliance is the norm, as is the case in the poorest parts of the world, most of the working population labors mightily to grow paltry amounts of food. All available land that lends itself well to agriculture is cultivated, as are many fields that are poorly suited to farming. Relative to everyone's meager earnings, food is expensive; for a typical family, the cost of a minimally adequate diet represents a serious financial burden. In poor places, people devote most of their resources to obtaining nourishment.

Precisely the opposite circumstances are observed where specialization and trade flourish, as these do if a robust market economy is in place. Most people work outside of agriculture. As a rule, farming is confined to settings well suited to crop production, and non-agricultural uses of land, including nature reserves, are extensive. Rather than being very scarce, food is sufficiently available that its expense is well within the reach of the vast majority of households.

Food economies of the kind that exist in the United States do not emerge spontaneously. To unleash market forces, the state needs to establish an enabling institutional environment – an environment expressed by the term, rule of law, and characterized by transparent and even-handed enforcement of contracts and property rights. Government also facilitates commercial exchange by developing the infrastructure needed for transportation and communications. Without this infrastructure, trade is impeded and people find it difficult to specialize. Furthermore, gains from specialization are enhanced by productivity-enhancing investment in human capital and new technology – investment often made possible by government, alone, or by initiatives involving both the public and private sectors.

As market forces are unleashed, specialization reaches an advanced stage within the food economy. Rather than being self-reliant, farmers purchase agricultural inputs and marketing and processing services from agribusinesses. The ultimate beneficiaries of the cost savings created by this exchange are consumers. In the United States, food expenditures, including payments for restaurant meals, amount to 10 percent of total household income (ERS-USDA, 2003). The food ingredients supplied by farms make up about one-fifth of these expenditures (Gardner, 2002, p. 155), or 2 percent of consumers' earnings. The other four-fifths of what Americans spend to feed themselves are payments for the inputs and services provided, with great efficiency, by agribusinesses.

These firms are also a source of technological innovation. Over the years, major advances in machinery and seeds have come from the private sector. Today, agribusinesses are dominant players in biotechnology. They are no less important, then, than farmers and the government. Indeed, it is useful to think of the food economy as a tripod, with each leg supporting and being supported by the other two legs. Just as this economy could not exist without crop and livestock producers, its efficiency would be seriously impaired if either government or agribusiness were absent.

As the food economy develops, the impacts reverberate far and wide. With edible commodities growing less scarce, people not only eat better, but their purchases of things other than food rise. In addition to buying more cars, electronic appliances, and other consumer goods, they spend more on education, health care, and other services. These changes in household expenditures contribute to economic diversification, which is an integral feature of development. Cheaper food also leads to an increase in savings, which are then channeled into investment. As a result, economic growth accelerates.

Profound changes occur as living standards rise, thanks in no small part to efficient performance of the food economy. People stop thinking of childbearing merely in terms of an addition to the labor pool – the family's agricultural workforce, to be specific – and more consideration is given to the rewards captured by people with education and training. Since the formation of human capital is costly, the number of children in each household falls in order to accommodate the large investment that most parents hope to make in the skills and capabilities of each son and daughter. Family size also declines because women enjoy improved opportunities for education and employment. Their economic empowerment, which is part and parcel of development, raises the trade-offs associated with rearing children. Among the consequences of the decline in family size that results from rising prosperity is, obviously, decelerating growth in the demand for food.

Compared with the pessimistic assessments of the adequacy of food supplies that grabbed headlines in the late 1960s and early 1970s, today's concerns about the global food economy seem less urgent. More than 80 percent of humankind is adequately nourished today and will continue to be so for the foreseeable future (FAO, 2003b, p. 7). Moreover, the steps that knowledgeable observers say are needed to ease the suffering of the hungry minority of the world population are not prohibitively expensive, but rather are well within our reach. If the specter of global famine truly loomed, any and all measures to boost agricultural production would be universally applauded. The very strength of current opposition to some of these measures – including the biotechnological process of modifying one agricultural species by inserting a gene from an entirely different part of the plant or animal kingdom – is evidence of the progress made to date in supplying people with enough of the food they desire at prices they can afford.

Just as pessimism about the adequacy of food supplies is not warranted, there is no call for complacency. Much of the world invests too little in agriculture and its technological improvement. Moreover, governmental interference with market forces diminishes incentives for food production in many places. These are some of the reasons why hundreds of millions of people eat too little.

1.1   Our Focus

Given the stakes involved, recent progress in the world's food economy merits careful examination. In this book, trends in demand and supply are analyzed, those during the recent past as well as those anticipated in the years to come. While properly acknowledging the decline in food scarcity that occurred during the second half of the twentieth century, we emphasize that future progress is not guaranteed and that fully alleviating hunger remains a challenge.

To assess what the future holds in store for producers and consumers of food, one may suppose that trends from the recent past will continue into the future. However, simple extrapolation of this sort would be very misleading. During the past 100 years, the number of mouths to be fed skyrocketed, from a little more than 1.5 billion in 1900 to a little less than 2.5 billion in 1950 and ultimately surpassing 6.0 billion by 2000. But even before the turn of the century, there were clear signs that population growth was decelerating and, hence, that food demand will not increase as quickly during the next few decades as it did in the 1900s. Likewise, the pace of supply growth is changing. In particular, the rise in crop yields that has coincided with gains in agricultural productivity is slackening. There is a real possibility that demand growth, slower though it is sure to be, will exceed increments in supply during the first part of the twenty-first century. If so, the pronounced decline in food scarcity that occurred during the second half of the twentieth century will stall. Scarcity, and thus food prices, might even go up for a while.

For much of the human race, a departure from recent trends would hardly be noticeable. As already mentioned, the value of unprocessed agricultural commodities is roughly equivalent to 2 percent of total household income in the United States. Even a 50 percent increase in the former value, which is far more pessimistic than most available forecasts, would only oblige the typical American family to reallocate 1 percent of its budget. But for hundreds of millions of people in South Asia, Sub-Saharan Africa, and other impoverished regions, any increase in scarcity would create real hardship. Food expenditures by a Zambian family, for instance, claim three-fifths of its meager earnings (ERS-USDA, 2003). Since agribusiness is poorly developed in the country, specialization and trade are limited in the national food economy. Raw commodity values can easily comprise two-thirds of what the family spends to feed itself, in which case a 20 percent increase in commodity prices leads to an 8 percent reallocation of household income. For people not eating well to begin with, the belt-tightening that results is severe. Indeed, some of these people could starve.

Aside from the suffering that poor people endure if the things they eat become more expensive, increased food scarcity has a number of other impacts. Some of these are environmental. For example, a rise in commodity values, which is to be expected if gains in agricultural productivity are not enough to drive down food scarcity, causes agricultural land use to expand. As natural habitats are cleared to make way for new cropland and pasture, the benefits of undisturbed habitats, including the conservation of biological diversity, are lost. Other consequences of rising food prices are more narrowly economic. It bears repeating that diminished food scarcity has contributed to growth and development by increasing savings and investment and accelerating economic diversification. If food scarcity goes up in a poor country, the economy will diversify at a slower pace and there will be less investment and growth. As the country struggles without success to escape poverty, it will not be the only loser. Its trading partners, actual as well as potential, will forego the benefits of commercial exchange with a country that could be more efficient and more prosperous.

To understand why food becomes more or less scarce as well as the various consequences of such a change, the conceptual framework that economics provides has no substitute. As emphasized at the beginning of any introductory textbook or course on the subject, economics examines how people respond to scarcity, which comes about whenever supply is limited relative to demand. Economists give special attention to specialization and trade as well as investment and technological improvement as ways to deal with scarcity.

The global food economy being as complex as it is, each and every one of the responses to scarcity cannot be analyzed thoroughly in a single volume. On the supply side, this book focuses much more on production agriculture than on agribusiness. The former is a distinctive sort of enterprise, not least because of its being inseparable from the natural environment. In contrast, agribusinesses have a lot in common with firms that are unrelated to the food economy. In this book, overall trends in the availability of farm commodities are examined. We also analyze the impacts that changes in land use and other inputs as well as improvements in productivity have had on these trends. Certain elements of supply – cereal grains consumed directly by people, fed to livestock, or both – are examined more closely than others (e.g., the output from marine fisheries).

On the demand side, the focus is likewise on overall trends, with special attention paid to population growth and improved living standards as primary drivers of changes over time in food consumption. Given this emphasis, food safety, public acceptance of genetically modified organisms (GMOs), and related topics are not examined in great detail. Also touched on lightly are some important aspects of human nutrition. One of these is education aimed at informing people about what is good to eat and what is not. Another aspect is sanitary improvements needed to eradicate intestinal parasites, which prevent many people from receiving full nourishment from the food they consume. When it comes to economic development, our overriding concern is to clarify linkages between the agricultural sector and the rest of the economy, and to analyze how these linkages are altered as an economy grows and diversifies, as opposed to exhausting the topic of economic development.

In this book, we do not duplicate the content of more specialized contributions to the economics literature. In a comprehensive study of food consumption, one that addresses all aspects of human nutrition and food safety, consumer behavior is modeled in sophisticated ways. Likewise, texts on agricultural economics contain a thorough analysis of such topics as farmers' choices among inputs and production methods, their response to risks created by variable weather, and the performance of markets in which they purchase inputs and sell outputs. By the same token, books on the economics of development provide a comprehensive treatment of topics in that field. Recognizing the availability of specialized volumes, we have written this book precisely so that the overlap among food consumption, agricultural production, and development can be better understood. The pages that follow have more to say about agricultural production and development economics than a book focused on consumption alone. Similarly, we look more carefully at population growth and its impact on food demand than the authors of agricultural economics texts normally do. In addition, while agriculture is sometimes relegated to a single chapter of a book on development economics, linkages between agriculture and the rest of the economy are, to repeat, one of our major interests.

Aside from addressing the overlap among three interrelated topics, we try to derive maximum insights from minimal economic theory. Fundamental economic concepts, which are reviewed in various chapter appendices, are used to make sense of trends in demand for the things we eat, trends in supply, as well as changes in the markets in which food is traded. The same fundamental concepts can be put to good use in an analysis of the economy-wide impacts of a rise or fall in food scarcity.

A particular benefit of not trying to duplicate volumes that use more advanced theory and have a narrower topical focus is that we are able to examine the availability of affordable supplies of food not just at a global level, but at the regional and even national levels as well. This allows for clarification of differences in the food economy from one country or part of the world to the next. In particular, differences in undernourishment among regions and countries can be highlighted.

1.2 Chapter Outline

Our examination of the world's food economy begins, in the next chapter, on the demand side. During much of the twentieth century, death rates plummeted in poor countries without simultaneous and equivalent reductions in birth rates. Consequently, human numbers have swollen throughout Africa, Asia, and Latin America. However, there is clear evidence that death and birth rates are now converging – more quickly in some places than in others, to be sure – and it is reasonable to expect that the world population will peak midway through this century, perhaps within the lifetimes of young people assigned to read this book in a college class. As population growth decelerates, other factors influencing the demand for food, most notably changes in income, will become more important.

Responses to demand growth are analyzed in Chapter 3. Attention is paid to how the production increases of the twentieth century were accomplished as well as how output growth is likely to be achieved during the decades to come. The role of expanded agricultural land use is acknowledged, as are its costs. So is the role of increased use of commercial fertilizers and other non-land inputs. Emphasis is placed on the impacts of productivity growth and the contributions that government and agribusinesses have made to this growth. We stress that, for agricultural production to keep pace with demand in the future, investment that raises productivity must continue.

How markets bring consumption into line with production, thereby enhancing economic well-being, is demonstrated in Chapter 4. We describe the changes observed in food markets as demand goes up because of growth in population, income, and other variables, supply increases because of technological innovation and other reasons, or both. As is also made clear, commodity prices are affected by government policies. In addition, we examine the downward trend in inflation-adjusted commodity values observed during the twentieth century as well as the prospects for continuation of this trend, which to repeat is not guaranteed.

For the most part, commercial exchange that is competitive and unregulated is enormously beneficial. However, the marketplace often leaves environmental problems unresolved. Many of these problems have to do with governmental interference with market forces – subsidizing water and other agricultural inputs, for instance. Others are a consequence of what economists refer to as market failure, a standard example being output that is too high and prices that are too low because producers are not compelled to pay for the harm they do to the environment. In Chapter 5, we survey the damage done in the food economy to natural resources because of market failure. Also examined is the environmental cost of misguided governmental intervention in the marketplace. Of special concern is the habitat loss that corresponds to an expansion of agricultural land use. Similarly important is land degradation, resulting from erosion and other processes. This problem is especially severe in Sub-Saharan Africa, not in spite of but because that region has benefited less than other parts of the world from advances in agricultural technology.

As stressed in Chapter 6, an important part of the world's food supply is traded internationally. Also recapitulated is the theory of comparative advantage, now nearly 200 years old, which demonstrates the benefits of specialization and commercial exchange among different countries. Enhanced food security is one of these benefits. Reviewed as well in the chapter is the damage that has resulted as agricultural trade has been suppressed – damage that is particularly acute for poor countries that can produce crops efficiently for international markets.

Chapter 7 focuses on relations between agriculture and the rest of the economy. The literature on the contributions that the former sector makes to general development is surveyed. We also assess the performance of different countries in which governmental treatment of agriculture varied markedly during the twentieth century. An important lesson of this assessment is that governmental investment in education, science, and information services to raise agricultural productivity has had much higher economic pay-offs than governmental manipulation of prices and incomes. Also highlighted are the synergies between productivity growth in agriculture and dynamism in other sectors.

Lest a review of broad trends in demand, supply, and commodity values incline one toward complacency, we acquaint the reader in Chapter 8 with the dimensions of food insecurity around the world. These dimensions are staggering. Of the 6 billion people alive today, approximately 844 million, or 14 percent, lack economic access to the food they require for healthy and productive lives (FAO, 2003b, p. 7). This deplorable situation has much less to do with scarce supplies than with the penurious incomes of those who are hungry. Economic development of the kind that lifts people out of poverty, then, must be the centerpiece of a strategy for alleviating food insecurity.

The rest of the book is regionally focused. Food economies differ in terms of climate and natural resources, history and culture, and recent economic development. Interregional differences are summarized in Chapter 9. Each of the six chapters that follow is about one part of the world: the world's affluent places (Western Europe, North America, Japan and South Korea, and Australia and New Zealand); Asia; Latin America and the Caribbean; the Middle East and North Africa; Eastern Europe and the former Soviet Union; and Sub-Saharan Africa. Data on principal features of the food economy at the national level are presented in each of these regional chapters, in which patterns of economic growth, population dynamics, agricultural development, and food consumption are analyzed.

As we look to the future, neither complacency nor panic about the global food economy is warranted. If policies are put in place that encourage the proper utilization of the world's resources and current and emerging technologies, it is entirely possible for the economic expansion of recent decades to continue and, moreover, for the benefits of this expansion to be widely distributed. In particular, the food-insecure portion of the human population will keep on declining – not just proportionately, but in terms of absolute numbers as well. As the specter of hunger recedes, other concerns about the food economy will grow more prominent. Among these are the environmental impacts that might result as new technologies for crop and livestock production are developed and applied, thereby making food less scarce. Also of mounting concern is the poor health that results from overeating – a phenomenon observed not just in the United States and other wealthy countries, but also in China and other places where famine threatened just a generation or two ago. The only safe predictions that can be made about the twenty-first century are that age-old challenges to the food economy will persist and that novel problems will emerge.

Study Questions

1. In the late 1960s and early 1970s, Paul Ehrlich, the Club of Rome, and others forecast imminent famine throughout the world. Describe what has happened and what has not happened in the food economy to avert this disaster.
2. What are the three main elements of the food economy and how are these elements interrelated?
3. Describe the economy-wide impacts of agricultural development.
4. Compare and contrast the impacts of scarcer agricultural commodities in a poor country, such as Zambia, and in an affluent place, such as the United States.
5. Compare and contrast the subject matter of this volume with those of textbooks on agricultural economics and development economics.

2

The Demand Side:
How Population Growth and Higher Incomes Affect Food Consumption

Not so long ago, predicting trends in food consumption could be treated as a straightforward exercise. Little attention was paid to things like the impacts on demand of higher incomes since the vast majority of people were very poor and variations in their living standards were negligible. Instead, population growth was the only driving force that really mattered. If the population of a country or the world was going up by 10 percent, it was safe to conclude that food consumption was rising by 10 percent as well.

Estimates of demographically driven trends in demand were positively alarming whenever it was supposed that human numbers were on an exponential trajectory. If the population were compounding steadily at the same annual percentage rate, one could readily identify the number of years required for it to double in size – in particular, by dividing 70 by the yearly growth rate. For example, Ecuador's population was increasing by 2.9 percent per annum during the early 1980s (World Bank, 1987, p. 254), which implies a doubling time of just under 25 years (70 ÷ 2.9 = 24.14). Had that rate not diminished, as indeed it has, then the country's population, which was approximately 9 million in 1985, would have reached 18 million in 2010, 36 million in 2035, 72 million in 2060, and so forth. It would not have taken long for Ecuador's capacity to feed itself from domestic production and imports to be overwhelmed.

However, no exponential trend continues indefinitely. One possible brake on human numbers – the purported tendency of food supplies always to increase more slowly than the population – was posited 200 years ago by Thomas Malthus, the author of a classic treatise on demography. Still influential, his observations comprise a suitable point of departure for this chapter's discussion of trends in food demand. But as we shall see, profound demographic changes have occurred around the world since the late 1700s. These changes, which Malthus and his contemporaries could not have conceived of, have much to do with economic progress and the empowerment of women, which happen to go hand in hand. Much of this chapter is devoted to describing and analyzing recent trends in population – trends that could lead to a stabilization of human numbers midway through this century. We also emphasize that changes in living standards are bound to affect food demand more and more as population growth slows.

2.1 Classic Malthusianism, Its Modern Variants, and Its Critiques

An ordained clergyman in the Church of England, Thomas Malthus, was also the first professor in Great Britain of what was then called political economy (Pullen, 1987). Born in 1766, he witnessed the dawn of the Industrial Revolution, when the steam engine and other technology were harnessed in the mass production of textiles and other goods. Nevertheless, farming dominated the British economy for much of Malthus's lifetime. The portion of the country's work force employed in agriculture, for example, was nearly 50 percent in 1770 (Cipolla, 1965, p. 67) and stayed high after the turn of the nineteenth century. Agriculture therefore commanded Malthus's attention. The same was true of David Ricardo, who gave us the theory of comparative advantage (Chapter 6), and other classical economists, who were active during these years.

If Great Britain's economy was predominantly agricultural, it was by no means stagnant. Increased trade in the 1600s and 1700s coincided with economic expansion. During this same period, the country's population increased steadily, as did the populations of a number of countries on the European continent. A sensible question to consider on the eve of the nineteenth century was how long this demographic expansion could continue (Pullen, 1987).

In addressing this question, Malthus (1963) formulated a "principle of population" that rests on a very simple characterization of human behavior. Strikingly at variance with modern views, to say the least, this characterization is best expressed by two words: vice and misery. Vice, by which was meant profligate sex and reproduction, arises whenever food supplies exceed what bare subsistence requires. That is, any positive difference between living standards and the basic necessities of life not only keeps hunger and illness at bay, but also causes people to breed with abandon and the population to grow exponentially (e.g., from 3, to 9, to 27, etc.). But if consumption levels are below bare subsistence, perhaps because people lacking foresight have reproduced beyond the number that can be fed, then misery is experienced. Disease and starvation, and perhaps war, cause deaths to exceed births, thereby inducing demographic contraction. Malthus contended that the only way to avoid either exponential growth or a rapid decline in human numbers is for living standards to be at bare subsistence, nothing more and nothing less. Otherwise, vice sets in or its opposite, misery.

This is not to deny that population growth over the long term is impossible. Malthus believed that it is possible, although he stressed that growth is inevitably constrained by food availability. He also contended that, unlike human numbers, agricultural output rises linearly (e.g., from 7, to 9, to 11, etc.) over time, not exponentially. This conviction would be reasonable if technological advances in agriculture were insignificant.

As Malthus emphasized, an exponential trajectory for population cannot be reconciled with linear increases in food availability. Furthermore, the collision between the two trends can be wrenching if rapid reproduction continues even as the threshold of bare subsistence is being passed. This possibility is illustrated in Figure 2.1, in which a positively sloped line depicts increases in agricultural output. The other curve represents population (which varies nonlinearly as the years pass) multiplied by a minimally adequate diet. At an initial date, 0, food supplies exceed what is required for bare subsistence. According to Malthus's principle of population, this abundance induces "vice" and an exponential increase in the number of people and the amount of food they must eat. But by year t^*, this increase has caused the basic dietary requirements of the population to draw even with food availability. At this point, an overshoot is imminent. As human numbers continue growing a little while longer, average consumption falls below what is required for bare subsistence. Soon afterward, the population contracts because of disease and starvation. As shown

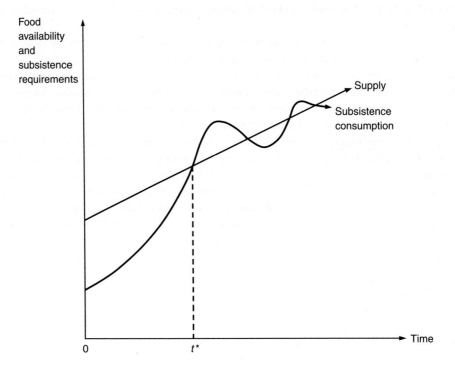

Figure 2.1 Malthusian trends in food demand and supply.

in Figure 2.1, the amplitude of the alternating cycle of overshoot and collapse may diminish over time, with consumption levels eventually settling in at bare subsistence.

According to the classic Malthusian model, humankind's lot is to experience periodic episodes when the population contracts due to food shortages. Even the good times between these episodes only serve to set the stage for subsequent demographic collapse. No wonder that early critics of political economy, mindful of the principle of population, referred to the new scholarly discipline as "the dismal science"!

Neo-Malthusianism

To recognize that the implications of Malthus's ideas are unpleasant does not, of course, constitute a refutation. As far as some are concerned, these ideas have not been refuted at all. Indeed, the classic Malthusian model has proven resilient, its variants cropping up regularly during the past two centuries.

In the early 1970s, for example, the Club of Rome – an influential group of civic, governmental, and business leaders concerned about long-term consequences for the environment and human well-being of demographic and economic expansion – commissioned a team of systems engineers to develop a model of the global economy and the natural resources it uses (Meadows *et al.*, 1972). The model features a series of feedback loops, both positive and negative. For example, economic growth begets investment, which in turn stimulates growth. Also, growth causes environmental quality to deteriorate, which leads to increased mortality and thus to slower growth. But the model's core is Malthusian. Trends in population, industrial output, pollution, and so on are all exponential. Moreover, these trends, if allowed to run unchecked, always overshoot limits imposed by the world's fixed endowments of fossil fuels, agricultural land, and other natural resources.

Without comprehensive birth control, strict limits on current resource use, and other measures, Meadows *et al.* (1972) predicted that a few decades of exponential growth in the human population and its economy would exhaust available resources. The ensuing collapse would be permanent. Rather than foreseeing oscillations of the human population around a linear, though positively sloped, trend in food availability, as depicted in Figure 2.1, the simulation modelers predicted that the current age of rapid growth and industrialization would be succeeded by conditions similar to what preceded it. As shown in Figure 2.2, the world's population would experience a precipitous and irreversible decline. So would per capita supplies of food and industrial products.

As economists are quick to point out, the model developed for the Club of Rome ignores an intrinsic feature of economic life – namely, markets. This is a critical omission because markets constitute an important feedback mechanism for dealing with resource scarcity. Consider, for example, how buyers and sellers of petroleum respond to its growing more scarce, due possibly to demographic or economic expansion. Obviously, sellers raise prices. This, in turn, accelerates the search for new supplies as well as the development of alternative energy sources and conservation

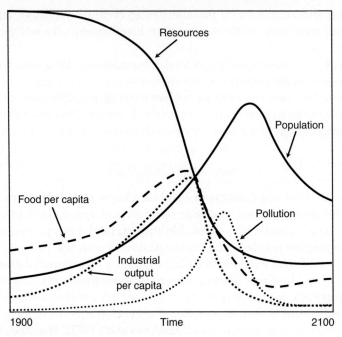

Source: Medows et al. (1972), p. 124.

Figure 2.2 Standard run of Club of Rome model.

technology. Even if exploration proves fruitless and technological progress is not achieved, higher prices induce customers to cut back their purchases, thereby delaying the exploitation of previously identified deposits of oil.

The omission of markets by Meadows *et al.* (1972) can be rationalized by observing that markets do not exist for various environmental resources, including the air we breathe and a number of others of vital importance. As is explained in Chapter 5, this absence of markets has to do with these resources not being owned by anyone, which is required for commercial transactions. While ownership of something like atmospheric oxygen is far-fetched and may even seem objectionable, the absence of property rights means that no one has to pay for using, or abusing, the resource. Likewise, the growing scarcity of clean air never induces the sort of price rise that would encourage better control of pollution.

Granted that property rights do not exist for various natural resources, the complete neglect of markets by Meadows *et al.* (1972) causes them to ignore all the substitution, conservation, and development of new resource supplies that take place as economically motivated firms and households respond to rising environmental scarcity. Something else is neglected in the classic Malthusian model and its later variants. This is the simple human desire for self-improvement and the results observed when, as always happens in a market economy, people are individually empowered to act on that desire. Lasting improvements in living standards cannot be reconciled with people thriving or perishing pretty much as animals do, abandoning themselves to vice when there is temporary respite from bare subsistence and plagued by misery at other times. By the same token, individual desires for a better life are all but impossible to factor into the Club of Rome model since the best setting for acting on these desires – the marketplace – is lacking. By going to the trouble to write a book, Meadows *et al.* (1972) implicitly conceded that at least some human beings can recognize an unsustainable trend when presented with convincing evidence and can be counted on to support measures needed to avoid a catastrophic collapse. But individual adaptation to mounting scarcity in a market setting is outside the purview of their model. Instead, coercive, regulatory action, inevitably carried out by a central authority that is benevolent as well as capable, is emphasized. Mandatory birth control would be an example of such action.[1]

Interestingly, Malthus himself did not entirely discount what can happen as people strive individually for self-improvement. In a cogent survey of trends in population, agricultural development, and economic progress during the past two centuries, D. Gale Johnson quotes the passage that follows from the second edition of Malthus's book, published 5 years after the first edition, which suggests that the principle of population can be beneficially tempered.

> "(Though) our future prospects respecting the mitigation of the evils arising from the principle of population may not be so bright as we could wish, yet they are far from entirely disheartening and by no means preclude gradual and progressive improvement in human society" (Johnson, 2000, pp. 1–2).

Is the World Dismally Malthusian?

During the past two centuries, as Johnson (2000) makes clear, human existence has improved dramatically because better ways to harness available resources have been found. For one thing, technological innovation in agriculture has driven down the cost of food and released productive factors to non-farming sectors of the economy. Human nutrition has improved, and not just by a small amount and not just in a few places. Furthermore, people are considerably healthier today than they were 200 years ago, as indicated by increases in life expectancy at birth. Finally, people around the world have responded to these changes not by stepping up reproduction, but rather by doing the opposite. On average, women today have fewer children than their female ancestors did.

Agricultural development and its consequences are examined in the next chapter and elsewhere in the book. For now it is important to note something that Malthus never could have imagined, which is the increase in the world's population that has coincided with significant improvements in the average person's diet. Human

---

[1] Ehrlich offers a specific proposal in *The Population Bomb*. "One plan often mentioned involves the addition of temporary sterilants to water supplies or staple food. Doses of the antidote would be carefully rationed by the government to produce the desired population size" (Ehrlich, 1968, p. 122).

numbers did not rise above 1 billion until 1820 or so. Since then, a 500 percent increase has occurred, the global population reaching 6 billion at the turn of the twenty-first century. Malthusian analysis would suggest that this increase should have been accompanied by a drop in per capita consumption of food. However, precisely the opposite has happened. During the 1600s, long before the principle of population was posited, the typical European ate no more than 1,700 calories per day – about the same as average daily consumption throughout the world in the middle of the twentieth century. By the early 1960s, the global average had risen to 1,940 calories per day. During the 1990s, the global daily average was approaching 2,600 calories and less than 10 percent of the world's people lacked reliable access to 2,200 calories per day of food (Johnson, 2000).

Health improvements have been at least as significant. In 1900, infant mortality rates in Europe varied from 88 per thousand live births in Norway to 221 per thousand in Austria (Cipolla, 1965, p. 83). China's rate 60 years later was in this same range: 157 per thousand. Since then, it has gone down to 62 per thousand (Johnson, 2000). Thanks mainly to improved survival of infants and young children, human longevity has increased. Two hundred years ago, English and French newborns had a life expectancy of 30 years. By 1900, this number was approaching 50 in industrialized nations (Cipolla, 1965, p. 86). Increases in life expectancy at birth in India during the twentieth century have been quicker and larger – from 23 years in 1900, to 32 during the 1940s, to 43 in 1960, to 62 years in 1996 (Johnson, 2000).

Living standards have shot up as well. Between 1500 and 1820, growth in income per capita was barely noticeable, amounting to just 0.04 percent per annum. From 1820 to 1992, the yearly rate of increase was much higher: 1.21 percent, to be precise. During the nineteenth century, growth in average incomes was concentrated in Europe and the places Europeans had colonized. But during the past 100 years, growth has been more universal, leaving few parts of the world unaffected, and if anything has accelerated in recent decades (Maddison, 1995, pp. 20–21).

According to Malthus's model, better diets, improved health, and higher incomes ought to have caused a skyrocketing of human fertility. However, his principle of population has proved to be a very poor guide to recent trends. In some places, most notably China, strict controls on family size, of the sort advocated by Meadows *et al.* (1972) and other modern Malthusians, have been applied. But many other parts of the world have experienced declines in fertility comparable to China's without resorting to coercive measures. The latter result can be explained only in terms of people, especially women, deciding that their individual interests are best served by having fewer children and also being able to act accordingly.

2.2 Demographic Transition

The desire for fewer children has arisen and been acted on only after the beginning of the demographic transition, which gets under way at the start of a sustained

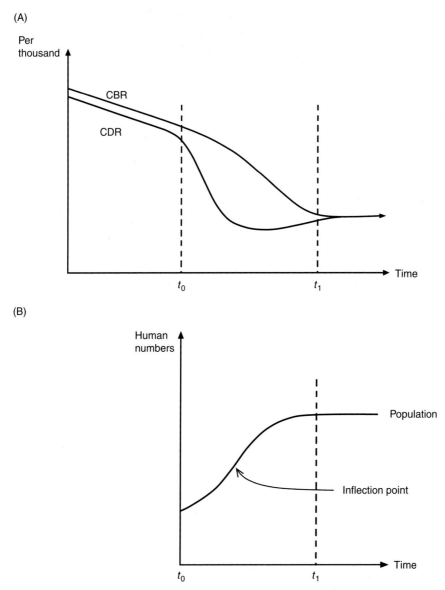

Figure 2.3 The demographic transition. (A) Changes over time in birth and death rates. (B) Changes over time in population.

and historically unprecedented decline in mortality. This decline occurs after date t_0 in Figure 2.3A, in which the transition is depicted abstractly. Demographic adjustment is entirely over once the age composition of the population has stabilized, with old people comprising a large share of total numbers (Lee, 2003). While each and every part of the world has experienced mortality decline, there is no country where age composition has reached an equilibrium. Thus, demographic transition is still happening everywhere. However, quite a few nations have reached the point, represented by t_1 in Figure 2.3A, at which the birth rate has fallen into line with the death rate. Once this demographic milestone is passed, of course, human numbers stop increasing.

By definition, a population is in demographic equilibrium – defined in terms of equivalent birth and death rates as well as a stable age composition – both before and after the transition. However, the two equilibria, pre- and post-transition, differ sharply. Characteristic of the impoverished, rural communities in which practically all humankind dwelt for millennia, the pre-transition state features annual crude death rates (CDRs) that are extremely high by the standards of the modern world: 40 per thousand or higher.[2] If there is no net immigration, then a community with a CDR in this range must have a correspondingly high annual crude birth rate (CBR) in order to avoid population contraction. For example, suppose a village loses 4.5 percent of its residents to illness every year. This means that there must be 45 live births annually per thousand villagers if demographic equilibrium is to be maintained. With males comprising about half the population, 9 percent of the female population must deliver a child each year. Of course, the percentage for those females who are of childbearing age is higher still. Prior to the demographic transition, then, the total fertility rate (TFR), which is the median number of children that women deliver during their lifetimes, is elevated. For example, TFRs 20 years ago in Honduras, Kenya, and a number of other countries where the transition was not far along exceeded seven children per woman (World Bank, 1987, p. 256).

CDRs, CBRs, and TFRs all being high before the demographic transition begins – again, at t_0 – these fundamental indicators converge to low levels as the transition proceeds. In particular, convergence between the CDR and CBR, which occurs at t_1, means that natural increase, which is found by subtracting the former rate from the latter, has ended. This state has been achieved in Sweden, where the CDR is low (10 per thousand in 2001) and equal to the CBR. In addition, the country's TFR, 1.6 births per woman, is a little under the replacement level – approximately 2.1 births – needed to maintain a stable population (World Bank, 2003b, pp. 40 and 102). Sweden is clearly in the demographic transition's last stage, which begins at t_1 and during which the impacts of previous changes in fertility and mortality on age composition gradually fade.

Conceivably, demographic transition could occur without any natural increase. All that would be needed is for the CBR and CDR to fall in unison between t_0 and t_1. However, this never occurs. Instead, a positive gap inevitably opens up between the two rates because mortality always drops off first. This reflects people's enthusiasm for making full use of health care and public sanitation once all of this becomes available, as generally happens with development. In contrast,

---

[2] The convention among demographers is to state most rates as a number per thousand, rather than as a percentage. A CDR of 15 per thousand is equivalent to 1.5 percent annual mortality.

contraceptive technology is adopted more gradually, which prevents the TFR from falling as rapidly as the CDR. Even after family size shrinks, in response to diminished mortality, improved standards of living, and better educational and job opportunities for women, the CBR declines at a leisurely pace. This is largely a delayed consequence of better infant and child survival. Many more of the babies born early in the demographic transition, when the TFR is close to its pre-transition level, live past their fifth birthday. Hence, large numbers of females are reaching sexual maturity 15–20 years after the transition is set in motion by a drop in the CDR. Each of these women would have to deliver far fewer children than her mother or grandmother did for the CBR to decline as rapidly as the CDR. Since this never occurs, natural increase is an intrinsic feature of the demographic transition.

The sizable female cohort reaching sexual maturity 15–20 years into the demographic transition creates demographic momentum, which explains the relatively slow pace of CBR declines. Of course, demographic momentum, which accounts for a large share of natural increase after human fertility has fallen from the pre-transition range, slackens as TFRs stay at or near the replacement level. Accordingly, an inflection point is always passed in the time-path of total human numbers, which is represented in Figure 2.3B. Past this point, the rate of natural increase diminishes, eventually reaching zero at t_1. At t_1, the population, though no longer growing, is at a peak.

Nicholas Eberstadt, of the American Enterprise Institute, succinctly explains why natural increase always has occurred once the demographic transition began: "Rapid population growth commenced not because human beings suddenly started breeding like rabbits but rather because they stopped dying like flies" (Eberstadt, 1995, p. 21). People respond to the welcome achievement of diminished mortality, especially for the very young, by cutting back on family size. However, holding the grim reaper at bay creates demographic momentum, as large numbers of females reach childbearing age. Accordingly, natural increase stays positive. Once this momentum is spent, as it eventually must be, birth and death rates come together at a low level. As shown in Figure 2.3B, however, human numbers are much higher once this equivalence is achieved than what they were before the whole adjustment began.

The Modern Transition in the Developing World

Regardless of where and when the demographic transition has been experienced, some basic features of the process are always observed, such as the lag between CDR declines that are sustained and historically unprecedented and falling birth rates. However, specific trends in CBRs and CDRs vary considerably from place to place. By the standards of our own time, the transition in northern and western Europe, which was under way 250 years ago, proceeded at a leisurely pace, with natural increase never exceeding 15 per thousand annually. In contrast, natural increase rose to and above 3 percent per annum in a number of developing countries during the second half of the twentieth century, thanks mainly to precipitous drops in mortality. Although CBRs have fallen as well, the wide gaps that have opened up between birth and death rates in Africa, Asia, and Latin America explain two fundamental demographic facts. One is that human numbers have gone up faster since the Second World War than at any other time in history. The other is that most of the increase has occurred in developing countries.

Few parts of the world have longer demographic records than England and Wales and the Scandinavian nations (Table 2.1). By the 1750s, CDRs in these places had fallen to or below 30 per thousand, while CBRs were around 35 per thousand – except for Finland, where the birth rate was 45 per thousand (column 1). During the next two centuries, the two rates went down slowly. Indeed, the decline in mortality halted at times. The CDR in England and Wales remained at 23 per thousand from the early to middle 1800s (columns 2 and 3). One reason for this is that people were moving to cities, where industrial employment was increasing and yet where living conditions were insalubrious, especially for the working classes. Kremer (2002) points out that achieving lower mortality during this period, long before the discovery of antibiotics and other pharmaceutical products for combating disease, required a sizable investment in potable water systems and sanitary infrastructure; as the numbers in Table 2.1 show, progress was slow. But after the middle of the nineteenth century, much of this infrastructure was in place, which allowed for an appreciable drop in CDRs. Not coincidentally, natural increase peaked in England and Wales and Scandinavia during the late 1800s and early 1900s (column 4). CDRs did not stabilize at 9–12 per thousand until the middle 1900s, more than 200 years after these began to fall, and CBRs reached this level even more recently (column 5).

If the demographic transition were proceeding as slowly today in Africa, Asia, and Latin America as it did in northern and western Europe in times past, the prospects for a stable population during the next few decades would be very remote. However, the transition has been much faster in the developing world than in places like Britain and Scandinavia. Consider mortality trends. Different from what occurred in Europe and North America, other parts of the world have not had to discover factors contributing to the spread of mortal illness on their own before

Table 2.1 Birth and death rates in Great Britain and Scandinavia, 1750s to 2001

Country	1751–1755		1801–1805		1851–1855		1905–1909		1950		2001	
	CBR	CDR	CBR	CDR	CBR	CDR	CBR	CDR	CBR	CDR	CBR	CDR
	(1)		(2)		(3)		(4)		(5)		(6)	
England and Wales*	35	30	34	23	34	23	27	15	16	12	11	11
Finland	45	29	38	25	36	28	31	18	25	10	11	9
Norway	34	25	28	24	33	17	27	14	19	9	13	10
Sweden	37	26	31	24	32	22	26	15	16	10	10	10

*2001 data are for Great Britain as a whole.
Source: Cipolla (1965, p. 81) for 1750s through 1950; World Bank (2003b, pp. 38–40) for 2001.

dealing with these factors. Instead, Africa, Asia, and Latin America have been able to apply knowledge and technology originally developed elsewhere. Containing mortality mainly with pharmaceuticals has turned out to be much cheaper and quicker than relying exclusively on the improvement of water and sanitation systems, as Europeans had to do during the 1800s; as a result, CDRs have declined rapidly (Kremer, 2002). For example, death rates in Tanzania, Laos, and other impoverished nations 40 years ago (Table 2.2, column 1) were about the same as what these rates had been in northern and western Europe in the early 1800s. Due partly to acquired immune deficiency syndrome (AIDS), which is caused by the spread of the human immunodeficiency virus (HIV), Tanzania's CDR has stayed high. But elsewhere, the decline that Europe needed 100 years or more to accomplish has been achieved in three decades or so.

With CDRs falling a lot during the twentieth century, accelerated natural increase has been inevitable in the developing world. Further amplifying the gap between birth and death rates was that CBRs were high in Africa, Asia, and Latin America before the demographic transition began in those places. In six of the developing countries listed in Table 2.2, birth rates in 1960 (column 1) were at least as high as Finland's rate was in 1750. However, CBRs have dropped significantly in these countries, just as their CDRs have not taken long to decline. Once again, Tanzania is an exception. Although human fertility slackened there during the late twentieth century, demographic momentum has kept the birth rate high. The same is true in Laos. Elsewhere, the fall-off has been dramatic. Whereas 200 years were required for Finland's CBR to go down from the middle 40s to the middle 20s, Algeria, Bangladesh, Ecuador, and Thailand – not to mention many other countries – accomplished the same feat in about four decades (Table 2.2, column 4).

Table 2.2 Birth and death rates in selected countries: 1960s–2001

Country	1960–1965		1970–1975		1985–1990		2001	
	CBR	CDR	CBR	CDR	CBR	CDR	CBR	CDR
	(1)		(2)		(3)		(4)	
Tanzania	49	22	50	18	48	13	39	18
Laos	45	23	45	22	46	16	37	13
Bangladesh	47	21	48	20	40	14	28	9
Bolivia	46	21	45	18	42	14	30	8
Ecuador	45	13	39	10	32	7	24	6
Algeria	50	18	46	14	37	8	23	5
Thailand	41	10	33	8	22	7	15	5
Chile	34	11	25	8	23	6	17	6
Poland	17	7	19	9	16	10	10	9
Sweden	16	10	13	11	13	12	10	10

Source: World Bank (1991, pp. 6, 22, 36, 60, 88, 174, 248, 296, 304, and 316) for 1960s through 1990; World Bank (2003b, pp. 38–40) for 2001.

The schematic representation of the adjustment from high CBRs and CDRs to low rates contained in Figure 2.3A corresponds to the demographic transition that began in most of the developing world during the twentieth century. This transition has started with a CDR plunge that is so sudden that death rates have gone below the level observed at t_1 (i.e., around 10 per thousand) for a while. A rate of 5 or 6 per thousand reflects the low average age of the population (resulting from diminished mortality of infants and young children), which also accounts for demographic momentum. For some time after t_0, the CBR does not change very much, because the TFR remains elevated, the effects of diminishing fertility are largely offset by demographic momentum, or both. Obviously, natural increase accelerates as long as declines in the CBR do not come close to matching those in the death rate. However, this part of the transition is over once human fertility has fallen toward and stayed near the replacement level. Once this happens, natural increase is being driven mainly by demographic momentum. This momentum is not spent until t_1 in Figure 2.3, when human numbers have reached their peak.

The Revolution in Human Fertility

While declining mortality catalyzes the demographic transition, the transition's duration is governed by how quickly human fertility falls from its pre-transition level. The numbers reported in columns 1 and 2 of Table 2.3 show that a large part of humankind truly experienced a revolution in human fertility as the twentieth century was drawing to a close. In little more than one generation, TFRs in places like Algeria and Bangladesh halved. The median number of children for female Ecuadorians was five in 1980; for their daughters, 21 years later, the number was three. In Thailand, where the TFR was 3.5 in 1980, births per woman have actually fallen below the replacement level. Even Tanzania and Laos have experienced reductions in human fertility.

Needless to say, these declines in family size could not have occurred without the use of contraception, and the inverse relationship is strong between TFRs and the prevalence of birth control (Table 2.3, column 3). This prevalence is low in Tanzania and Laos, where there is high fertility. But where women have fewer children, they invariably use contraceptives. This is true even where the majority of them belong to religious faiths led by men who oppose birth control – for example in Ecuador, where most people are Roman Catholic, and in Algeria and Bangladesh, which are Muslim nations. The World Bank (2003b) does not report contraceptive prevalence for Chile, Poland, and Sweden. But these are countries where women routinely do what is required to avoid having children, so the data are not worth collecting. To quote Cole Porter, "Everyone's doing it!"

That birth control is common in places with low TFRs does not prove that the former is a fundamental, as opposed to a proximate, cause of the latter. Instead, contraception is something used by women who have decided to limit fertility. This decision is influenced by various factors, some of which – like religious affiliation – are non-economic. However, diminished childbearing is mainly related to the

Table 2.3 Fertility decline between 1980 and 2001 and causal factors in 2001

Country	1980 TFR (births per woman) (1)	2001 TFR (births per woman) (2)	Percentage of females 15–49 years old that use contraception (3)	Urbanized percentage of population (4)	GNI per capita (PPP $) (5)	Illiterate percentage of adult females (6)	Infant mortality rate (deaths per 1000 births) (7)
Tanzania	6.7	5.2	25	33	520	32	104
Laos	6.7	4.9	25	20	1,540	46	87
Bangladesh	6.1	3.0	54	26	1,600	69	51
Bolivia	5.5	3.8	49	63	2,240	20	60
Ecuador	5.0	2.9	66	63	2,960	10	24
Algeria	6.7	2.9	51	58	5,910	42	39
Thailand	3.5	1.8	72	20	6,230	6	24
Chile	2.8	2.1	Unavailable	86	8,840	4	10
Poland	2.3	1.3	Unavailable	63	9,370	0	8
Sweden	1.7	1.6	Unavailable	83	23,800	Unavailable	3

Source: World Bank (2003b), pp. 100–102 for TFR and contraceptive prevalence; pp. 156–158 for urbanization; pp. 14–16 for GNI per capita; pp. 88–90 for female illiteracy; and pp. 112–114 for infant mortality.

economic circumstances of a woman and her family. All interrelated, these circumstances include urbanization, living standards, female literacy, and infant mortality.

As is evident in Table 2.3, there is an inverse relationship between fertility and urbanization (column 4). The reason why rural families tend to be larger, in the developing world and elsewhere, is that there is always a need for an extra pair of hands to fetch water or firewood, tend livestock, and so forth. Incentives are different in urban settings, where fewer people raise large families merely out of a desire to have a large pool of family labor. Of course, access to contraception is superior in urban settings than in the countryside, as are opportunities for education and employment.

In the 10-country sample, Laos is clearly a place where low urbanization helps to explain a high TFR. In contrast, three out of five Ecuadorians live in cities, which is a reason why the median number of children per woman has fallen to three. Bangladesh and Thailand seem to be outliers. But the former country, where three-quarters of the population are categorized as rural, is a little misleading. As humorist P.J. O'Rourke has gone to the trouble to find out, average population density in Bangladesh, where 133 million people live (World Bank, 2003b, p. 14), is a little above 1,000 per square kilometer. This happens to be comparable to the population density of Fremont, California, a suburban community located along the southeastern shore of the San Francisco Bay. In other words, much of the Bangladeshi countryside is about as crowded as metropolitan areas in affluent parts of the world are (O'Rourke, 1994, p. 53), which has an effect on human fertility. Thailand is more interesting, a place where other factors driving down the TFR more than offset the impacts of limited urbanization.

Like urbanization, living standards, as gauged by either gross national income (GNI) or gross domestic product (GDP) per capita (Table 2.3, column 5),[3] rises as development takes place and is inversely related to human fertility. Quite poor in the 1970s, Thailand now has a GNI per capita of more than $6,000, which helps to explain why its TFR is below the replacement level. In Laos, just to the northeast, urbanization is comparable, though living standards are more than 75 percent lower. As a result, Lao women have nearly five children on average.

No factor affects human fertility more than female empowerment, as indicated by educational attainment. Since women with more years of schooling enjoy a wider range of employment options outside the home, there is a corresponding rise in the opportunity cost of time spent raising children. Chile and Thailand are two places where female illiteracy practically does not exist, which facilitates women's participation in the labor force (Table 2.3, column 6). The latter country is especially interesting because its TFR is comparable to that of China, where draconian policies have been applied to limit fertility. In contrast, one of every two Laotian women can neither read nor write. This circumstance, along with the armed conflict and economic stagnation their country has gone through in recent decades, explains why

---

[3] In this chapter, the purchasing-power-parity (PPP) measure of GNI per capita is used. This is a good indicator of average living standards since it takes into account whether prices in any given country – especially for non-traded services – are higher or lower than prices in the United States (Perkins *et al.*, 2001, pp. 30–33).

they have many children on average. Something a little more subtle to observe is that Algeria's TFR, 2.9 births per woman, is actually a little high in light of the country's living standards – GNI per capita of nearly $6,000. This is largely because of the low status of women, as indicated by elevated female illiteracy.

To recapitulate, women who are empowered, because they are educated and have access to jobs outside the home, are generally in a much better position to decide how many children to have than are women with neither schooling nor jobs. Decisions at the individual level depend on where a woman lives (i.e., in a city or in the countryside), how much she and other members of the household earn, and other economic circumstances. Also having an effect is the frequency with which children die. In fact, the inverse relationship between the infant mortality rate (Table 2.3, column 7) and the TFR comprises clear proof that childbearing is, indeed, a conscious decision, for empowered women at least. As the chances that a newborn will live past his or her fifth birthday go up, such women are less inclined to go through an extra pregnancy or two merely so that they will end up with the desired number of children. There is no better evidence that, once impediments to female empowerment are overcome, human fertility is driven by choices made by people desiring a better life, not Malthus's principle of population.

2.3 Trends in Human Numbers, Past and Present

As the data presented in Tables 2.1 and 2.2 indicate, the course of demographic transition varies considerably from country to country. Specific observations about population growth and its determinants in Africa, Latin America, and other parts of the globe are offered later in this book. The focus in the paragraphs that follow is on long-term trends for the world as a whole.

The most obvious thing to notice about these broad trends is how quickly the population has shot up in recent centuries. Ten thousand years ago, our ancestors, who took advantage of the end of the last Ice Age by starting to cultivate crops, only numbered 5 million (Deevey, 1960, cited in Matras, 1977, p. 34). By Roman times, eight millennia later, about 250 million were alive (Clark, 1967, cited in Matras, 1977, p. 36). During the next 1,500 years, Asia, Europe, and other parts of the world suffered occasionally from devastating plagues. Human numbers grew slowly, only reaching 427 million in 1500 (Table 2.4, column 7). The only place to experience demographic contraction since then is Latin America, where most of the indigenous population succumbed quickly to smallpox, measles, and other diseases that were unknown in the region before European colonization (column 3). Otherwise, unprecedented growth has occurred during the last half-millennium, human numbers having increased 14-fold between 1500 and 2000 (column 7).

In addition to multiplying, the population has changed its distribution around the planet since 1500. When Columbus landed in the Caribbean, just 18 percent of all human beings lived in Europe (including Asiatic Russia), North America, and

Table 2.4 Estimated world population by region since 1500 (millions)

Date	Africa (1)	Asia* (2)	Latin America (3)	Europe* (4)	North America (5)	Oceania (6)	World (7)
1500	85	225	40	74	1	2	427
1750	106	498	16	167	2	2	791
1800	107	630	24	208	7	2	978
1850	111	801	38	284	26	2	1,262
1900	133	925	74	430	82	6	1,650
1950	217	1,335	162	572	166	13	2,465
1975	400	2,262	318	822	243	21	4,066
2000	780	3,460	513	960	314	30	6,057

*Russia's entire population is included in European totals, not Asia's.
Source: Matras (1977, p. 36) for 1500 through 1950; UNDP (2002b, pp. 162–165) for 1975 and 2000.

Oceania (Table 2.4, columns 4–6). More than four-fifths of the population was in Africa, Asia (excluding Russia), and Latin America (columns 1–3). The latter share declined slightly during the next 250 years or so. It then fell sharply, as Europe and other parts of the world with predominantly European populations experienced demographic transition. Annual growth of the combined populations of these places, which was slightly below 0.50 percent in the second half of the eighteenth century, accelerated to nearly 0.75 percent between 1800 and 1850. Even faster growth, 1.00 percent per annum, occurred during the second half of the nineteenth century. By the early 1900s, more than three in every ten people lived in Europe, Russia, Canada, the United States, Australia, and New Zealand (Table 2.4).

With two world wars and the genocides and mass murders directed by Hitler and Stalin, increases in population in Europe and its "offshoots" slowed, although did not turn negative, during the first half of the twentieth century. After the Second World War, growth picked back up, although by this time natural increase was higher in the developing world. This pattern has held ever since, which means that Africans, Asians, and Latin Americans have come to comprise a larger share of the total population. Equal to 73 percent in 1975, that share is now 78 percent, exactly what it was 250 years ago. Given current trends, more than four-fifths of the human race will soon be living outside of Europe and its offshoots, as was the case before the demographic transition began in these places.

As emphasized in the preceding discussion of the demographic transition, natural increase is not about people reproducing more. Rather, it is mainly a consequence of delayed mortality. Asian, Latin American, and now African women have responded to diminished mortality by having fewer children. Even with demographic momentum, the revolution in human fertility has pulled down CBRs, dramatically so during the waning years of the twentieth century (Table 2.2).

People whose knowledge of demographic facts is dated, including anyone who has not examined CBRs and CDRs during the 1990s, are bound to be surprised by

Table 2.5 Projections of the world population in 2025 and 2050

	2025 (billions) (1)	2050 (billions) (2)
Low	7.47	7.87
Medium	7.94	9.32
High	8.39	10.93
With constant fertility	8.65	13.05

Source: UNPD (2001), pp. 38–39.

how close human numbers are to stabilizing in places like Chile and Thailand. In such countries, demographic momentum is now the only reason why there is still a gap, albeit a modest one, between birth and death rates. Likewise, many are unaware that Algeria, Bangladesh, Ecuador, and many other countries, where natural increase equaled or exceeded 2.5 percent as recently as the late 1980s (Table 2.2), could soon be in the same position that Chile and Thailand are in now.

Popular misunderstanding of current population trends cannot be faulted too severely since those trends – especially the recent slide in TFRs and CBRs – have caught professional demographers a little by surprise. That confidence in their predictions has been shaken as a result of this is indicated by the recent disappearance of certain types of demographic forecasts. Estimates of doubling time – obtained, as indicated at the beginning of this chapter, by dividing 70 by the growth rate – used to be a prominent feature of the literature. Such estimates are now very hard to find. Likewise, *World Development Reports*, issued annually by the World Bank, used to contain forecasts of stabilization dates and levels for every country's population. These forecasts were dropped in the early 1990s.

There is now universal agreement that, for humankind as a whole, the inflection point in the S-shaped curve relating population to time during the demographic transition (Figure 2.3B) has been surpassed. Demographers now concern themselves mainly with predicting how quickly, not whether, growth in human numbers will decelerate. As time passes, their projections of future growth are becoming smaller.

One standard source of information is the Population Division of the United Nations (UNPD). During the early 1990s, for example, it offered three forecasts of human numbers for 2025: 7.85 billion (low), 8.47 billion (medium), and 9.08 billion (high). Also provided was a forecast of what the population would be if human fertility never varied: 10.44 billion (UNPD, 1993, pp. 284–285). Mainly because TFRs have indeed declined, and more than was anticipated, these forecasts have had to be revised downward. The current estimates for 2025 are 7.47 billion (low), 7.94 billion (medium), and 8.39 billion (high), as well as the constant-fertility figure of 8.65 billion (Table 2.5, column 1). Note the 17 percent decline that has occurred in just 8 years in the constant-fertility estimate, which should be treated only as a benchmark. It is also significant that the newer constant-fertility estimate, 8.65 billion, is just a little higher than the medium forecast from the early 1990s, which was 8.47 billion. Clearly, TFRs are falling, with far-reaching demographic consequences.

In the middle 1990s, Avery (1996) predicted that the human population would peak in 2045, at 9.00 billion. At the time, this forecast was considered outside the mainstream, well below the consensus estimates. Now, just a decade later, it does not look implausible at all. The UNPD's medium forecast is for human numbers to reach 9.32 billion in 2050 (Table 2.5, column 2). The low forecast is for the world's population to rise at an annual rate of 0.2 percent between 2025 and 2050, reaching 7.87 billion halfway through the twenty-first century. There is a very real possibility that, rather than underestimating increases in population, Avery (1996) was overestimating them. The chances are good that, because of the revolution in human fertility that was well under way in the late 1900s, today's college students will see the day when human numbers stop growing.

2.4 Food Consumption and Income

As observed at the beginning of this chapter, predicting the future consumption of food used to be a straightforward exercise. Standards of living changed very little from one year to the next. Aside from fluctuations caused by bad harvests every now and then, commodity prices were similarly stable. Accordingly, forecasting how much would be eaten in a country or throughout the world at some later date involved little more than extrapolation of trends in population.

With human numbers growing more slowly, trends in food consumption are coming to be driven more and more by other factors. Some of these are demographic. For example, the incidence of pregnancy has an effect, as has distribution of the population among different age cohorts. As the demographic transition proceeds, fewer women are pregnant at any given time, which cuts down on food requirements. Also adults, who eat more than children do, come to comprise a progressively larger portion of the population, which leads to a rise in per capita consumption (Foster and Leathers, 1999, pp. 173–174). On the other hand, people of retirement age generally eat less on average than adolescents and young and middle-aged adults. Patterns of food consumption in Japan, for instance, reflect the fact that old people comprise a large part of the country's population, as they do wherever the demographic transition is far advanced.

Economic factors also have gained importance as a driver of consumption trends. As indicated in the next chapter and other parts of the book, agricultural development puts downward pressure on commodity prices. As food becomes cheaper relative to other goods and services, household budgets are reallocated, with more food consumed though with food purchases claiming a smaller share of total earnings. In addition, changes in income matter.

The very poor, whose earnings are not enough to cover the cost of a minimally adequate diet, subsist almost entirely on starchy carbohydrates and often consume food (and water) that is contaminated with microbes and other harmful things. As living standards rise, per capita consumption of staple carbohydrates increases. Furthermore, people opt for diets that are safer and more varied. In affluent settings, diets are remarkably diverse, with a wide range of fruits and vegetables, livestock products, as well as grain-based foods being eaten. In addition, aversion to products that may threaten human health is very strong, even when the threat is miniscule.

Income growth does not cause purchases of each and every food item to rise. As people put destitution behind them, for example, they often give up the cheapest starchy carbohydrates, like cassava (also called manioc), in favor of more palatable substitutes, including rice as well as bread, noodles, and other products made from wheat (Timmer *et al.*, 1983, pp. 22–35). Things like cassava are inferior goods, to use an economic term, in the sense that people consume less of these as incomes rise. At the opposite end of the earnings spectrum, some people refuse to eat products that are genetically modified or may contain minute pesticide residues. In light of the negligible health risks posed by these products (Ames and Gold, 1989), this choice reflects just how strong the demand for food safety is in affluent settings.

There seems to be no threshold beyond which income growth coincides with diminished consumption of milk, eggs, and other livestock products. To the contrary, demand for these goods consistently increases with income. As indicated in Chapter 11, livestock products account for slightly over half (in cereal equivalents)[4] of the diet in Laos, which is very poor. This share of the diet in Thailand, which has experienced much more development, is nearly two-thirds. In South Korea, which is now one of the world's affluent nations, livestock products account for more than 70 percent of all cereal equivalents consumed.

There are other reasons why aggregate food expenditures exhibit the fundamental property of a normal good, which is that these expenditures rise with earnings. One is that, as income goes up, the opportunity cost of time increases. This causes food preparation at home to be replaced by food preparation elsewhere. In-home consumption of frozen and microwaveable meals grows. Of course, demand for these products is linked to refrigerators, microwaves, and other appliances, which definitely can be categorized as normal goods. Another such good is entertainment, which arguably is part of what one enjoys (besides the convenience of staying out of the kitchen) at many restaurants. As people experiencing better living standards demand more food preparation and other services, the value of raw ingredients as a portion of total food expenditures diminishes. In the United States, this portion has fallen to 20 percent (Gardner 2002, p. 155). Moreover, because livestock products and food services of various sorts are all normal goods, food expenditures never stop responding positively to increases in income.

The sensitivity of food purchases to a rise or fall in earnings is expressed by the income elasticity of food demand. This measure is analogous to own-price elasticity, which is discussed in the Appendix to this chapter. Just as the latter compares relative change in consumption of a particular good to relative change in that good's

---

[4] The cereal equivalents of livestock products are found by multiplying the calorie content of such products by their respective conversion rates – that is, the calorie content of plant products required to produce 1 calorie's worth of livestock. The conversion rates for eggs, poultry, and pork are 4-to-1. For milk, the rate is 8-to-1. Eleven calories' worth of plant products are required to produce 1 calorie of beef or mutton (Foster and Leathers, 1999, p. 177).

price, income elasticity equals relative change in food expenditures divided by relative change in earnings. If a 10 percent change in income causes purchases of food to go up by 4 percent, for example, then the income elasticity of food demand is 0.40.

Unlike own-price elasticity, which is negative because higher prices lead to diminished consumption, income elasticity for a normal good is positive. Furthermore, the elasticity of overall food expenditures with respect to income is generally below 1.00. True, there are individual goods – for example, gourmet vegetarian dishes consumed in expensive restaurants – for which income elasticity of demand is above this threshold. Also, very poor families spending most but not all they have on food may decide to buy food with any incremental money that comes their way, in which case income elasticity will definitely be greater than 1.00.[5] However, these are exceptional instances. As a rule, the proportion of income spent on food declines as income rises. This relationship, called Engel's law, implies that income elasticity of demand for food is generally less than 1.00 and falls as income per capita increases. Where poverty is rampant, a 10 percent increase in earnings might cause food purchases to rise by 6 percent or more, which implies an income elasticity of 0.60 or higher.[6] In contrast, this elasticity is 0.35 or lower in wealthy places, such as the United States.

International patterns of earnings spent on food as well as income elasticity of demand are entirely consistent with Engel's Law. ERS-USDA (2003) reports estimates of both these variables for 117 countries around the world. Algeria and Laos are not in this group, although the other eight nations listed in Tables 2.2 and 2.3 are included. Reported in Table 2.6 are food purchases as a share of total expenditures (column 2) as well as income elasticities of food demand (column 3) for the reduced sample. Each measure declines as living standards rise, of course. An extreme case is Tanzania, where GNI per capita is only $520, spending on food accounts for nearly three-quarters of all expenditures, and income elasticity is 0.80. At the other extreme is Sweden, where everybody eats better and food purchases amount to a much smaller share of total expenditures. Income elasticity of food demand is still positive, although much smaller than the same elasticity in Tanzania.

International variations of the sort on display in Table 2.6 are matched within many individual countries, especially where disparities between the rich and poor are acute. In places with great inequality, one finds a small minority living and eating as well as or better than the average Swede, while large numbers are no better off than the typical Tanzanian. Being an average measure, GNI per capita does not capture this sort of difference. Examination of the distribution of income (Chapter 7) is also required.

---

[5] Define the incremental income, which is spent entirely on food, as Δ. If $F < I$, then $[\Delta \div \frac{1}{2}(F + \Delta + F)] \div [\Delta \div \frac{1}{2}(I + \Delta + I)] > 1.00$.

[6] Among poor households, income elasticity and own-price elasticity of the dietary staple (e.g., cassava or rice) that dominates the diet have similar absolute values. Since there are no close substitutes for the staple and expenditures on it comprise a large share of earnings, the effects of a price change on poor households' purchasing power and their consumption of the staple are comparable to the effects of a change in their incomes.

Table 2.6 Living standards, food purchases, and income elasticity of food demand in selected countries

Country	GNI per capita (PPP $) (1)	Food purchases as share of total expenditures (%) (2)	Income elasticity of food demand (3)
Tanzania	520	73	0.80
Bangladesh	1,600	56	0.73
Bolivia	2,240	43	0.71
Ecuador	2,960	29	0.71
Thailand	6,230	29	0.65
Chile	8,840	23	0.59
Poland	9,370	31	0.58
Sweden	23,800	13	0.36

Source: World Bank (2003b, pp. 14–16) for GNI per capita and ERS-USDA (2003) for food expenditures and income elasticities.

2.5 Demand Trends and Projections

In this chapter, we have examined population growth, past and present. How income and food consumption are related also has been considered. We now turn our attention to current trends in food demand. Presented as well in this section are estimates of what global consumption will be later in the twenty-first century.

Let us begin with current trends. Barring major variation in food prices or tastes, these trends are driven by changes in human numbers as well as per capita consumption. Suppose the population is growing by X percent. Suppose as well that growth in per capita income is Y percent and that income elasticity of food demand is E, which implies an increase in per capita consumption equal to the product of the two. The overall trend in food demand, reflecting demographic expansion as well as better living standards, is:

$$(\{[100 + X] \times [100 + Y \cdot E]\} \div 100) - 100. \tag{2.1}$$

For example, population growth of 2 percent combines with a 6 percent increase in per capita income and an income elasticity of 0.67 to yield an overall trend of 6.08 percent:

$$([100 + 2] \times [100 + 6 \times \tfrac{2}{3}] \div 100) - 100$$
$$= (102 \times 104 \div 100) - 100 = 6.08 \text{ percent}.$$

Let us consider growth trends in a pair of countries discussed earlier in this chapter. One of the poorest places in the Western Hemisphere, Bolivia, is experiencing population increases of 1.8 percent per annum as well as 1.0 percent annual declines in GDP per capita (World Bank, 2003b, pp. 14 and 38). With GNI per capita barely above $2,000, the income elasticity of food demand is 0.71 (Table 2.6, column 3), which is very high. Thus, the yearly increase in food demand is:

$$[100 + 1.8] \times [100 + (0.71 \times \{-1.0\})] \div 100 - 100$$
$$= 101.8 \times 99.3 \div 100 - 100 = 1.1 \text{ percent.}$$

With living standards deteriorating, demand is not growing as fast as the Bolivian population.

A distinct situation presents itself nearby, in Chile. With the country very far along in its demographic transition, population growth is lower: 1.0 percent a year. Although not rising as rapidly as in the recent past, GDP per capita is still going up, by 1.5 percent per annum (World Bank, 2003b, pp. 14 and 38). Average income in Chile, $8,840 in 2001, is appreciably higher than Bolivia's, which means that income elasticity is lower: 0.59 (Table 2.6, column 3). The annual trend in demand for food is:

$$[100 + 1.0] \times [100 + (0.59 \times 1.5)] \div 100 - 100$$
$$= 101.0 \times 100.9 \div 100 - 100 = 1.9 \text{ percent.}$$

Chile is a place where growth in food demand has nearly as much to do with improved living standards as with increases in human numbers.

Moving from measures of current trends in food demand to estimates of demand at a later date is a challenge. One reason, of course, is that projecting world population in the future is problematical. As stressed earlier in this chapter, recent declines in human fertility have caught demographers by surprise, which has obliged them to make significant adjustments every few years in their predictions. Similarly difficult to forecast are living standards in the future. Yet another reason why pinpointing future demand is not easy is that changes in income affect income elasticities. If living standards are going up, for example, then income elasticity is declining. This interaction needs to be taken into account.

In an analysis of global food demand during the twenty-first century, Tweeten (1998) considers the interaction between GDP per capita and income elasticity. Conceding that the latter is being driven down in various parts of the world by income growth, he also points out that ever larger portions of the increase in human numbers are occurring in poor countries, where income elasticities are higher, as the demographic transition winds down in richer places. Due to this latter reality, growth in per capita consumption will hold steady during the next few decades. In particular, average annual growth of 0.3 percent will continue through 2050, although this rate could be as low as 0.2 percent or as high as 0.4 percent.

Since precise estimates of future population, living standards, and annual growth in per capita consumption of food are not available, various projections of food demand in 2025 and 2050 are presented in Table 2.7. Every single figure corresponds to a specific forecast of population during the first quarter or half of this century as

Table 2.7 Growth in food demand (%), from 2000 to 2025 and from 2000 to 2050

	2000–2025 (1)	2000–2050 (2)
Low UNPD population forecast*		
0.2% yearly growth in consumption per capita	30	44
0.3% yearly growth in consumption per capita	33	51
0.4% yearly growth in consumption per capita	36	59
Medium UNPD population forecast*		
0.2% yearly growth in consumption per capita	38	70
0.3% yearly growth in consumption per capita	41	79
0.4% yearly growth in consumption per capita	45	88
High UNPD population forecast*		
0.2% yearly growth in consumption per capita	46	99
0.3% yearly growth in consumption per capita	49	110
0.4% yearly growth in consumption per capita	53	120

*See Table 2.5.

well as a specific assumption about annual increases in consumption per capita. Given the UNPD's medium demographic prediction and 0.3 percent growth in per capita consumption, for example, demand in 2025 will be 41 percent above 2000 levels. If average consumption were static, then demand would increase by just 31 percent, entirely due of course to the rise in human numbers. By the same token, demand would grow by 54 percent during the first half of the twenty-first century if everyone's diet in 50 years were exactly the same as what it is now and the UNPD's medium forecast proved to be correct. But since consumption per capita will be going up, demand will actually be four-fifths above current levels for this same demographic estimate. The only way that demand will not increase by one-half during the next half-century is if average consumption rises slowly (i.e., by 0.2 percent per annum) and the UNPD's low forecast turns out to be true. On the other hand, demand will be at least double its current level if the high forecast is correct, irrespective of the growth in average consumption (Table 2.7).

2.6 Summary and Conclusions

Accustomed as many people are to thinking about trends in human numbers in Malthusian terms, the nature and consequences of the demographic transition now happening in Latin America, Asia, and Africa is not fully appreciated. Rather than increasing because TFRs are going up, population has risen mainly because death rates have fallen dramatically. People have not responded instantaneously to this happy change by cutting back on family size. However, there has been a revolution in human fertility, with women around the developing world having far fewer children on average than their mothers and grandmothers. Due to this revolution, growth is decelerating and there is a very good chance that human numbers will peak well before the end of this century.

This is not to say that food demand will stabilize any time soon. As living standards continue to improve, average consumption will increase. Demand in the middle of the twenty-first century will be at least 50 percent above the current level. Indeed, with faster demographic expansion than the UNPD currently is forecasting, there could be a doubling of food demand before the current century is halfway finished (Table 2.7). Clearly, feeding the human race will still be a major challenge for many more years.

Study Questions

1. Describe Thomas Malthus's principle of population and the view of human behavior on which this principle rests.
2. Compare and contrast overshoot and collapse in Malthus's model of trends in food demand and supply and overshoot and collapse in the systems-engineering model developed for the Club of Rome.
3. Identify critical missing elements of the Club of Rome model and assess the implications of these omissions.
4. What happens at the beginning of the demographic transition? When is the transition at an end?
5. Why does natural increase always occur during the demographic transition?
6. Explain why mortality has declined faster in the developing world during the twentieth century than it did in places like Britain and Scandinavia during the 1800s.
7. What factors have caused human fertility to decline throughout the developing world in recent decades?
8. Describe demographic momentum and its contribution to natural increase.
9. Explain Engel's Law and its implications for trends in food demand in an economy experiencing a rise in average income.
10. Describe likely trends in global food demand during the first half of the twenty-first century.

Appendix:
The Fundamental Economics of Demand

Economic analysis of changes in food consumption is a complicated task given the host of factors that need to be considered. To make the undertaking tractable, some abstraction from complex reality is desirable. With respect to consumption of any particular commodity – corn, for example – it is convenient to suppose, as economists routinely do, that everything influencing consumers' purchases is unchanging – everything, that is, aside from the commodity's price. This assumption – referred

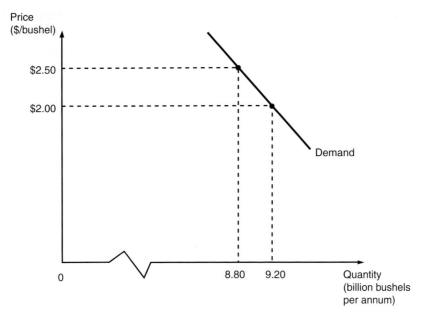

Figure 2.4 US corn demand.

to with a Latin phrase, *ceteris paribus* (other things being equal) – allows for the transformation of a general consumption function, in which all factors affecting people's purchases are represented, into the simple abstraction of a demand curve, which relates consumption of a good solely to its own price. With such a curve identified, growth over time in demand can be examined. This growth is caused by changes in precisely those things, including population size and household incomes, that must be held constant so that a demand curve can be traced out in the first place.

Measuring price on the vertical axis and tracking yearly consumption of corn on the horizontal axis, we can plot out a demand curve for corn in the United States (Figure 2.4).[7] The curve has a downward, or negative, slope, which implies an inverse relationship between price and quantity. For example, a rise in price from $2.00 to $2.50 per bushel causes yearly consumption to fall from 9.20 billion to 8.80 billion bushels, *ceteris paribus*. In contrast, quantity demanded increases if the price declines, again barring any change in other variables affecting consumption.

---

[7] In the standard representation of a relationship between X (the independent variable) and Y (the dependent variable) – $Y = a + bX$, for example – the former is measured horizontally and the latter vertically. If price is influencing consumption rather than the other way around, then mathematical custom is being flouted in the conventional representation of the demand curve, which can be confusing for students being introduced to economics. Price could be tracked on the horizontal axis and quantity demanded on the vertical axis. However, measuring market value vertically, as is done in Figure 2.4, makes sense if one thinks about the maximum price that producers can charge and still sell all their output. That maximum price is a decreasing function of output. This alternative view of the demand curve makes sense if one treats output as predetermined by past prices and other exogenous variables in the supply function, as some agricultural economists do when they model equilibrium in commodity markets.

That the demand curve slopes downward reflects a fundamental characteristic of consumers' preferences. While people are generally happy to receive more of all sorts of things, the satisfaction (or utility) they derive from another unit of any particular item grows smaller as their consumption of that item increases. This characteristic of consumers' preferences, called diminishing marginal utility (MU), explains why the marginal value of a good or service – that is, the maximum amount of money that is offered for the very last, or marginal, unit – falls as the total quantity of the good or service rises. The demand curve's negative slope is a manifestation of this declining marginal value. Diminishing MU also explains why, from the perspective of suppliers, getting people to buy more of something generally requires that its price be lowered.

Own-Price Elasticity

While it can be taken for granted that a demand curve slopes downward, the sensitivity of quantity demanded to price changes varies considerably from product to product and from consumer to consumer. This sensitivity, called own-price elasticity, is found by dividing relative change in consumption by relative change in price.

Since quantity and price always move in opposite directions, own-price elasticity is never positive. If dividing the relative quantity change by the relative price change yields an absolute value less than one, then the elasticity of demand is low. To say the same thing, demand is inelastic, which means that consumption is not very responsive to price changes. Suppose, for example, that a 50 percent decline in price causes consumption to go up by 10 percent; dividing the latter by the former yields an own-price elasticity of -0.20, which is smaller in absolute magnitude than -1.00 (i.e., unitary elasticity). In contrast, the relative change in quantity may exceed the relative change in price, in which case the elasticity of demand is high. For example, a 20 percent increase in price might cause consumption to go down by 40 percent. With own-price elasticity equal to -2.00, which is obviously greater in absolute magnitude than the unitary benchmark of -1.00, demand is elastic. Consumption changes quite a lot in response to lower or higher prices.

As long as two points on a single demand curve have been identified, own-price elasticity can be estimated. Let P and Q_D represent one combination of price and quantity consumed on a particular curve. If the price rises to P', then there will be movement to another point on that curve, with consumption falling to Q_D'. Own-price elasticity is:

relative quantity change ÷ relative price change
$$= [\Delta Q \div \text{average of } Q_D' \text{ and } Q_D] \div [\Delta P \div \text{average of } P' \text{ and } P]$$
$$= [(Q_D' - Q_D) \div \tfrac{1}{2}(Q_D' + Q_D)] \div [(P' - P) \div \tfrac{1}{2}(P' + P)]. \qquad (2.2)$$

In Figure 2.4, P and Q_D are $2.00 per bushel and 9.20 billion bushels per annum, respectively, and P' and Q_D' are $2.50 per bushel and 8.80 billion bushels per annum, respectively. The change in prices, ΔP, is positive, $0.50, while ΔQ_D is negative,

−0.40 billion bushels. Since P and Q_D move in opposite directions, own-price elasticity of demand is less than zero:

$$[-0.40 \text{ billion} \div \tfrac{1}{2}(8.80 \text{ billion} + 9.20 \text{ billion})] \div [\$0.50 \div \tfrac{1}{2}(\$2.50 + \$2.00)]$$
$$= [-0.40 \text{ billion} \div 9.00 \text{ billion}] \div [\$0.50 \div \$2.25] = -0.20.$$

Since the absolute value of elasticity is less than the unitary cutoff, one concludes that US demand for corn is inelastic.

Two demand curves, with very different own-price elasticities, are presented in Figure 2.5A and B. In the case of Jimmy's demand for jam (Figure 2.5A), own-price elasticity happens to be low, which is manifested by his curve's steep slope. In contrast, Betty's consumption of peanuts is highly sensitive to the price of the good, as is reflected by the gentle slope of her demand curve (Figure 2.5B). When the own-price elasticity of Jimmy's demand for jam is calculated, the number we come up with has a low absolute value:

$$[-2 \div \tfrac{1}{2}(8 + 10)] \div [3 \div \tfrac{1}{2}(5\tfrac{1}{2} + 2\tfrac{1}{2})] = [-2 \div 9] \div [3 \div 4] \approx -0.30.$$

With Betty and peanuts, the absolute value of own-price elasticity is high:

$$[40 \div \tfrac{1}{2}(60 + 100)] \div [-0.10 \div \tfrac{1}{2}(1.05 + 0.95)] = [40 \div 80] \div [-0.10 \div 1.00] = -5.$$

Comparing absolute values of these estimates to unitary elasticity, one concludes that Jimmy's demand for jam is inelastic and Betty's demand for peanuts is elastic.

Changes in Demand

As long as there is no variation at all in the number of consumers in a market, their incomes and tastes, and the prices of other goods and services, then consumption of the good being exchanged in that market changes only because that good's price rises or falls. However, any adjustment in the former set of factors brings about an entirely new relationship between consumption and price. In other words, there is a new demand curve, one that is either closer to or farther from the vertical axis.

As shown in Figure 2.6, falling demand is represented by movement of the demand curve toward the vertical axis, which means that less is consumed at any given price. Demand can fall for various reasons. A contraction in the population of consumers is one reason. Another is lower prices of substitute goods. For example, demand for beef declines if chicken becomes cheaper. There are also complementary goods. If these grow more expensive, then less will be consumed at any given price. A case in point is hot dogs and buns; if the price of the former goes up, then demand for the latter falls. In addition, income influences demand. If consumers have diminished earnings, then demand for most goods falls. Finally, economists recognize that tastes – a catch-all category that encompasses everything

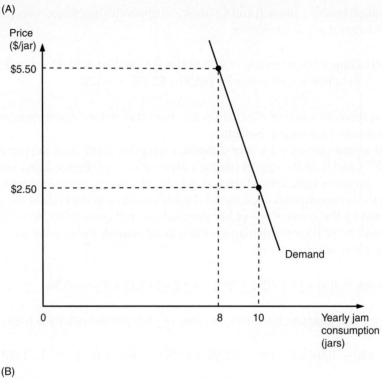

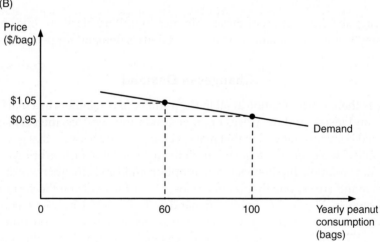

Figure 2.5 Demand elasticity. (A) Jimmy's demand for jam. (B) Betty's demand for peanuts.

from whims influenced by passing fashion to dietary requirements related to age and other demographic variables – influence demand. Diminished food consumption caused by an aging of the population would be regarded by economists as a fall in demand caused by changing tastes.

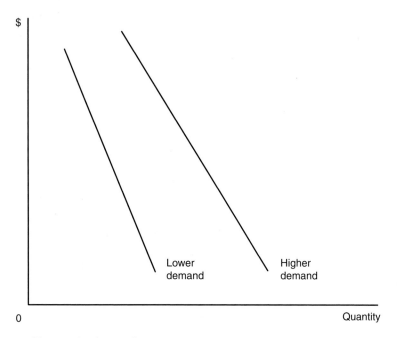

Figure 2.6 Changes in demand.

Just as it can decline, demand can increase. As indicated in Figure 2.6, the latter is represented by movement of the demand curve away from the vertical axis, which means that more is consumed at any given price. The causes of this sort of shift are a mirror image of the causes of falling demand. One of these is population growth, which of course is a major focus of this chapter. Demand also increases because of higher prices of substitute goods and lower prices of complementary goods. It can also go up because of changing tastes. Likewise, income growth causes consumption of normal goods to be higher at any given price.

Just as own-price elasticity indicates the sensitivity of consumption to price changes, *ceteris paribus*, income elasticity indicates how much demand for some good changes in response to higher or lower earnings, all else remaining the same. Suppose, for example, a 5 percent rise in earnings leads the Smith family to increase its spending on food by 2 percent. Under these circumstances, its income elasticity of food demand is 0.40 (2 percent divided by 5 percent).

If spending on food is known at a pair of earnings levels, then income elasticity can be estimated. Let F and Y represent one combination of food expenditures and income, respectively, and F' and Y' represent another combination. Income elasticity is:

relative change in food purchases ÷ relative income change
$$= [\Delta F \div \text{average of } F' \text{ and } F] \div [\Delta Y \div \text{average of } Y' \text{ and } Y]$$
$$= [(F' - F) \div \tfrac{1}{2}(F' + F)] \div [(Y' - Y) \div \tfrac{1}{2}(Y' + Y)]. \tag{2.3}$$

For the sake of illustration, consider a family of six in Bangladesh, where GNI per capita is $1,600. Its earnings increase, from $9,500 to $10,500. This causes its food expenditures to rise from $4,815 to $5,185. Its income elasticity of demand is:

$$[(5{,}185 - 4{,}815) \div \tfrac{1}{2}(5{,}185 + 4{,}815)] \div [(10{,}500 - 9{,}500) \div \tfrac{1}{2}(10{,}500 + 9{,}500)]$$
$$= [370 \div 5{,}000] \div [1{,}000 \div 10{,}000] = [7.4\%] \div [10.0\%] = 0.74.$$

3

The Supply Side
Agricultural Production and Its Determinants

With human numbers and living standards continuing to rise, the amount of food people eat is nowhere close to cresting. For demand growth not to outstrip increases in supply, production will have to go up by at least one-half during the next 50 years. In fact, output may have to double by 2050 if the trend of recent decades toward diminished food scarcity and lower commodity prices is to be sustained.

One way to raise production is to make wider use of farming's main natural resource, which is land. Another is to increase the application of other agricultural inputs, such as labor, machinery, fertilizers and pesticides, and water. Also possible are changes in the technology that farmers use to convert inputs into edible output. Each of these responses to demand growth has been an important feature of agricultural progress in various parts of the world during the last century. Extensification (i.e., using more land for crop and livestock production) and intensification (resulting from increased application of non-land inputs, technological change, or a combination of the two) will both contribute to increased food supplies during the next 100 years.

This chapter addresses increases in food production and how these are achieved. Special attention is given to the second half of the twentieth century. The consumption growth that occurred during this period, when human numbers increased from 2.47 billion to 6.06 billion (Table 2.4), was unprecedented. However, increases in supply exceeded those in demand – not by a small margin, but significantly so. Continuation of the trend toward diminished scarcity during the next 50 years is certainly possible, though not guaranteed. In this chapter, we assess prospects for further extensification of agriculture. Also examined is the role of intensification, especially the yield increases resulting from technological innovation. How much people eat and how much they pay for food during the next few decades hinge on these responses to demand growth. Indeed, whether some of us survive depends on how well supply keeps up with demand.

Before trends in global agricultural output and the driving forces that underlie them are described, we highlight some special features of farming – its sensitivity to environmental conditions and its close resemblance to the economic ideal of perfect competition, to be specific. The rest of the chapter is about trends in food supply, those in recent decades as well as those expected in the years to come.

3.1 The Nature of Agriculture

For most people in affluent places, farming is a vaguely old-fashioned activity, something engaged in long ago by grandparents or great-grandparents. Yet agriculture represents something quite new in human experience, a development that is no more than 10,000–12,000 years old. For innumerable millennia before people raised crops and cared for livestock, our distant ancestors fed themselves exclusively by hunting and gathering.

To be sure, hunter-gatherers came up with countless incremental improvements. Before the agricultural revolution, people undoubtedly figured out that removing weeds would promote the growth of food-bearing plants. Simple observation also would have revealed that plant growth is more robust where some manure or a discarded fish has been left to decay. But from innovations based on observations like these to agriculture, with domesticated crops and livestock, a great leap was required.

Until that leap occurred, human existence itself was tenuous. Even in the most luxuriant setting, such as a rainforest, extensive tracts of land were (and are) required to support a small, nomadic group of hunter-gatherers. Food storage was rare, and survival was constantly jeopardized by drought, floods, insect plagues, frost, and disease, not to mention the predations of rival human bands. None of these threats were very predictable and the total population was miniscule. As indicated in the last chapter, approximately 5 million people were alive 10,000 years ago, not long after farming began. Moreover, urban culture and all its arts and industry were inconceivable as long as a few hunter-gatherers, each sustained by a substantial amount of natural resources, were scattered across the face of the Earth.

The problem people faced before the dawn of agriculture was that they exercised almost no control over the biological processes that provided them sustenance. Today, these same processes remain the underlying source of food, although biological nature has come to be manipulated intensively by farmers. Due to this manipulation, far fewer resources are needed to feed one person. The agricultural revolution also allowed human numbers to rise. By the standards of the age, there was a population explosion in various parts of the Western Hemisphere as corn farming, which originated in southern Mexico and Central America, spread northward and southward. As Diamond (1997, pp. 86–92) emphasizes, the existence of an agricultural population that could produce more food than it required for its own survival made incipient urbanization possible. It is significant, for example, that the most ancient towns excavated in the Middle East – such as ancient Jericho, which is 9,000 years old – were founded after the agricultural revolution (Cipolla, 1965, pp. 18–25).

Farmers' manipulation of biological processes takes myriad forms. Domesticated animals are protected from parasites and pathogens, and their reproduction is controlled. The soil in which the seeds of crops are planted is managed, both to prevent its being eroded by wind and rain and to maintain its fertility. Fertilization traditionally meant the application of manure, which contains the three macronutrients required for plant growth – nitrogen, phosphorus, and potassium. These days, it is much more common to replace the macronutrients used up during the course of crop production by applying chemical fertilizers, which are an industrial product. Insecticides are another such product. Ever a concern, weed control has long been accomplished by cultivation (e.g., hoeing). Here again, many farmers instead now use an industrial input – herbicides, to be specific. Finally, just as fertilization replenishes soil fertility, rainfall can be augmented by irrigation, the additional water applied to farm fields being extracted from underground aquifers or diverted from rivers, lakes, or streams.

Although nature is managed in various ways on a farm or ranch, crop and livestock production remains essentially biological, which has a number of ramifications. One of these is that large volumes of organic residuals are produced along with edible commodities. For example, the rice or corn we eat comprises a very small portion of a rice or corn plant. Domesticated animals may be fed some of the stalks and leaves. But much more of this material is left in the field to decompose, thereby adding to the organic content of soil. This is largely beneficial since decayed organic matter holds water and nutrients, which crops utilize. Beneficial use is also made of the organic residuals of livestock production. To repeat, livestock wastes contain the macronutrients that feed plants. However, these macronutrients are not utilized exclusively by the crops farmers raise. Whenever the rate at which manure (or fertilizer) is applied to fields exceeds plant uptake, there is a good chance that nutrients attached to eroded soil or dissolved in run-off water will find their way into waterways or aquifers, which has various consequences. For example, algae and other aquatic organisms proliferate, which deprives fish of dissolved oxygen.

The water pollution resulting from excessive concentrations of waste byproducts is an impediment to the industrialization of livestock production, which involves confining large numbers of animals to small spaces.[1] As a matter of fact, wide expanses of rangeland continue to be grazed by cattle and other domesticated species. Crop farming is similarly extensive, although selected horticultural products are raised intensively in greenhouses and other controlled settings.

Agriculture, Soils, and Climate

Agriculture's massive land requirements mean that the sector is very exposed to the elements. A sustained downpour and associated flooding can wash away an entire harvest. In addition, extreme heat is a problem, as is excessive cold. Far from

---

[1] Also arousing much concern are the conditions in which confined livestock live (Pollan, 2002).

the equator, in places like northern Russia and Canada, the growing season can be curtailed by heavy frosts in the late Spring or early Autumn. This problem is less serious farther from the polar regions. Indeed, temperate-zone winters are ideal since these are strong enough to induce significant mortality among insects and other pests. Something else that benefits agriculture in the middle latitudes is the availability of soils that are fertile and not too susceptible to erosion and other forms of degradation (Sanchez and Logan, 1992). Also favorable are the many hours of daylight during the growing season, which promote photosynthesis and therefore rapid plant growth. It is no mystery, then, that temperate settings with adequate rainfall (or ready access to irrigation sources) and fertile soils are important sources of corn, wheat (which was originally cultivated in the Middle East), and other staple grains. Among these areas are the Midwestern United States and the Argentine *pampas*.

Farming nearer the equator is more challenging. Land quality is one problem. Stewart *et al.* (1991) estimate that approximately 57 percent of tropical soils are Alfisols, Oxisols, and Ultisols, each of which tends to suffer from chemical limitations (e.g., low fertility, aluminum toxicity, etc.), have undesirable physical properties (e.g., high susceptibility to erosion or compaction), or both. Entisols, Inceptisols, and Mollisols, which lend themselves well to farming, comprise 32 percent of all soils within 23½ degrees of the equator, as compared to 46 percent in the temperate zone (Sanchez and Logan, 1992). Another problem in the tropics and subtropics is that solar radiation is intense, which accelerates evaporation and transpiration. Conditions in many parts of the low latitudes are arid or semi-arid, which means that irrigation is a prerequisite for farming. One such dry setting with irrigated agriculture is the Punjab, which straddles the Indian–Pakistani border and where the Green Revolution (see below) has had a substantial impact.

Other parts of the tropics and subtropics are rainy, which makes plant growth profuse. One problem, though, is that precipitation tends to occur in heavy storms, which wash away soil and nutrients rapidly. Also, organisms that harm people or the species they consume thrive in places that are always warm and humid. Crude death rates as recently as 40 years ago in places like Laos and Tanzania were still well above the rates in Scandinavia 250 years ago (Tables 2.1 and 2.2). No doubt, winters that were long and hard cut back on longevity in the past. However, the lack of a period when sub-freezing temperatures kill off mosquitoes and other vectors that transmit parasites and pathogens is a clear disadvantage. Diseases such as malaria are responsible for widespread morbidity and mortality in the tropics and subtropics. Likewise, black sigatoka, which attacks bananas, is symptomatic of the toll taken by plant diseases. Furthermore, constant labor is required to control weeds, which always threaten to dominate farm fields at the expense of crops in the low latitudes.

Two agricultural species of major importance are typical of the tropics and subtropics. One is sugarcane, which grows so rapidly that it usually out-competes weedy species in the fields where it is planted. Originally domesticated in Asia, this crop gradually migrated westward. Sugar's introduction in the Caribbean and

Brazil, where the indigenous population collapsed during the 1500s (Chapter 2), coincided with the beginnings of the trans-Atlantic slave trade. The other species is rice, which has long been the mainstay of Asian diets. Production of this grain is water-intensive, especially if rice seedlings are transplanted to flooded fields (i.e., paddies) as a weed-control measure. This transplanting also makes rice production labor intensive. Asian cultures were shaped in no small way a long time ago by the need to organize the hydrological and human inputs of rice farming.

The most pleasant climates in the world are found in another kind of tropical and subtropical setting – hills and mountain valleys. These places have permanently spring-like climates, with warm afternoons and cool nights, and have fewer insects and other pests than at lower altitudes. Conditions are ideal for the production of coffee and tea, which are the traditional cash crops of hilly areas in the low latitudes. Also, horticultural products, including cut flowers, are now finding their way to international markets from upland farms in Meso America and the Andes, Southeast Asia, and East Africa.

Specialization and Diversity

Due to variations in temperature, rainfall, and soil fertility, agriculture is extremely heterogeneous. However, diversity is not just the result of environmental differences. From place to place, the economic forces and public policies that influence the level and mix of farm inputs and outputs vary considerably.

Consider, for example, how the same crop, corn, is grown in two very different places. One of these is Tanzania, which is one of the poorest places in the world, and the other is the United States, which of course is one of the richest. The typical farm in the former country occupies a mere 10 hectares[2] of un-irrigated ground. Capital (i.e., the manufactured portion of productive capacity) is similarly meager, comprising hoes, sickles, and other equipment worth about $50. It is also exceptional to have draft animals (e.g., oxen), so agricultural equipment is wielded with muscle power provided by the farm household itself. Human muscle power being as limited as it is, the household never cultivates more than two of its 10 hectares at any time. The harvest from these two hectares is barely enough to satisfy the nutritional needs of the farming family, and oftentimes less than that. Without a marketable surplus, there is no cash for buying improved seeds and commercial fertilizers, which are needed to raise yields. Even if there were a surplus, trade would be severely hampered by the poor state of roads and other infrastructure.

The situation could hardly be more different in the Midwestern United States, where a farming family typically possesses 500 hectares as well as $500,000 worth of tractors, combines, and so forth. Fossil fuels and electricity power all this equipment and machinery, which allows household members to work the entire holding on their own. Practically all output is sold and the use of improved seeds,

---

[2] There are 2.471 acres in 1 hectare, which is the area of a square with 100-meter sides.

commercial fertilizers, and other purchased inputs is routine. Production technology is up to date, with inputs converted into grain with great efficiency. Transportation and communications infrastructure is similarly advanced, which allows enormous amounts of inputs and output to move to market at minimal cost.

Disparities of comparable magnitude also exist between irrigated rice production in the Mississippi Delta of the United States and upland rice farming in West Africa. The former is characterized by large, capital-intensive operations that take advantage of the latest technology and rely heavily on purchased inputs. In contrast, subsistence farmers in Africa tend to practice shifting cultivation.[3] This process begins with the clearing of bush forest with saws, axes, and cutlasses (machetes) and the burning of residue, which releases nutrients into the soil. Rice is then grown for a year without additional fertilizers, followed by a couple of years of producing vegetables, root crops, or tubers. To restore soil fertility after this cycle, the land should be left undisturbed for 15–20 years. Thus, shifting cultivation requires extensive tracts of land per farmer to be sustainable, since 15–20 hectares must lie fallow for every cultivated hectare. However, this system, which makes no use of chemical inputs, becomes unsustainable if the fallow period is curtailed, as frequently occurs because of population growth.

Environmental variation creates differences among small, impoverished farms in the developing world. Yet these farms have much in common with one another. In particular, they are self-sufficient to a high degree, which is not coincidental at all. Economic progress requires specialization and trade, with each agent (be it a household or an entire nation) concentrating on the activity in which it has a comparative advantage (Chapter 6) and making exchanges with other agents with comparative advantage in different activities. In poor, rural areas, little distinguishes one household from another, which means that commercial activity among them yields modest gains. Also, trade is impeded by severe deficiencies in infrastructure. Thus, much of the African, Asian, and Latin American countryside is mired in isolation, poverty, and hunger.

Specialization and trade are salient features of the food economy in affluent parts of the world. Farms themselves are diverse. Those with rich soils produce crops while livestock grazing occurs where land quality is low and water is scarce. A rice farm in Japan, which may encompass as little as 1 hectare, is highly mechanized so that the operator can spend most of his time earning non-agricultural income. In contrast, a sheep station in inland Australia can be 1,000–100,000 times larger and, due to its being located far from major cities, the operator ranches full time. Of greater importance is that an evolution has occurred from a food economy comprising self-reliant operations to one made up of farms focused on crop and livestock production proper as well as a wide array of firms specializing in the production of inputs and services used on and by those farms. In times past,

---

[3]Another term for bush-fallow farming of this sort is swidden agriculture. Also synonymous is slash-and-burn farming, which can have pejorative connotations.

on-farm processing of food (e.g., pickling, canning, and smoking) was routine. Likewise, it was not unusual for an individual operator to produce various crops and livestock, and sell directly to consumers, in public markets for example. Nowadays, many farmers raise just one or two commodities. Rather than producing inputs like seeds and fertilizer on their own, they buy these from agribusinesses, which enjoy economies of scale and other advantages. Most farmers also rely on specialized firms for marketing services, even though self-reliance for storage and transportation is an option. This reflects the growing predominance of services throughout the economy (Chapter 7).

Populist suspicion of agribusiness can be traced back to the nineteenth century. During the 1870s and 1880s, for example, local monopolization of rail transport was the rule in the farm country west of Chicago; as a result, freight rates were high and service was limited, much to the detriment of grain producers (Cochrane, 1979, pp. 93–94). Likewise, sanitation was poor in a number of meatpacking plants, as Upton Sinclair and other muckrakers let the US public know in the early 1900s. A legal and regulatory framework now exists to curb such excesses, thanks to political action by populist farmers and progressive reformers around the turn of the twentieth century. In part because of anti-trust legislation, pure-food laws, and other such arrangements, there is little evidence today of monopolization and other abuses. Nevertheless, populist suspicion remains in some quarters. The critics of agribusiness pay little heed to the rural poverty that persists, in Sub-Saharan Africa and other places, where there are no specialized firms to provide inputs and marketing services to farmers. By the same token, they do not concede that farms in Europe, North America, and other prosperous settings are as prosperous as they are precisely because agribusiness is robust.

The specialization and trade among farms and agribusinesses that create rural prosperity in places like the United States have been made possible thanks to sizable public investment. Government constructs and maintains the roads and navigable waterways along which trade flows. Likewise, commerce would be very hard to conduct without courts and related institutions that enforce contracts and property rights – what can be called the legal and institutional infrastructure that makes the whole market system possible. In addition, public support has been given to agricultural research, which has dramatically improved technology for crop and livestock production, and extension (or outreach), which transfers this technology to producers. Public education, including agricultural extension as well as general schooling, creates the foundation for the application of science to food production, in the sense that educated farmers are better able to learn about and apply technological advances. Also, the spread of irrigated crop production has been enabled by dams and canals built by the public sector.

The reason why government usually takes the lead in building transport and irrigation systems, establishing the legal and institutional infrastructure that markets require, and supporting research, extension, and education is that many of the services yielded by these investments are what economists call public goods. Two textbook illustrations are military defense and environmental quality, both of which share a critical characteristic – the high cost of excluding non-paying consumers. If a paying subset of the population were to be defended, for example, then denying the same protection to those who do not pay would be impossible. The same holds for clean air. Since some or all the beneficiaries of defense, environmental quality, and other public goods cannot be made to pay – to repeat, because of high exclusion costs – private firms have little interest in providing these goods. Accordingly, government has to be the main supplier.

To summarize, a well-developed food economy is made up of three elements. One of these, of course, is production agriculture, which comprises farms and ranches where crops and livestock are raised. A second element is agribusiness, the emergence of which reflects the increased specialization and trade which create prosperity. The third is government, which furnishes infrastructure, technology, and other public goods used by production agriculture and agribusiness. None of the three elements is superfluous and, if any one were to disappear, the remaining two would suffer.

Notwithstanding the interdependence among farms and ranches, agribusinesses, and the public sector, emphasis is placed in this chapter and throughout this book on production agriculture. Government as a source of public goods, as well as laws and regulations that influence market performance, is an important topic for us. But other than to compare the benefits and costs of, say, agricultural research supported by the public sector, we do not address the motivation of governmental investment or policy-making. Neither does this book focus very much on agribusiness, for the simple reason that little distinguishes the enterprises that supply agricultural inputs and marketing services from firms in other parts of the economy.

Some enterprises engaged in agricultural production are very similar to firms outside the food economy. In a laying hen enterprise, for example, trucks deliver feed at one end of a building complex that houses 5,000 birds or more. This feed is distributed throughout the building on conveyor belts to hens that spend their lives in cages with tilted floors. The eggs they lay roll down these floors onto another network of conveyor belts that lead to a part of the facility where mechanized washing and packaging occur. Clean, packaged eggs are then loaded on trucks, to be transported to market. Pointing to concentrated animal feeding operations (CAFOs) of this sort, some critics decry "factory farming." They have a point since these businesses are organized much like any other industrial enterprise, with ownership by public corporations, routine use of debt financing, and assembly-line production.

However, factory farms remain exceptional in production agriculture. Independent operators are the norm in crop farming even in places like the United States. Apparently, no one has figured out a profitable way to use large sums raised in equity markets to purchase enormous tracts of rural real estate and employ a hired workforce to grow corn, soybeans, wheat, and other commodities. Perhaps the difficulty of supervising a small number of workers across a large area is the main impediment to assembly-line production of crops. Maybe business risks (related to variable commodity prices) and weather and biological risks are too daunting.

Regardless, public corporations largely steer clear of activities like farming, thereby leaving the field open to individual operators.

A word or two about these operators is in order. For one thing, the majority of them do not rely very much on debt, probably because of agriculture's environmental and business risks. Thus, most of the value of their land and capital comprises equity. Moreover, their great numbers – tens of thousands in the United States, alone – mean that no single farmer is in a position to influence input or output prices unilaterally, by varying factor employment or production on his own. If ever a sector resembled the economic ideal of a perfectly competitive market, in which no single agent exercises market power, production agriculture is it. Accordingly, the conceptual framework that economists have developed to describe competitive supply serves splendidly for the analysis of trends in food production.

3.2 Increases in Agricultural Supply

As explained in the Appendix, in which the fundamental economics of supply are reviewed, supply goes up for various reasons. One of these is investment, which causes more to be produced at any given price. A decline in input prices also lowers production costs and therefore increases supply. In addition, supply growth results as better ways are found to transform inputs into output. As indicated in the pages that follow, the discovery and spread of new agricultural technology has had profound impacts around the world.

These impacts are in line with the contributions that technological improvement has made over many years throughout the US economy. A framework for accounting for these contributions was suggested in the 1950s by Nobel laureate Robert Solow, who found that increases in capital per worker explained just 12 percent of the rise over time in gross domestic product (GDP) per capita while the other 88 percent resulted from technological improvement (Solow, 1957). Within a few years, economists were focusing on growth in total factor productivity (TFP) – that is, output expansion beyond what can be explained by increases in conventional inputs, such as capital and labor – and finding that GDP growth results mainly from TFP increases in the United States and a number of other countries.[4]

The origins of the public goods that have created TFP growth in US agriculture date back nearly 150 years. In 1862, when prospects for the United States emerging whole from the Civil War seemed dim, Congress passed the Morrill Act, which created land-grant colleges that provided the general population with opportunities for higher education – opportunities formerly reserved for a small, privileged elite. The Hatch Act of 1887 established state agricultural experiment stations, which were similar to counterpart institutions that had been established beforehand

---

[4] Recent analysis suggests, however, that East Asian countries have experienced more GDP growth because of the accumulation of capital, which formerly was quite scarce, than because of increases in TFP (Young, 1995). As these countries continue to develop, TFP growth is likely to become more important.

in Europe and Japan, for carrying out research benefiting the farm economy. The Smith-Lever Act of 1914 created an extension service partially funded by the federal government to educate farmers about management and marketing as well as technological advances.

Passage of the Morrill Act was followed by a sizable investment in education. As stressed by Theodore W. Schultz, who was both an agricultural economist and a Nobel laureate, education is a critical element of human capital formation, which is in turn needed for TFP growth (Schultz, 1961). Likewise, agricultural productivity was enhanced by improvements in the biology of crops, which were accelerated by the Hatch Act, as well as the dissemination of these improvements, thanks to the Smith–Lever Act. Corn is a case in point. There was no discernible change in corn biology between the end of the Civil War and the eve of Second World War, national yields averaging 25 bushels per acre[5] throughout this period. But once farmers started using hybrid seeds, after the middle 1930s, average yields began to climb, rising above 125 bushels per acre as the twentieth century came to an end. There have been similar trends in average yields of wheat, and other commodities (Gardner, 2002, pp. 19–22).

Yields have not gone up in response to a rise in commodity prices. To the contrary, increased output caused real prices to fall dramatically between 1950 and 2000 (Chapter 4). Nor can yield growth be explained by increased labor use since the agricultural workforce has declined steadily since the 1930s. Increases in per-hectare production have coincided with increased use of purchased inputs – improved seeds, fertilizer, etc. But since the added expense of these inputs has been offset by diminished expenditures on labor and other factors, real input costs have not gone up very much. In contrast, the inflation-adjusted value of US farm output has risen dramatically. Thus, TFP growth has occurred.

The substitution of other inputs for farm labor is plain to see in the changing composition of agricultural production costs. In 1949, wages amounted to 48 percent of these costs and expenditures on purchased inputs accounted for another 26 percent. During the next 36 years, these shares were nearly reversed, labor's share of total costs falling to 27 percent and purchased inputs' share rising to 40 percent. Capital costs as a portion of the total rose from 11 to 14 percent while land's share did not vary at all, being a little more than 18 percent in 1949 and a little less than 19 percent in 1985 (Craig *et al.*, 1994, cited in Alston and Pardey, 1996, pp. 103–105).

Comparable to these changes in the mix of agricultural inputs have been the increases in agricultural TFP. Until the late 1930s, the value of aggregate output was about equal to total costs, with no variation aside from the fluctuations resulting from an unusually good or bad harvest. Since that time, the gap between aggregate output and aggregate input values has grown wide. Inflation-adjusted production costs rose by just 7 percent between 1949 and 1985, as the real value of output was doubling. The real story of US agricultural development during the twentieth century,

---

[5]A bushel of corn weighs 56 pounds, or 25.397 kilograms, and, as mentioned in footnote 2, there are 2.471 acres in a hectare. Thus, a yield of 25 bushels per acre is equivalent to 1.569 metric tons per hectare and a yield of 100 bushels per acre is the same as 6.276 metric tons per hectare.

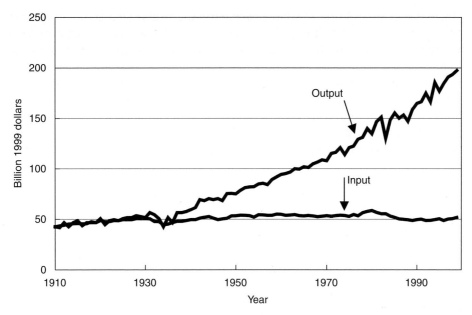

Figure 3.1 US real agricultural output and input from 1910 to 2001.
Source: US Department of Agriculture (USDA) (2002).

then, has been one of tremendous productivity growth, with TFP going up by 80 percent between the late 1940s and middle 1980s (Craig *et al.*, 1994, cited in Alston and Pardey, 1996, pp. 103–107). As indicated in Figure 3.1, the inflation-adjusted value of US agricultural output has quadrupled since the 1930s as real costs have held steady.

Since productivity gains have been high, it comes as no surprise that pay-offs on the investment in agriculture's scientific base making these gains possible have been generous. In a landmark study, Griliches (1958) found that the internal rate of return on the research and development that produced hybrid corn was 35–40 percent. What this means is that $100 spent on coming up with this innovation subsequently led to a recurring annual benefit (consisting of additional output and cost reductions) of $35–40. Another perspective on this internal rate of return is that financing this research and development with a loan would have had a positive pay-off even if the interest rate on that loan had been 35–40 percent. More recently, Huffman and Evenson (1989) found that, from 1950 to 1982, the internal rate of return for expenditures on applied research at state agricultural experiment stations and other public institutions exceeded 40 percent. Empirical research indicates that support for agriculture's scientific base has produced comparable benefits in Australia, Canada, Japan, New Zealand, and Western Europe (Alston and Pardey, 1996, p. 208). Likewise, Tweeten and McClelland (1997) have found that public investment in research, infrastructure, education, and other assets for the rural economy has yielded large returns in a number of developing countries.

Increases in agricultural TFP have been particularly spectacular, not to mention very well documented, in the world's wealthy nations. Elsewhere, a larger share of the growth in food supplies has resulted from increased use of land and other inputs. In the rest of this section, previous trends in farmed area are examined, as are the prospects for future extensification. Also analyzed is the role that yield growth (i.e., intensification) has come to play in the growth of food supplies around the world. With respect to intensification, we are particularly interested in distinguishing between the impacts of technological improvement and those resulting simply from increased application of fertilizer and other non-land inputs.

Extensification

As US experience illustrates, augmenting the supply of food by using science to manipulate the biology of agricultural species (plant as well as animal) is a very recent phenomenon. The science of genetics, which provides a basis for modern crop and livestock breeding, did not exist before Father Gregor Mendel's experiments with garden peas in the middle 1800s. Truth be told, these experiments, which revealed how dominant and recessive genetic traits are passed from generation to generation, had no immediate impact on agricultural yields. Instead, per-hectare output started to go up in Germany, Britain, and other parts of Europe as farmers started to apply commercial fertilizer. Mendel's genetics had to be rediscovered after his death (Pardey and Beintema, 2001) and one of the first examples of yield growth resulting from scientific breeding comes from Japan, where the adoption of improved varieties of rice began in the late 1800s (Hayami and Ruttan, 1985, pp. 232–237).

Relative to the impacts of incipient intensification, extensification remained the primary driver of supply growth during the early part of the twentieth century. This was certainly true in the United States, where as already mentioned there was no appreciable growth in crop yields between the late 1860s and middle 1930s. During the 1800s, American farmers, who had started to venture west of the Appalachian Mountains as the eighteenth century drew to a close, pushed the agricultural frontier (i.e., agriculture's extensive margin) across the continent. The western frontier vanished in 1890 or so and the area used for crop and livestock production reached an all-time peak around 1920 (Cochrane, 1979, pp. 37–102).

Undertaking intensification only after the major opportunities that existed in the United States for extensification had been exploited was entirely sensible. In particular, this pattern of agricultural development is easily reconciled with the hypothesis of induced innovation, by which is meant that changes in technology respond primarily to the scarcest factor of commodity production (Hayami and Ruttan, 1970). During the nineteenth century, land was abundant and labor in short supply in North America. Since the enhancement of labor productivity was much more rewarding than raising land productivity, technological improvement took the form of mechanization. Due to the invention and adoption of the reaper, the steel plow, and (subsequently) the tractor, the area that a single farmer could cultivate went up dramatically. Along with facilitating agriculture's geographic expansion, this change allowed labor to be released from farms and reallocated to other parts of the economy, in which employment opportunities were increasing due to the Industrial Revolution (Cochrane, 1979, pp. 189–208).[6]

One hundred years ago, when the human population was little more than one-fourth its current size (Table 2.4), there were just a few places where extensification was not a satisfactory response to growing food demands. One of these was Japan, where land was scarcer than labor; in that country, induced innovation was aimed at enhancing the productivity of the former factor, in particular by raising yields through fertilization and the scientific improvement of crop varieties. After the middle of the last century, this sort of response to demand growth became predominant throughout the world. Thus, the geographic expansion of agriculture has abated significantly.

Global agricultural land use, including fields planted to crops as well as pastures and other land grazed by livestock, went up by a little more than one-tenth between 1961 and 2001 (Table 3.1, column 4). This growth was modest compared to increases in population in recent decades (Table 2.4). There was a larger increase, 21 percent, in the area planted to the six major crop groups, which do not include cotton, sugar, and commercial tree crops (e.g., coffee and cocoa). Land used for the production of oil crops like soybeans has almost doubled since 1961, largely because demand for soy products has gone up considerably. Also, the area planted to fruits and vegetables (including melons), which comprised a tiny fraction of farmed area 40 years ago, has nearly doubled, reflecting the dietary improvement and diversification that has coincided with improved living standards (Chapter 2). On the other hand, there has been very little change in the land used to produce the commodities that supply most of the calories that people consume. For roots and tubers (e.g., potatoes), this expansion amounted to 10 percent between 1961 and 2001. For rice, corn, and other cereals, which account directly or indirectly for two-thirds of total caloric intake, as well as pulses (e.g., peas and lentils), the increases have been especially small.

Just as different patterns of agricultural land use reflect changes in the human diet, the global totals reported in Table 3.1 have been influenced by regional trends, which have been far from uniform. This is true of cereal area, for example. Between 1961 and 1981, the global total went up from 648 to 729 million hectares (columns 1 and 2) – an increase of 13 percent. Subsequently, a sizable contraction occurred, leaving the area in 2001 (column 3) little changed from what it had been 40 years earlier. Some of this latter contraction took place in Western Europe, where there was a partial reform of government policies encouraging excessive production of grain and other commodities (Chapter 10). The decline was even steeper in the Former Soviet Union, where socialized agriculture began to fail

---

[6] By no means was this the first time in history when extensification had been facilitated by technological change. Between 1000 and 1500, for example, farmers in northwestern Europe adopted deep-cutting iron plows in place of wooden implements originally introduced by the Romans. This change made possible the tillage of heavy soils in river valleys and other locations. As a result, agricultural land use increased substantially north of the Loire River between the eleventh and fifteenth centuries (Landes, 1998, p. 41).

Table 3.1 Agricultural land use (million hectares) in 1961, 1981, and 2001

Crop group	1961 Area	1981 Area	2001 Area	Percentage change, 1961–2001
	(1)	(2)	(3)	(4)
Cereals	648	729	676	4.3
Oil crops	114	164	221	93.9
Pulses	64	62	68	6.3
Roots and tubers	48	46	53	10.4
Fruits	25	33	49	96.0
Vegetables and melons	24	26	47	95.8
Subtotal	923	1,060	1,114	20.7
Other crops	434	382	421	−3.0
Total cropped area	1,357	1,442	1,535	13.1
Grazing land	3,149	3,298	3,491	10.9
Agricultural total	4,506	4,740	5,026	11.5

Source: FAO (2004).

before the 1980s and where food demand has been contracting due to population shrinkage and falling standards of living (Chapter 14). Obviously, the reductions in cereal area in Europe and Russia between 1981 and 2000 have exceeded the increases in other parts of the world.

Global or regional declines in farmed area cannot really be ascribed to urban encroachment on the countryside. Land-use conversion of this sort provokes concern in a number of places, such as the northeastern United States and along China's Pacific Coast, and policies have been adopted in various countries to halt the loss of farmland. However, well under 5 percent of the world's land is urbanized, which is small compared to all the world's farms and ranches. Even significant expansion of urban areas would still leave plenty of real estate for crop and livestock production. Soil erosion and other forms of degradation comprise another threat to the natural resources needed for agriculture. As indicated in Chapter 5, land degradation cannot be taken lightly, although it is usually not severe enough to render farm fields entirely useless.

Could extensification be stepped up if this were needed to increase food supplies? Almost certainly, yes. One obvious candidate for this sort of geographic expansion would be those temperate settings where agricultural land use has declined in recent years. Along with Europe and Russia, these include North America. As Rudel (2001) documents, forests have taken root in many parts of the United States east of the Mississippi River that were farmed 50 or 100 years ago. While much of the land that has been reforested (mainly because of natural succession rather than tree planting) does not have great farming potential, the ebbing of the agricultural frontier could be reversed. In contrast, the prospects for extensification are very limited in South Asia, the Middle East, and North Africa. In these regions, virtually all land suited to crop production, not to mention some that is not, is cultivated already.

Table 3.2 Projected extensification in the developing world from 1990 to 2050

Region	2050 cropland as a share of 1990 cropland (%) (1)	Share of increase in production between 1990 and 2050 resulting from extensification (%) (2)	Share of new cropland obtained from forest or wetland conversion (%) (3)
Africa	196	29	61
Asia	119	10	73
Latin America	149	28	70
All developing countries	147	21	66

Source: Fischer and Heilig (1997).

In the developing world, the best opportunities for extensification are in Africa and Latin America. Drawing on prior natural resource assessments, Crosson and Anderson (1992) conclude that, of all the potentially cultivable land in the developing world that is not currently being farmed, one-half is in South America and 45 percent is in Africa. However, they also stress that extending agricultural frontiers on these two continents is likely to have limited benefits, carry high costs, or both. Much of Africa's unfarmed land, for example, is semi-arid or nearly so. Without irrigation, which requires a sizable investment, crop yields are low. Elsewhere, production is hampered by erosion, chemical limitations of soils, and other problems. Even if the costs of clearing land and preparing it for farming are not too high and if output is satisfactory, getting that output to market is a challenge, especially in the hinterlands where much of the unused land in Africa and South America is located.

Like Crosson and Anderson (1992), Fischer and Heilig (1997) see limited prospects for extensification in Asia. Using FAO data on the agricultural capabilities of unfarmed land as well as UNPD forecasts of population from the middle 1990s, they predict that farmed area throughout the continent in 2050 will be less than one-fifth above agricultural land use in 1990. In contrast, the relative increases in Africa and Latin America during those same 60 years are projected to be 96 and 49 percent, respectively (Table 3.2, column 1). Since the demographic forecasts on which these projections rest have since been revised downward (Chapter 2), Fisher and Heilig (1997) probably have overestimated farmed area in the future. Regardless, their findings are thought-provoking. In particular, wide discrepancies exist between relative increases in cropland expected during the twenty-first century and the contributions of this expansion to output. In Asia, using 19 percent more land for crop production will account for just 10 percent of the output increase anticipated between 1990 and 2050 (column 2). In Latin America, expanding farmed area by one-half is expected to account for a little less than 30 percent of that region's increased harvests. This share is about the same even in Africa, where intensification has proven to be a great challenge and agricultural land use is expected to double. Throughout the developing world, much more food will be produced by intensifying the use of the natural resource base that already exists for agriculture. Likewise, the discrepancies between the numbers in the first two columns of Table 3.2 strongly suggest that, by and large, the quality of land yet to be farmed is inferior to the resources currently harnessed for crop production.

As is also indicated in Table 3.2, most of agriculture's projected expansion in Africa, Asia, and Latin America will come at the expense of forests and wetlands. This is the case with 70 percent of the extensification in Latin America and nearly three-quarters of the geographic extension in Asia (column 3). As explained in Chapter 5, various impacts are suffered as natural habitats are lost. Deforestation, for example, often causes the local climate to grow drier and warmer, especially in the tropics and subtropics, and some contend that it accelerates global warming. The clearing of tree-covered land to make way for new cropland and pasture often leads to increased erosion and sedimentation, and more variability in stream flow (e.g., more flooding during and right after major storms). In addition, deforestation causes some species to disappear, which by definition diminishes biological diversity. Left to his own devices, an individual farmer has little reason to consider these impacts. However, their containment is in the interest of society as a whole. Recognizing this, governments may well decide to arrest agricultural land clearing by subsidizing forest conservation or applying other measures.

To summarize, increasing the supply of agricultural commodities through extensification, which used to be the rule throughout the world, continues to be an option in some places. But it is a costly option. To prepare a new field, a farmer must remove trees and other vegetation, improve drainage, and carry out related tasks. Whether or not this preparatory work is warranted depends on the returns coming his way. Especially if the additional output won from newly cleared land is modest, as would be the case where an agricultural frontier traverses low-quality land, extensification is rewarding for the farmer only if the market value of output is rising, which happens if farm goods are becoming more scarce. Even with rising scarcity, expanding agricultural land use may be undesirable because of the adverse environmental impacts of habitat loss.

Intensification

In light of the costs, environmental and otherwise, associated with extensification, it is fortunate that intensification emerged as the primary mode of agricultural supply growth after the middle of the twentieth century. In developing countries, alone, yield increases accounted for 69 percent of overall growth in food production from 1970 through 1990; extensification's contribution amounted to just 31 percent (Table 3.3).

To be sure, the relative contributions of intensification and extensification have varied considerably from one part of the developing world to another. During the 1970s and 1980s, yield growth was responsible for nearly three-quarters of increased production in the Middle East and North Africa, and more than four-fifths

Table 3.3 Sources of increased crop production in the developing world from 1970 to 1990

Region	Share of output growth from higher yields (%) (1)	Share of increased output from extensification (%) (2)
Sub-Saharan Africa	53	47
Middle East and North Africa	73	27
East Asia (not including China)	59	41
South Asia	82	18
Latin America	52	48
All developing countries (not including China)	69	31

Source: Alexandratos (1995), p. 170.

Table 3.4 Yield increases (annual percentages) for six major crop groups in 1961, 1981, and 2001

Crop group	Yield growth in 1961 (1)	Yield growth in 1981 (2)	Yield growth in 2001 (3)	Increase in farmed area, 1961–2001 (4)
Cereals	3.26	1.97	1.41	0.11
Oil crops	3.53	2.07	1.46	1.66
Pulses	0.92	0.78	0.67	0.15
Roots and tubers	1.39	1.21	1.07	0.25
Fruits	0.49	0.45	0.41	1.68
Vegetables	1.91	1.38	1.08	1.68

Source: FAO (2004).

of all output gains in South Asia. In East Asia – outside of China, where yields have gone up markedly (see below) – this share was 59 percent. In contrast, intensification and extensification had comparable impacts in Sub-Saharan Africa and Latin America (Table 3.3).

Yield growth and increases in farmed area also have had varied impacts among crop groups (Table 3.4). For example, the land used for fruit production has gone up faster than yields of that commodity. The same relationship has applied to vegetables in recent years. However, the combined area planted to fruits and vegetables has yet to rise above 10 percent of total farmland (Table 3.1). Yields of crops like peas and lentils have not gone up very much, but then again the area used for pulse production has increased even more slowly and is now exceeded by the combined size of all the world's fruit and vegetable farms. For roots and tubers, yield growth has been limited. This has an adverse impact on food security in places like West Africa, where many of the poor subsist largely on yams, cassava, and other starchy roots. However, even this modest yield growth has exceeded

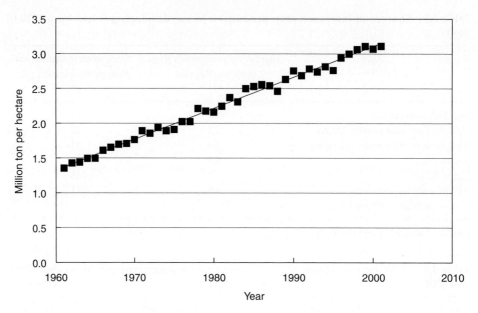

Figure 3.2 Average global cereal yield from 1961 to 2001.
Source: Tweeten (1998).

increases in the area planted to these crops by a wide margin. In the case of cereals, yield growth has been substantial and much greater than extensification. As mentioned already, the area used for the production of soybeans and other oil crops, which was much smaller than cereals area in the early 1960s, increased a lot in the late 1900s, as production expanded first in the United States and later in Brazil and other tropical and subtropical settings. In recent years, annual yield growth, which still is above 1 percent, has fallen a little below the annual rate of extensification for this crop group. However, this represents a modest exception to the general trend of supply growth being driven mainly by intensification.

As the figures in Table 3.4 clearly show, annual percentage increases in crop yields have been trailing off for a number of years. What this reflects is linear growth in per-hectare output, which has been sustained year in and year out since the early 1960s. Shown in Figure 3.2 is the linear trend for cereals, which continue to be planted on more land than the other five crop groups combined. Of course, dividing a constant annual increment by average yields, which are rising over time, produces an ever smaller ratio as the years pass. Thus, percentage yield growth, which is what this arithmetical exercise gives us, declines.

Just as the development of new crop varieties started to drive up per-hectare output and add to agricultural TFP in Japan during the late 1800s and in the United States before World War Two, the agricultural intensification experienced in various parts of the developing world during and since the Green Revolution has been a consequence largely of scientific improvement – improvement of a sort that has permitted output to climb much more dramatically than factor employment.

Research that was to lead eventually to the Green Revolution was under way before the middle of the twentieth century. In 1944, the Rockefeller Foundation recruited a young plant scientist from the Midwestern United States, Dr. Norman Borlaug, for a group of crop breeders in Mexico intent on developing a variety of wheat that could withstand the rust virus. After a few years of concentrating its efforts on disease resistance, the research team became primarily concerned with yield improvement, and there were achievements in this latter area during the 1950s. In the 1960s, the Rockefeller Foundation established the International Maize and Wheat Improvement Center (CIMMYT) near Mexico City as the Ford Foundation was founding the International Rice Research Institute (IRRI) just outside Manila. These two institutions subsequently became the leading two elements of the Consultative Group on International Agricultural Research (CGIAR), which has worked with national agricultural research institutes and extension agencies to develop and disseminate improved crops (Dalrymple, 1985). The success of this venture was commemorated by the Nobel Peace Prize awarded to Dr. Borlaug in 1970.

Green Revolution varieties have various advantages. For one thing, they are not as tall as their traditional relatives, which is why they are often described by the adjective, semi-dwarf. As can be appreciated, shorter plants are less apt to lodge – that is, to be bent to the ground by wind and rain. Another merit of Green Revolution varieties is that their grain-bearing panicles are recessed, which increases exposure of the leaves (where photosynthesis occurs) to sunlight. This enhances cereal production. The beneficial trait of greatest importance, however, is that the varieties coming out of the CGIAR yield much more grain than traditional strains when supplied with ample nutrients and water. Accordingly, increased use of the former varieties has been accompanied by major growth in fertilizer use and irrigation (Dalrymple, 1985). Also, since fertilizer, improved seeds, and other purchased inputs must be paid for months before crops are harvested, demand for credit has grown.

Especially in Asia, farmers were quick to exploit the advantages of the Green Revolution. Introduced in the middle 1960s, semi-dwarf varieties accounted for 30 percent of all the wheat planted in Bangladesh, India, Nepal, and Pakistan by 1970; little more than a decade later, that share had risen above 70 percent. During the same period, the share of total rice area in South and Southeast Asia planted to varieties developed by the CGIAR rose from 10 percent to nearly 40 percent (Figure 3.3). It is also noteworthy that the Green Revolution did not greatly exacerbate income inequality. Larger, wealthier farmers comprised a disproportionate share of those making an early shift from traditional varieties. Within a very few years, though, small and medium farmers were about as inclined to use semi-dwarf crops. Benefits even accrued to landless people, who are among the poorest of the rural poor, because the Green Revolution reduced food prices while strengthening demand for their labor (Anderson *et al.*, 1985).

The linkage between farmers' adoption of improved varieties and increased fertilization and irrigation underlies trends in yields and input use during the 1980s and 1990s in the developing world's 15 most populous countries (Table 3.5). Each

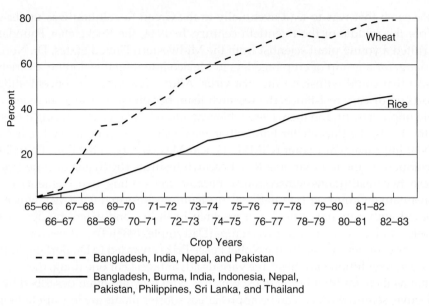

Figure 3.3 Area planted to high-yielding varieties of rice and wheat in South and Southeast Asia from 1965–1966 to 1982–1983.
Source: Dalrymple (1985, p. 1071).

of these countries has at least 60 million inhabitants and, as a group, they have 60 percent of the world's 6.13 billion people and nearly four-fifths of the combined population of Asia (outside of the Former Soviet Union), Africa, and Latin America. Needless to say, agricultural trends in China, India, and other large developing nations matter a great deal in terms of how well the entire human race eats.

In South Asia, average cereal yield went up by more than 70 percent in Bangladesh and India and two-fifths in Pakistan (Table 3.5, column 1). Historically very high in the third country, the irrigated portion of agricultural land increased substantially in the first two (column 4). Likewise, fertilizer application rates rose dramatically: by 150 percent in Pakistan, more than 200 percent in India, and nearly 250 percent in Bangladesh (column 3). Similar changes have taken place farther to the east in Asia. High cereal yields in China – nearly 5 tons per hectare, after 61 percent growth during the last two decades of the twentieth century – have come about largely because of improved varieties and a doubling of fertilizer application rates. In addition, irrigated area has gone up, although the irrigated portion of cropland has declined due to recent extensification in rain-fed areas. Crop production has intensified as well in Indonesia, the Philippines, and Thailand. Particularly impressive has been Vietnam's emergence as a major producer of rice after decades of warfare. Since the late 1970s, output per hectare has nearly doubled as fertilizer use went up 10-fold and the irrigated portion of cropland climbed from 26 to 41 percent.

Table 3.5 Agricultural intensification and its underlying factors for the 15 most populous developing nations

Country	Cereal yield (kg/ha)		Per-capita cropland (ha)		Fertilizer use (kg/ha)		Irrigated % of Aarable land		Average value added per worker (95$)	
	79–81	99–01	79–81	99–01	79–81	99–01	79–81	99–01	79–81	99–01
	(1)		(2)		(3)		(4)		(5)	
Bangladesh	1938	3322	0.10	0.06	45.9	159.3	17.1	47.6	217	311
Brazil	1496	2665	0.32	0.32	91.5	109.9	3.3	4.4	2048	4356
China	3027	4879	0.10	0.10	149.4	291.1	45.1	39.0	161	321
Egypt	4053	7015	0.06	0.05	286.4	404.3	100.0	100.0	721	1240
Ethiopia	n.d.	1141	n.d.	0.16	n.d.	15.5	n.d.	1.8	n.d.	138
India	1324	2299	0.24	0.16	34.5	105.8	22.8	33.6	272	397
Indonesia	2837	3915	0.12	0.10	64.5	141.5	16.2	15.5	609	736
Iran	1108	2030	0.36	0.25	43.0	64.7	35.5	39.8	2197	3756
Mexico	2164	2604	0.34	0.26	57.0	70.6	20.3	23.8	1482	1772
Nigeria	1265	1206	0.39	0.23	5.9	6.1	0.7	0.8	414	672
Pakistan	1608	2261	0.24	0.16	52.5	126.1	72.7	81.7	394	630
Philippines	1611	2434	0.11	0.07	63.6	131.5	12.8	15.5	1347	1328
Thailand	1911	2659	0.35	0.25	17.7	111.3	16.4	26.0	626	904
Turkey	1869	2196	0.57	0.38	52.9	83.1	9.6	15.8	1860	1886
Vietnam	2049	3955	0.11	0.07	30.2	317.9	25.6	41.3	n.d.	240

Source: World Bank (2003b), pp. 124–126, 128–130.

Intensification has had similar effects in the Middle East and North Africa. The yield impacts of increased fertilization and irrigation may have been modest in Turkey. However, per-hectare output in Iran, which was very low two decades ago, has almost doubled since then thanks to a 50 percent increase in the fertilizer application rate and a rise in the irrigated portion of farmed area. Of all the countries listed in Table 3.5, none has higher average cereal yields than Egypt, where all agricultural land has long been irrigated. Output per hectare has risen from four to seven tons as applications of fertilizer have increased from a little less than 300 to a little more than 400 kilograms per hectare.

Elsewhere, intensification has had less effect. The growth in yields, fertilization, and irrigation experienced in Mexico, which along with the Philippines was the cradle of the Green Revolution, has been of limited magnitude. In Brazil, relative increases in per-hectare output have exceeded those in fertilizer application rates. In that country, as in China, increases in irrigated area do not show up in the data of Table 3.5 on the irrigated portion of agricultural land because of extensification in places where farming is rain-fed. Finally, the changes affecting agriculture in other parts of the developing world have largely bypassed Sub-Saharan Africa. In Nigeria, where fertilizer use and irrigation are rare, average yields have actually declined. Twenty years ago, Ethiopia was ruled by a Marxist dictatorship, which did not bother to keep agricultural records as it made war against the opposition. Even by the standards of the developing world 20 years ago, current agricultural conditions in the country are backward.

The lack of a Green Revolution in Sub-Saharan Africa relates in part to opportunities for extensification in that region. In addition, intensification has been held back by geographic impediments to irrigation development – in some places, a lack of good reservoir sites. But as is explained in Chapter 15, numerous African governments have neglected agriculture or applied policies injurious to the sector, thereby delaying the spread of improved production technology.

Since fertilizer application rates have risen and irrigation has expanded at least as rapidly as yields in a number of countries, should one conclude that intensification during and since the Green Revolution has not constituted technological improvement of the sort that began to occur in the United States before the Second World War? The answer to this question is an unequivocal no. The clearest proof that agricultural TFP has increased in countries embracing the Green Revolution is revealed by an analysis of value added per agricultural worker. The numerator of this indicator equals the difference between the value of agricultural output and the costs of fertilizer, capital goods, and all other inputs aside from labor. The denominator comprises the entire agricultural workforce, including farmers as well as the people they hire.

As indicated in column 5 of Table 3.5, value added per member of this workforce has gone up in the late twentieth century, not least in those parts of the developing world where the Green Revolution has had the greatest impact. True, an increase happened in Nigeria. However, this reflects a decline in the agricultural workforce rather than a change in average yields, which have not grown. Conversely, value added per agricultural worker has stagnated as output per hectare has increased in the Philippines. Elsewhere, the positive impacts on TFP are plain to see: a 43 percent rise in value added per member of the agricultural labor force in Bangladesh, a doubling of this indicator in China, a 72 percent increase in Egypt, and so forth.

Clearly, major gains in productivity underlie the intensification that, since the beginnings of the Green Revolution, has created the lion's share of additions to the world's food supply.

3.3 Has Intensification Run Its Course?

In light of the Green Revolution's enormous benefits and the modest dimensions of some of its adverse impacts (e.g., on income inequality in rural areas), environmental consequences have been the focus of much of the debate over agricultural intensification. For example, one empirical study undertaken in the Philippines suggests that the value of the additional rice harvested due to the application of pesticides may be exceeded by the costs of illness suffered by farmers and other people exposed to those pesticides (Pingali *et al.*, 1995). Since it is rare for these costs to be fully taken into account in estimates of agricultural value added, the net benefits of intensification, made possible by heavy use of chemical inputs, tend to be exaggerated. Under-counting of the economic damage that results when agricultural chemicals find their way into rivers, lakes, and streams likewise distorts our view of the true pay-offs of intensification.

Another concern about intensification has to do with the natural resources needed for yield growth. Of these resources, none is more important than water. Of course, pollution from industrial and municipal sources circumscribes irrigation development, just as agricultural run-off impairs water quality to the detriment of society as a whole. Moreover, the capital costs of irrigation are sizable. For a medium- or large-scale system, constructing canals and other basic infrastructure costs $8,000 per irrigated hectare; if roads, power grids, and other complementary infrastructure must be built up, then the per-hectare expense rises above $18,000 (FAO, 1992). Once this infrastructure is in place, operating and maintaining it are by no means free. In addition, costs are incurred for the technology transfer directed at farmers accustomed to rain-fed production, who need training in the effective use and management of water distributed to their fields.

Thanks to the investment of tens of billions of dollars, irrigated area grew during the second half of the twentieth century, from a little under 140 million hectares around the world in 1961 to more than 260 million hectares today; amounting to little more than one-sixth of the world's cropland, this irrigated area is the source of approximately one-third of all agricultural output (FAO, 2004). More irrigation development is possible. In a 1990 study, the World Bank and the UN Development Program (UNDP) estimated that 110 million rain-fed hectares could be irrigated, thereby causing agricultural water use around the world to go up by approximately 30 percent; the resulting production would be sufficient to feed up to 2 billion people (World Bank and UNDP, 1990). To put this impact on food output into perspective, keep in mind that human numbers are expected to go up by 2–3 billion during the next 50 years (Table 2.5).

Although accelerated irrigation development may be feasible, it seems unlikely to happen any time soon. For one thing, competition for water from non-agricultural consumers has been growing rapidly and promises to continue doing so. As Rosegrant *et al.* (2002) point out, domestic and industrial demand quadrupled between 1950 and 1995, while demand from crop and livestock producers went up by a little more than 100 percent. At the end of the twentieth century, agriculture accounted for 80 percent of water consumption around the globe and for 86 percent in developing countries. Given the importance of satisfying unmet needs for clean drinking water in poor countries, Rosegrant *et al.* (2002) anticipate that, between 1995 and 2025, uses of water other than irrigation will rise by 62 percent and diversions to farmland will go up by just 4 percent. This is obviously a small fraction of the feasible growth of 30 percent estimated by the World Bank and UNDP (1990). Furthermore, irrigation subsidies, which governments around the world offer farmers to boost agricultural production but which create economic inefficiency and harm the environment (Chapter 5), are bound to come under closer scrutiny as competition over water grows more intense with each passing year.

There are other impediments to irrigation development aside from greater competition over the liquid input. Large dams, which create substantial environmental and social impacts, tend to arouse fierce opposition – in affluent places and even in poor settings. When these impacts are evaluated and combined with capital and other costs, many projects begin to look prohibitively expensive. This helps to explain why Rosegrant *et al.* (2002) are forecasting that irrigation will expand slowly. The only way to justify the costs, including damages to the environment and local communities, would be if agricultural commodities were becoming much scarcer, thereby causing a major run-up in food prices.

While expanded irrigation often jeopardizes environmental quality, biotechnological innovation generally has the opposite impact. To date, moving genes from one species to another has yielded corn, cotton, and other crops that can defend themselves against insects and other harmful organisms as well as crops that tolerate herbicides. The environmental advantages of pest resistance are obvious, having to do with diminished use of chemicals with toxic properties. Likewise, the planting of herbicide-resistant commodities is advantageous since doing so allows farmers to cut back on deep plowing and other tillage practices that accelerate soil erosion and other forms of land degradation (Chapter 5).

It is lamentable that many self-styled environmentalists are staunchly opposed to the GMOs that agricultural biotechnology yields (Rauch, 2003). This opposition helps to explain why there are worrying trends in public funding for agricultural research and development, which have driven intensification and productivity growth for decades. In 1995, this financial support totaled $22 billion, which was nearly double the level in 1976. However, practically all the increase happened before 1990. During the last decade of the twentieth century, governmental budgets for the technological improvement of agriculture held steady in affluent nations. In many developing countries, especially in Sub-Saharan Africa, these budgets actually declined (Pardey and Beintema, 2001).

In terms of stimulating some though not all types of technological change, increased private spending on agricultural research and development can compensate for stagnating governmental budgets in this area. Over the years, individual firms and entrepreneurs have come up with a number of improvements in farm machinery. Likewise, biotechnology is currently dominated by the private sector. However, profit-oriented enterprises are interested in technological improvement only if it generates something for which a price can be charged, because exclusion costs are low. A better kind of tractor fits this description. With patent laws and other protections, the innovating firm can easily deny access to a tractor to anyone unwilling to pay the going price. Genetically modified varieties that are hard to reproduce on a farm comprise another product of technological improvement that is excludable and therefore marketable. New crops that are reproducible will be developed by the private sector only if these can be patented and if courts and other institutions stand ready to enforce patents. Otherwise, development of these crops, such as the semi-dwarf varieties of the Green Revolution, amounts to the production of a public good. As pointed out earlier in this chapter, this sort of good has high exclusion costs, which diminishes its marketability and therefore makes it uninteresting to the private sector.[7]

In light of the understandable reluctance of profit-oriented firms to produce public goods, private spending on agricultural research and development in the United States and other rich countries, which surpassed $11 billion in 1995 and continues to rise, is not a perfect substitute for governmental spending, which to repeat has not gone up recently in the same countries. Concerns are very different in the developing world. Potential substitution of investigation in the private sector for publicly funded work is not a pressing issue. Not more than 7 percent of all private research throughout the world occurs in Africa, Asia, and Latin America; by the same token, biotechnological innovation in these places is at an incipient stage of development (Pardey and Beintema, 2001). The primary problem in poor countries is one of limited and declining support for any kind of invention and innovation. Indeed, Pardey and Beintema (2001) have identified a widening gap between developed and developing countries in the intensity of research investment – that is, budgets for agricultural research and development divided by the value of agricultural output. As they also observe, the pay-off from research often mounts as more research is done, in the sense that the gains from any single innovation or discovery are enhanced by prior or contemporaneous advances and findings. At times, a developing country can overcome its research gap by drawing on basic and applied science from wealthier nations. However, there is always a need to adapt agricultural technology to local conditions. If allowed to persist and widen, the gap in research intensities between the developed and developing worlds will inevitably cause growth in agricultural productivity to lag precisely where productivity gains are most sorely needed because food demand is going up rapidly and the environmental costs of agriculture's geographic expansion are high.

From time to time, there is speculation that limits on yield growth have been reached or are being approached. If this were true, the prospects for tropical forests and other habitats in the developing world would be gloomy. Just as a lot of deforestation has been avoided during and since the Green Revolution, continued yield increases are required if widespread extensification of agriculture is to be avoided during the twenty-first century. No doubt, the environmental risks of developments like agricultural biotechnology, which so far appear to be very small, must be monitored and contained. Regardless, clear statistical evidence that growth in per-capita

---

[7] For example, hybrid corn that is bioengineered to resist rootworms and corn borers cannot be reproduced. This is not true, however, of soybeans that have been genetically modified for resistance to glyphosphate herbicides. In the United States, firms that produce and sell the latter item enforce their patent rights vigorously, by prosecuting farmers who purchase and sow bioengineered seeds one year and then use a portion of their harvests as seeds the next year. In countries that do not recognize or enforce patents, the private sector does not supply bioengineered though reproducible seeds. Government, then, is the sole source of supply – not counting seeds reproduced without the consent of the original innovator.

production is not abating (Lomborg, 2001, pp. 97–101)[8] indicates that African, Asian, and Latin American farmers will not be impelled any time soon to use each and every square meter of available land in a desperate struggle to stave off hunger. Great news for farmers. Great news as well for the natural environment.

3.4 Trends in Per-Capita Production

Modern agriculture, of the kind found not just in wealthy nations but in many parts of the world, is a far cry from the kind of farming that our ancestors engaged in and that barely kept them fed. Muscle power, from farmers and draft animals, has largely been replaced by machinery driven by fossil fuels. Thus, the cost of this machinery now accounts for an appreciable share of total input expenditures. While growth of another category of inputs (i.e., land) has not ceased entirely, humankind's long experience with agricultural extensification is coming to an end. Instead, intensification, made possible by increased specialization and trade as well as investment in public goods, has proven to be a very efficient way to increase output. Farmers' embrace of improved technology is indicated by their adoption of improved varieties and increased use of fertilizer and other inputs. As they have done this, agricultural TFP has gone up dramatically.

Mainly because of intensification, increases in food supplies have exceeded the growth in food demand resulting from demographic expansion and rising standards of living. During the 1960s, agricultural output grew slightly faster than population in Africa and Asia, while there was no change at all in per-capita supplies in Latin America (Table 3.6, columns 1 and 2). Since then, the per-capita index has shot up in Asia and increased markedly in Latin America (columns 3–5). In contrast, human numbers rose faster than agricultural production in Africa during the 1970s and 1980s. All told, the per-capita index went up by nearly three-fourths in Asia and 44 percent in Latin America between the early 1960s and the turn of the twenty-first century. Meanwhile, per-capita output in Africa fell 10 percent. Due to the latter decline, the index for the entire world was prevented from climbing by more than one-fourth (Table 3.6).

Africa's problems aside, the improvements in food availability registered during the past 40 years comprise admirable performance. More than that, these improvements have been little short of heroic when compared to the predictions of impending worldwide famine made by people like Paul Ehrlich in the 1960s and early 1970s (Chapter 2). In light of this experience, no serious observer has any doubts about humankind's ability to feed itself during the twenty-first century – much more because of yield growth than because agriculture is expanding geographically at the expense of forests and other natural habitats.

---

[8] A statistician, Bjorn Lomborg, was motivated to test the hypothesis that yield growth is declining suggested by Lester Brown, an environmental activist from the United States. Neither Brown's claims nor his techniques for substantiating them turn out to be valid (Lomborg, 2001, pp. 96–97).

Table 3.6 Per-capita food production as a share of per-capita production in 1961–1965, various years

Continent	1961–1965 (1)	1971 (2)	1981 (3)	1991 (4)	2001 (5)
Africa	100	103	94	90	90
Asia	100	104	114	134	173
South America	100	100	115	118	144
World	100	107	112	115	126

Source: FAO (1976), p. 45; FAO (1983), p. 85; FAO (1993), p. 53; FAO (2003a), pp. 50–51.

If anything, we may now be taking it too much for granted that our growing demand for the products that nourish us will continue to be exceeded by supply increases. Ruttan (2002) emphasizes the environmental hurdles to continued intensification, which scientists must help to overcome. In addition, the bright promise of a few years ago that biotechnology would produce "better, more nutritious crops, which would be drought-resistant, cost-resistant, and salt-resistant" has yet to be fulfilled (Carr, 2003). If crops with these and other useful characteristics are to be made available to farmers, support for technological innovation must be adequate.

Complacency probably helps to explain why governmental funding of agricultural research and development and other public goods that the food economy relies on is flagging and why so many people are unwilling to put up with new irrigation projects and GMOs. Preferences like these must be respected, of course, although accommodating them entirely would undoubtedly slow, or possibly reverse, the long-term trend toward diminished food scarcity. In a world in which hundreds of millions of people still go hungry, the price to be paid for complacency about agricultural development would be enormous.

Study Questions

1. Why did human numbers grow and urban life arise after the agricultural revolution?
2. Compare and contrast temperate-zone agriculture and farming in the tropics and subtropics.
3. Compare and contrast the supply side of the food economy in impoverished settings in the developing world with the supply side of the food economy in affluent places.
4. In what ways are US farms different from the agribusinesses that supply inputs and services to these farms.
5. Describe the origins of productivity growth in US agriculture and compare this growth to productivity trends in the national economy.
6. Explain why US agriculture was mechanizing with little change in crop yields during the late 1800s and early 1900s, while yield enhancement was taking place with little mechanization in Japan.

7. Describe the relative contributions of increases in planted area and higher yields to growth in global production of various crops.
8. Describe the relative contributions of extensification and intensification to increased agricultural output in Asia, Latin America, and Africa.
9. Why were expanded irrigation and increased fertilization major elements of the Green Revolution?
10. What are the prospects for a major increase in irrigated area during the twenty-first century?
11. What are the current trends for agricultural research and development in rich countries as well as the developing world?

Appendix: The Fundamental Economics of Supply

Production agriculture is influenced much more by environmental conditions than is the rest of the economy. In addition, a host of non-environmental factors contributes to crop and livestock production. These include labor, machinery and other elements of capital (i.e., the manufactured portion of productive capacity), fertilizers and pesticides, as well as management inputs. The technology for transforming all these resources and inputs into agricultural output is also very important. Furthermore, how much farmers produce depends on market values, both the prices they pay in competitive markets for factors of production (i.e., inputs) and those they receive for output.

As explained in the Appendix of Chapter 2, growth in food demand is analyzed by first converting a general consumption function, in which everything that affects how much people eat is represented and freely variable, into a demand curve and then investigating shifts in that curve. Economists use an analogous approach to shed light on trends in food availability. That is, a general production function, in which everything affecting output is represented and freely variable, is converted into a supply curve and then shifts in that curve are analyzed. Of course, this sort of analysis rests on the assumption of *ceteris paribus*.

Ceteris paribus in production applies in what economists call the short run, which is not long enough either to effect an investment in productive capacity (comprising land and capital) or to apply new technology. Under these circumstances, output can be changed solely by increasing or decreasing other inputs (e.g., labor), which by definition are variable. A fundamental feature of the relationship between output and variable inputs is diminishing marginal product (MP), which means that the additional production that results from slightly increased employment of these inputs grows progressively smaller as employment rises. By the same token, the quantity of variable inputs needed to raise output a little grows ever larger as production increases.

Diminishing MP has implications for the marginal cost (MC) of production as well as supply in the short run. If the prices of variable inputs do not vary (another application of *ceteris paribus*), then the fact that ever larger amounts of variable inputs are needed to raise output marginally means that MC in the short run goes up as production rises. As is explained in any introductory economics text, profit-maximizing firms in a competitive industry, such as farming, seek out the production level (Q_S) at which output price (P), over which no single firm has any control, equals MC. Thus, the short-run MC function for either a single firm that maximizes profits or an entire industry made up of such firms is equivalent to the short-run supply curve for that same firm or industry.[9] Just as a demand curve's negative slope reflects diminishing marginal utility (Chapter 2), the upward slope of the supply curve corresponds exactly to the positive linkage between MC and output, which in turn reflects diminishing MP.

Unless stated otherwise, all supply curves presented in this chapter and the rest of this book are of the short-run variety, meaning that these are derived by holding productive capacity, technology, and input prices fixed. One such curve, the short-run relationship between price and output of corn in the United States, is depicted in Figure 3.4.[10] In the lower reaches of the curve, close to the vertical axis, variable inputs have a high MP, which implies that MC is low. As output rises, MP goes down, which drives up MC. Progressively higher prices are required for producers to justify using ever greater amounts of variable inputs to produce a little more output.

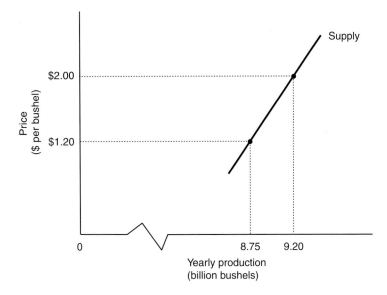

Figure 3.4 Short-run corn supply in the United States.

---

[9] To be more precise, the short-run supply curve comprises the upward-sloping portion of the MC curve above the minimum price a firm or industry needs to receive for its output in order to pay the costs of variable inputs in full. If price falls below this level, then the firm or industry produces nothing at all in the short run.

[10] Just as consumption (the dependent variable) is measured horizontally and variations in price (the independent variable) is tracked vertically in the standard representation of the demand curve, so too do quantity produced and price appear on the horizontal and vertical axes, respectively.

Supply Elasticity

Just as own-price elasticity of demand describes how sensitive consumption is to price changes, elasticity of supply, which is reflected in the supply curve's slope, expresses the sensitivity of output to rising or falling prices – that is, relative change in production divided by relative change in price. Since a higher price induces more output, this elasticity is always positive. Recall from Chapter 2 that own-price elasticity of demand is always negative because consumption goes down as price increases.

If relative change in output exceeds relative change in price, as would be indicated by a gently sloped supply curve, then supply elasticity is high (i.e., supply is elastic). As with own-price elasticity of demand, the threshold for categorizing supply as elastic is 1.00 (unitary elasticity). An example of high elasticity would be an 8 percent increase in output brought about because of a 5 percent rise in price – in other words, a supply elasticity of 1.60. In contrast, an elasticity of 0.80 – observed, for example because a 5 percent decline in price causes production to go down by 4 percent – means that supply is inelastic (i.e., the elasticity of supply is low).

As long as two points have been identified on a single supply curve, one can come up with a measure of supply elasticity. Let P and Q_S represent one such point and P' and Q_S' be another. Supply elasticity, then, is:

Relative quantity change ÷ relative price change
$= [\Delta Q \div \text{average of } Q_S' \text{ and } Q_S] \div [\Delta P \div \text{average of } P' \text{ and } P]$
$= [(Q_S' - Q_S) \div \frac{1}{2}(Q_S' + Q_S)] \div [(P' - P) \div \frac{1}{2}(P' + P)].$ (3.1)

This formula can be applied to the short-run supply of corn in the United States. As shown in Figure 3.4, cutting the price from $2.00 to $1.20 per bushel causes annual production to go down from 9.20 to 8.75 billion bushels. Supply elasticity is:

$[-0.45 \text{ billion} \div \frac{1}{2}(8.75 \text{ billion} + 9.20 \text{ billion})] \div [-\$0.80 \div \frac{1}{2}(\$1.20 + \$2.00)]$
$= [-0.45 \text{ billion} \div 8.98 \text{ billion}] \div [-\$0.80 \div \$1.60] \approx 0.10.$

This small positive number is consistent with actual estimates of supply elasticity in agriculture during the short run, which lasts 1 or 2 years. Beyond the short run, farmers can respond to higher or lower prices by varying more of their inputs and the elasticity of supply is higher. When farmers have 3–5 years to react to price changes, supply elasticity is in the range of 0.30–0.50. In the long run, which is a decade or longer in agriculture, unitary elasticity is typical. For example, permanently increasing the price of corn by 10 percent will eventually induce a 10 percent increase in output, as the new productive capacity created by investment comes on line (Askari and Cummings, 1976, pp. 87, 117, and 150; Henneberry and Tweeten, 1991).

Changes in Supply

By definition, any change in productive capacity or technology shifts the short-run supply curve, which means that there is a change in quantity produced at any given price. A rise or fall in the prices of variable inputs, which is another way to relax the assumption of *ceteris paribus*, has the same effect.

Since supply reflects MC, a rise in the latter causes supply to decrease. This is sometimes the outcome of disinvestment in productive capacity: switching land to another sector or industry, transferring mobile capital (e.g., vehicles), or gradually depreciating immobile capital (e.g., buildings and heavy equipment with no alternative use). MC also goes up if prices of variable inputs increase. Regardless of the reason, the supply curve moves toward the vertical axis, with less being produced at any given price (Figure 3.5).

MC can decline for various reasons. One of these is investment, either in land or capital goods, which raises the productivity of labor. MC can also be driven down by technological progress, which allows more to be produced with a given quantity of inputs (or, if one prefers, the same output to be produced with fewer inputs). In addition, lower costs can be the result of inputs being bought and sold at lower market values. If any of these events happen, the supply curve will move away from the vertical axis, thereby causing more to be produced at any given price (Figure 3.5).

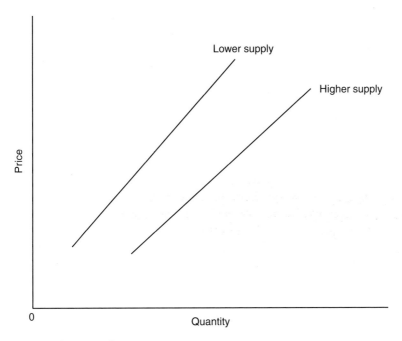

Figure 3.5 A shift in supply.

Just as a clear distinction needs to be made between movement along a stationary demand curve and a shift of that curve, care must be taken not to confuse movement along a stationary supply curve with a change in supply. As long as the *ceteris paribus* assumption holds, the short-run supply curve remains in place and output can go up or down only if price is increasing or decreasing. If there are changes in land, capital, technology, or the prices of variable inputs, then supply itself shifts, with output different at any given price. Clearly, treating supply (a relationship between price and output) as a synonym for production can create every bit as much confusion as treating demand as a synonym for consumption.

Finally, it needs to be acknowledged that analyzing changes in the supply of food can be a little tricky. Whereas a change in population or income obviously causes demand to go up or down, an increase or decrease in supply is sometimes difficult to distinguish from movement along a stationary curve. Consider an increase in crop yields resulting from increased use of fertilizer, which is a variable input. If the change happens only because this input is cheaper, then the *ceteris paribus* assumption has been relaxed and supply has increased. However, the change may be a consequence of a higher price for output, in which case movement along a stationary curve is happening. Unlike the different meanings of an increase in yields, however, there is no ambiguity at all about the consequences of technological improvement, which always lowers production costs and hence always amounts to an increase in supply.

Agricultural Supply: Real and Imagined Characteristics

Like the discussion of food demand in the Appendix to Chapter 2, this appendix's treatment of agricultural supply is orthodox. The supply curves derived when the assumption of *ceteris paribus* is applied have an upward slope. Also, these curves shift in and out precisely as expected in response to investment, technological improvement, and other changes.

Distinctly heterodox views exist of the supply of farm commodities. One thoroughly discredited notion is that agricultural supply curves bend backward, instead of sloping upward. Producers, it is alleged, respond to lower prices by increasing output, not reducing it, because they have mortgage payments and other fixed financial obligations. However, one cannot reconcile this view with the tendency of MC to rise with output, which suggests that defying the guideline for profit maximization (i.e., setting output where price equals MC) actually reduces the financial resources available for paying fixed costs. In any event, empirical investigation consistently yields positive estimates of supply elasticities.

More sophisticated thinking has gone into the explanation of low farm returns, as these are alleged to be. In his technological treadmill theory, Cochrane (1965) contends that improved inputs from science and industry are adopted quickly by farmers. As a result, agricultural labor requirements fall and commodity output rises, thereby driving down prices. Since redundant labor cannot move out of the agricultural sector fast enough, downward pressure is exerted on profits. To maintain economic viability in the face of low returns, individual operators continue to draw on new technology in the hope of lowering production costs. As everyone does this, technological improvement accelerates, output expands, and prices and profits fall. In short, the treadmill continues.

Related to the treadmill view is the fixed asset theory, which explains the alleged propensity on farmers' part to adjust sluggishly or not at all to chronically low returns (Johnson and Quance, 1972). Factors of production are imagined to be immobile – that is, not transferable out of agriculture – and are therefore used even if profits are extremely low. For example, milking machines and combines, which have little value outside the farm economy, are purchased when commodity prices are favorable but continue to be used to generate excessive amounts of output in the face of falling prices.

The treadmill and fixed asset theories do not explain chronically low returns in agriculture for a simple reason, which is that these returns are not chronically or systematically low on commercial farms. Numerous studies document that the profitability of these operations are neither abnormally high nor abnormally low (Hopkins and Morehart, 2002). Even people with small, less efficient farms find ways to take advantage of the tax code or even derive recreational or psychic value from their rural enterprises.

Although it cannot explain a nonexistent problem of chronically low returns on farm resources in developed countries, fixed asset theory yields insights into annual and cyclical instability that plagues agriculture, in rich and poor countries alike. Months and sometimes years separate the application of inputs from the realization of commodity output. Under these circumstances, a "cobweb" cycle can emerge after a drought or some other shock lowers output and raises prices. Responding to high prices, farmers during the next season increase input use so as to produce more. As a bumper crop is harvested, prices fall, which induces a cut-back on inputs and production. This variability, which needs to be distinguished from chronically low returns, can last quite a while.

4
Aligning the Consumption and Production of Food Over Time

Fundamental challenges facing the world food economy have been described in the two preceding chapters. On the demand side, human numbers are not increasing as rapidly today as they did during much of the twentieth century. However, population growth will continue for at least a few more decades. In addition, living standards, which have been improving in various parts of the world, will continue doing so this century. This will cause per-capita consumption to rise. As stated at the end of Chapter 2, aggregate demand ought to climb by at least 50 percent between 2000 and 2050 and, if human numbers and incomes go up enough, could easily double during this period.

Alternative responses to food demand growth are described in Chapter 3. Increasing the geographic domain of farming remains an option in some places. But elsewhere – including eastern and southern Asia, where well over half the human race lives – farmers have occupied most of the land that lends itself to crop production. Agricultural intensification, which has replaced extensification as the primary response to demand growth, is obviously still needed. Otherwise, the commodities that nourish us will grow scarcer, more people will experience hunger, and destruction of natural habitats will increase.

Significantly, no supreme agency exists to strike a desirable balance between food consumption and its availability. At a global level, there is no central bureau that decides on the production of food in various places or the allocation of available supplies among consumers worldwide. Even at the national level, this sort of central planning is very much the exception to the rule. Instead, decision-making is highly decentralized. On the demand side, countless households around the world make choices every day about how much of this or that product to buy based on earnings, individual tastes, and prices. Likewise, innumerable farmers decide what and how much to grow and sell entirely on their own, taking into account the resources they possess, the value of inputs they must purchase, as well as output prices. Governments exercise influence – sometimes a lot of influence – by manipulating market forces. However, neither consumers nor producers receive direct orders from above.

At first glance, the lack of central planning seems to be a formula for chaos. One might think that any match that may be struck between the production decisions of hundreds of millions of farmers and the consumption choices made by billions of consumers is totally serendipitous. Actually, precisely the opposite turns out to be true. Instead of preventing chaos, central planning creates it. This lesson was learned in brutal experiments with bureaucratic coordination that were conducted during the twentieth century in the Soviet Union, the People's Republic of China, and other totalitarian states. Secret police forces of enormous size were deployed in each of these nations to try to enforce planners' edicts and to suppress criticism when planning failed, as it did routinely. The Soviet Union, a nuclear power capable of killing everyone on the planet, collapsed largely because it was unable to accomplish the mundane task of matching the goods and services that people desire with production of these same goods and services. The Marxist mandarins of China phased out central planning before that system brought them down; this turning away from communism created the economic growth that so far has allowed these rulers to forestall political reform.

What totalitarian rulers sought to accomplish – at the expense of tens of millions of Chinese, Russian, Ukrainian, and other lives – was the elimination of markets. This turned out to be disastrous because markets are not just a setting – geographic, electronic, or otherwise – for exchange between businesses (including farms) with something to sell and their customers. Markets also comprise a mechanism for bringing production and consumption into line with each other. As explained in the Appendix, this coordination is accomplished with the price signals that markets generate. Prices are a remarkably effective way to guide the decisions of independent economic agents, not least when these decisions have to do with something as basic as food. Indeed, price signals are clearly superior to any alternative coordinating mechanism.

Along with bringing production into line with consumption, markets accommodate change with remarkable ease. This is especially important given how profoundly the global food economy is affected by population growth, technological improvement, and other trends examined in this book. How markets adjust to change is a major concern of this chapter.

Much of the discussion that follows, including an analysis of how government policies affect food markets, is conceptual, like the subject matter of a basic class in economics. The content of the latter portion of this chapter, though, is empirical. We examine the track record of food markets during the second half of the twentieth century, noting especially the downward trend in prices that came about because supply increased more than demand. Whether or not this downward trend will continue in the years to come is the final topic addressed in this chapter.

4.1 The Desirability of Competitive Equilibrium

As explained in the appendices of the two preceding chapters, economists analyze consumption and production in a competitive market, in which there are many buyers and many sellers, by deriving demand and supply curves – which express simple linkages between consumption and production of a good, respectively, and the good's price – and then examining shifts in these curves. In the case of demand, quantity consumed at any given price goes up or down in response to variations in income, consumers' tastes, and other factors that must be held constant in order to trace out a demand curve. By the same token, an increase or decrease in supply results from a change in productive capacity, technology for converting inputs into output, or input prices, all of which must remain fixed for a supply curve to be derived.

For any given combination of demand and supply, there is a price at which a competitive market is in equilibrium, in the sense that the quantity that consumers demand at that price equals the quantity that firms supply at the same price. The equilibrating tendencies of competitive markets are examined in the Appendix, as are changes in market equilibrium caused by shifts in demand, supply, or both. But there is something else to appreciate about competition. As is explained in Appendix, the difference between the value that consumers place on output – defined as their willingness-to-pay (WTP) for it – and the cost of labor, raw materials, and other variable inputs used to produce that output is maximized.

This difference, or net economic value (NEV), is distributed between consumers and producers. The former group's portion, which is called consumers' surplus (CS), equals WTP less the market value of output, or price multiplied by quantity. Producers' surplus (PS), which is the latter group's portion, equals the same market value less variable production costs. Both CS and PS can be identified by referring to a standard demand-and-supply diagram, such as the representation of the hypothetical market for caviar in Astoria-by-the-Sea in Figure 4.1. Likewise, interpretation of the same diagram indicates that the sum of CS and PS – which by definition equals NEV – is maximized if a market reaches competitive equilibrium.

First consider CS. Astoria-by-the-Sea happens to be inhabited by a wealthy individual with a strong craving for sturgeon eggs, and the maximum "bid" he would make for the first ounce coming his way every week is very high: $610. Since no one else offers more money for this first ounce, the rich individual's bid represents the first ounce's marginal value (MV), which is where the demand curve touches the vertical axis (Figure 4.1). By the same token, the MV of the second ounce supplied to the market comprises the maximum amount that someone offers for it; this bid is still high, though just a little under the MV of the first ounce: $606 (i.e., the second point on the demand curve). MV, which can always be found by inspecting the demand curve and in this example is a linear function of output,[1] continues to fall as more caviar is made available. This decline reflects the fundamental characteristic of consumers' preferences, which is diminishing marginal utility (see Appendix to Chapter 2).

---

[1] $MV_Q = 610 - 4Q$, where Q represents weekly output.

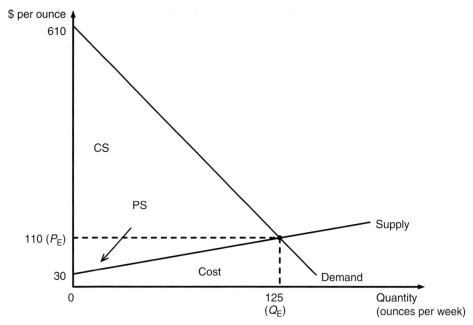

Figure 4.1 CS and PS.

As indicated in Figure 4.1, MV falls to $110, which is the competitive-equilibrium price (P_E), as weekly caviar purchases reach 125 ounces, which is the competitive-equilibrium level of consumption and production (Q_E). This makes perfect sense. What someone offers for the last, or marginal, unit consumed is precisely equal to its price, neither more nor less. Significantly, however, every buyer faces the same price, $110. There is obviously a $500 gap between the $610 that the rich caviar addict would pay for his first ounce and P_E, a $496 difference between the MV of the second ounce and P_E, and so on up to no gap at all between MV and price for the 125th ounce. Adding up all these differences, we obtain overall CS:

$$\begin{aligned} CS &= (\$610 - 110) + (\$606 - 110) + \cdots + (\$110 - 110) \\ &= \{\tfrac{1}{2} \times (\$610 + 110) \times 125\} - \{\$110 \times 125\} \\ &= \tfrac{1}{2} \times (\$610 - 110) \times 125 = \$250 \times 125 = \$31{,}250. \end{aligned}$$

As this calculation illustrates, CS is the difference between the sum of the 125 MVs and market value. The former sum is equivalent to WTP – represented in Figure 4.1 as the trapezoidal area under the demand curve (which is the same as the MV curve) between 0 and Q_E. Also, the market value of competitive output is represented in the same figure by the rectangle with a height of P_E and a length of Q_E. Accordingly, CS consists of the triangle bounded by the demand (or MV) curve, the vertical axis, and a horizontal line extending out of that axis from P_E.

The description of PS is analogous to that of CS. Just as there is a consumer in Astoria-by-the-Sea who values the first ounce of caviar he eats very highly, there is an unusually efficient firm that can supply 1 ounce to the same market at a cost that no other firm can match. To be specific, the marginal cost (MC) of this first ounce is $30, which is the supply curve's intercept on the vertical axis. While still low, the second ounce's MC is slightly higher: $30.64. MC, which like demand in this example is a linear function of output,[2] continues to be pulled up as output rises.

In competitive equilibrium, MC equals $110, which of course is P_E. Since every unit of output changes hands for this same price, regardless of its MC, PS accrues:

$$\begin{aligned}PS &= (\$110 - 30) + (\$110 - 30.64) + \cdots + (\$110 - 110) \\ &= \{\$110 \times 125\} - \{\tfrac{1}{2} \times (\$30 + 110) \times 125\} \\ &= \tfrac{1}{2} \times (\$110 - 30) \times 125 = \$40 \times 125 = \$5,000.\end{aligned}$$

Since the supply and MC curves are equivalent (as noted in Chapter 3), the sum of the first MC through the MC of Q_E is represented in Figure 4.1 as the trapezoidal area under the supply curve between 0 and Q_E. Subtracting this area from the rectangle that represents the market value of competitive output, one obtains PS: the triangle bounded by the horizontal line extending out from P_E, the vertical axis, and the supply (or MC) curve.

Having defined and described CS and PS, we point out that the sum of the two, which to repeat is NEV, is maximized if equilibrium is reached in the competitive market depicted in Figure 4.1. A clear indicator of efficiency, or NEV maximization, is that P_E, which equals MV, also equals MC. In other words, the value to the consumer who purchases the last unit supplied to the market (in this case, the 125th ounce of caviar) is equal to the cost some firm incurs to produce that same marginal unit. A lower level of output, at which MV exceeds MC, is undesirable in the sense that consumers' WTP for additional output exceeds the cost of raising production. The NEV losses associated with this kind of inefficiency, which arises if a market is monopolized by one or a few firms, are examined in the Appendix. In contrast, producing and consuming more than the efficient level means that the value of excess output (i.e., everything over and above Q_E) is less than the cost of same; in other words, NEV increases if output diminishes. As explained in the next chapter, excess production, at which MC is greater than MV, often happens because firms do not internalize environmental costs.

4.2 The Market Impacts of Commodity Programs

The equilibrating tendencies of competitive markets are unrelenting. Prices gravitate directly to levels that align consumption and production. Similarly, the

---

[2]$MC_Q = 30 + 0.64Q$, where Q represents weekly output.

marketplace's accommodation of changes in demand and supply is expeditious. As such an accommodation occurs, the sum of CS and PS is maximized.

But while desirable outcomes of the sort represented in Figure 4.1 are ubiquitous, inefficiency is also a regular occurrence. Often the problem is governmental interference with market forces. However, the state can also take action to reduce inefficiency. Anti-trust laws, for example, impede monopolization and anti-competitive business practices. Likewise, a fundamental purpose of environmental legislation is to oblige firms and households to take into account the impacts of their activities on natural resources.

Promoting competition and limiting environmental damages are two purposes of public policy for the food economy. But other purposes are at least as important. One of these is containment of the price instability often observed in commodity markets. As indicated in Chapter 3, agriculture is much more exposed to nature than are other parts of the economy. Thus, ideal climatic conditions and bumper harvests one year may be followed 12 months later by a shortfall in output resulting from drought, pest infestations, or heavy rains. If food demand were elastic, then positive or negative supply shocks would have modest effects on prices. But as a rule, these elasticities are low in the short run, which implies that prices vary a lot as supply expands or contracts.

Price variation in one country's food markets need not have a local environmental cause. There are also external sources of variation, such as unusual harvests in some other parts of the world. For example, disappointing production elsewhere will drive up the prices of imports, which are close substitutes for domestic output. This will, in turn, cause demand for local product to increase. If domestic supply is inelastic, as is usually the case, then higher prices will be the main consequence of the external shock.

Regardless of whether it is related to the local environment or something else, price instability can be contained in various ways. One of these is to maintain a stabilization fund, or even a stock of commodities that can be added to or sold in order to keep prices from falling or rising too much. When farm products are relatively abundant, withdrawals from the fund can be channeled to commodity purchases, thereby braking price declines. Conversely, the managing authority can enter the market as a seller during times of acute scarcity, which dampens price increases. Stabilization funds have been established by a number of governments, as have buffer stocks of commodities controlled by the public sector. However, it is debatable whether this intervention really creates more price stability than do periodic releases from and replenishment of private reserves (Williams and Wright, 1991, pp. 410–451).

Another way to reduce price volatility is to disseminate market information, including projections of output and demand around the world. Early warning of a supply glut, for example, allows farmers to alter production decisions in ways that stabilize prices and incomes. Information of this sort can be regarded as a public good – something that ought to be supplied by the state (Chapter 3). Aside from governmental action to prevent prices from going up or down excessively, yield-risk insurance, which rarely if ever is offered by private firms,[3] can be provided by the public sector to insulate farm incomes from the effects of poor harvests. Also, farmers can protect themselves from income variability by diversifying their output, rather than producing just one commodity and earning a lot when its price is high and doing poorly when its price is low.

Stabilization and enhancement of farm incomes, through yield-risk insurance and other means, have been fairly rare in the developing world. In much of Africa, Asia, and Latin America, farmers comprise a large share of the population and the financial resources of the state are limited. Thus, providing all farmers with income support is out of the question. If anything, governments of poor countries usually intervene in food markets to the detriment of producers. One such intervention is currency over-valuation, which makes imports of food and other goods cheaper and makes exporting farm products and anything else less profitable (Chapters 6 and 7). Price controls are another intervention, with a maximum price for food set below the equilibrium level. Of course, this policy creates shortages made up of the difference between consumption, which is stimulated, and production, which is discouraged. One way that governments controlling the price of food alleviate the resulting shortages is to draw on the excess production of nations where agriculture is subsidized. Thus, there is a connection between the extra income that producers in rich countries receive because of governmental interference with market forces and the losses suffered by farmers in poor countries due to market regulation (Chapter 6).

Over the years, farm earnings have been propped up in various ways in the United States, Western Europe, and other wealthy places. Although there has been an evolution toward policies that are supposed to create minimal economic inefficiency, each policy detracts from economic well-being.

One way to enhance agricultural income is to induce farmers to cut back on production. A potential benefit of this approach is environmental, in particular if output is reduced by taking erodible land out of production. But in commodity markets, supply management (assuming it actually works) has the same effect as monopoly. As discussed in the Appendix, output below the efficient, competitive level puts upward pressure on prices and reduces NEV.

Supply management, which was introduced during the New Deal of the 1930s (Cochrane, 1979, pp. 286–289), has been attempted on various occasions in the United States since the Second World War. In most of these initiatives, reducing farmed area has not had the desired impact on production, mainly because farmers have been able to compensate for diminished use of natural resources by applying more commercial fertilizer and other purchased inputs (Chapter 3). Also, demand

---

[3] The provider of any kind of insurance must deal with problems of asymmetric information. One such problem is adverse selection, which would arise with crop insurance if all farmers paid the same premium. Under these circumstances, low-risk farmers would not find the purchase of coverage to be worthwhile. Accordingly, the insurer would end up with a client pool comprised mainly of high-risk farmers. Another problem of asymmetric information is moral hazard, which would arise because farmers with coverage would be more inclined to run risks. Recognizing the costs they would incur because of asymmetric information, private companies generally do not insure all kinds of crop risks, although many cover hail damage (Wright and Gardner, 1995, p. 37).

grows more elastic as time goes by, which means that any reduction in output causes revenues to fall. In light of the limited effectiveness of supply management, alternative strategies for raising farm incomes have been pursued.

One alternative is commodity price supports – that is, a minimum market value (P_S) that exceeds the equilibrium price (P_E). This intervention creates a surplus of course, represented by the horizontal distance in Figure 4.2 between Q_S (i.e., the quantity farmers produce at the minimum price) and Q_D (i.e., the quantity demanded at that same price). As with the over-production that occurs if producers neglect environmental damages, the costs of producing Q_S instead of Q_E, which are represented by the trapezoidal area under the supply (or MC) curve between the two output levels, exceed the benefits to consumers of this extra output, which amount to the trapezoidal area under the demand (or MV) curve between Q_E and Q_S. In Figure 4.2, this difference between the costs and benefits of excess production is represented by area A.

For price supports to work, government must purchase all surplus output at the minimum price. The associated budgetary outlay is represented in Figure 4.2 as the rectangle lying between Q_S and Q_D and between 0 and P_S. Furthermore, the public sector must incur the costs of storing the surplus and disposing of it, although some offsetting revenues can be earned if commodities are sold (e.g., in a foreign country where food price controls are creating shortages). If the government prefers not to

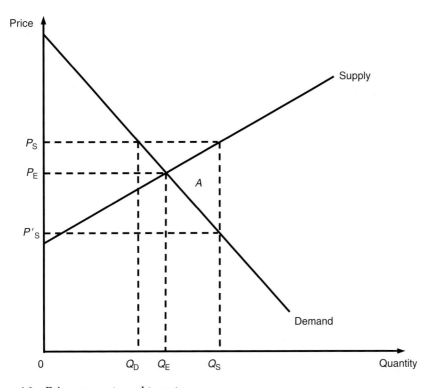

Figure 4.2 Price supports and targets.

get in the commodity business, it can still subsidize farmers by offering deficiency payments, which are equal to the difference between target values and prices at which farm products are actually sold. If the target price in Figure 4.2 is P_S, then the only way that producers will sell all their output, Q_S, is to offer it to consumers at P'_S. Overall deficiency payments, then, will consist of the rectangular area bounded by these two values, the vertical axis, and Q_S. While these payments may be less than those associated with a simple support price, there is no difference in NEV losses.

Policy-makers and others who want to avoid inefficiency while still propping up agricultural earnings currently recommend "decoupled" payments to farmers. The label here refers to the absence of a link between payments, which are financed by taxpayers, of course, and production. If decoupling were truly achieved, then prices and quantities would settle in at competitive-equilibrium levels and, at least in commodity markets, there would be no losses in NEV of the sort that result from supply management, support prices, or deficiency payments.

Perfect decoupling is unattainable. Since any payments that farmers receive from the public sector relax capital and credit constraints, production inevitably increases, which has a negative impact on NEV. Incentives to raise more crops are further strengthened because a linkage is perceived (correctly, quite often) between current output and future payments. Aside from production impacts, there are costs that decoupled payments share with other policies for enhancing agricultural income. One of these relates to the scarce time and resources that farmers devote to winning subsidies from government, instead of producing goods and services. Another consists of the opportunity cost of public bureaucracies needed to administer farm programs. In addition, payments from government do not materialize from thin air. Rather, this money must be extracted directly or indirectly from taxpayers, and some studies suggest that GDP contracts by at least $0.16 for every $1.00 collected by the government from the general public (Ballard *et al.*, 1985). This impact needs to be considered in a comprehensive evaluation of decoupled payments or any other farm policy.

Imperfect though decoupling undoubtedly is, this approach still creates less inefficiency than price supports and other governmental interventions in commodity markets. In particular, payments to farmers that are tied neither to prices nor output distort trade and production minimally (Tweeten and Thompson, 2002, p. 10). Decoupling is now being pursued by the United States and is being promoted in multilateral trade negotiations.

Governmental programs that support commodity prices, raise farm incomes, or both are sometimes rationalized as a means to stimulate farmers' adoption of improved technology, although no empirical evidence exists to support this claim (Gardner, 2002, pp. 257–260). To the contrary, Huffman and Evenson (1989) have found that 90 percent of the productivity gains in US agriculture have resulted from research and extension. A more likely impact of policy-induced distortions in markets for farm products is to drive up land prices, as David Ricardo – like Malthus, a luminary of classical economics – first pointed out in the nineteenth century (Ricardo, 1965, pp. 33–45).

To recognize that propping up commodity prices or farm incomes leads directly to the bidding up of real estate values is to appreciate that those who rent land gain little from farm policies. The same is true of individuals who buy land after such policies have been implemented – at elevated prices, of course. Instead, the main beneficiaries are those who happen to be landowners at the time of the policy change. Many of these people are able to sell their agricultural holdings and retire comfortably on the proceeds. The beneficial effects of this pay-off are negligible.

4.3 Historical Trends in the Scarcity of Agricultural Products

With agricultural earnings supported in a number of rich countries and farmers in quite a few developing countries suffering income losses because of governmental policies, one wonders whether the trend toward diminished scarcity of farm products has really showed up in the marketplace. Through the middle of the twentieth century, data limitations precluded a definitive answer to questions of this sort.

One of the first empirical studies of long-term trends in the scarcity of products from agriculture and other resource-based sectors made use of a time series of prices and production costs in the United States from 1870 through 1957. The two economists who carried out this study noted that prices of farm products had fluctuated markedly. Along with going down when harvests were good and increasing when output fell short, market values tended to fall during recessions and rise in wartime. Having declined after the Second World War, market values in 1957 were little changed from what these had been 87 years earlier (Barnett and Morse, 1963, pp. 211–212). But while the analysis of price trends indicated no major change in food scarcity, other parts of the study yielded evidence that scarcity was actually declining. It was found, for example, that the labor and capital required to produce a unit of farm output had gone down, not up, from 1870 to 1957. Evidently, mechanization – the substitution of capital for labor – had coincided with a decline in production costs. The study's two investigators also acknowledged the impacts of rising total factor productivity (TFP) – the difference between the value of crop and livestock output, on the one hand, and the costs of land, capital, labor, and other inputs, on the other (Chapter 3) – caused by the introduction of hybrid seeds and other innovations (Barnett and Morse, 1963, pp. 166–168, 197–198).

As reported in Chapter 3, TFP growth in agriculture was nowhere close to ending in 1957, which was the terminal date of Barnett and Morse's study. With agricultural productivity climbing steadily, food has become progressively less scarce, as indicated by declining commodity values during the second half of the twentieth century. Diminished scarcity is clearly evident in the United States, where markets are generally open to international competition and prices are consequently an accurate barometer of scarcity. As indicated in Figure 4.3, the inflation-adjusted price of corn fell by three-fourths between 1950 and 2000. For the other two staple grains, rice and wheat, the relative declines in real prices were even greater.

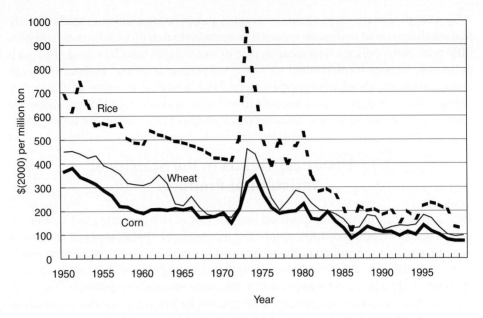

Figure 4.3 Real prices ($2000) of US rice, wheat, and corn from 1950 to 2000.
Source: US Department of Agriculture, prices deflated by implicit GDP.

As can be seen in Figure 4.3, exceptions to the general trend toward diminished scarcity have occurred. The most important of these was a spike in cereal prices during the early 1970s. Events in the Soviet Union were largely to blame. Reluctant to cut food supplies to consumers in the face of poor grain harvests and the expanded feeding of livestock, communist authorities abandoned their long-standing policy against imports from places like the United States. Thus, there was a positive shock to demand in international markets. At the same time, increased earnings from oil-exporting nations, which had benefited from a quadrupling of oil prices in 1973–1974, were being deposited in international banks, which in turn made loans in Latin America and other parts of the developing world. Some of these loans were used to buy grain on international markets, thereby augmenting demand.

The food crisis of the early 1970s – as it was called at the time – brought home the lesson that food availability cannot be taken for granted, even in international markets in which supply interruptions in one part of the world are normally cancelled out by unusual abundance somewhere else. Taken individually, none of the positive shifts in demand or negative changes in supply would have affected prices. But together, the combined demand and supply shocks were enough to affect market equilibrium dramatically, driving up grain prices by more than 100 percent in a little over a year.

There is another lesson to draw from the food crisis of the early 1970s, which is that markets recover quickly after an unusual shock. Just as prices climbed rapidly when a "perfect storm" of unusually strong demand coupled with supply shortfalls struck, market values declined without delay once the storm passed and conditions returned to normal. After modest run-ups in the price of rice later in the 1970s, grain values resumed their downward slide in the 1980s.

This latter lesson is not always fully appreciated, in part because an upward deviation from average prices receives more coverage in the press and from public figures than the subsequent return to normalcy. Furthermore, warnings from the Club of Rome about imminent over-population and resource exhaustion (Chapter 2) amplified the distress aroused by the food crisis 30 years ago. In particular, the doubling of grain values seemed to confirm that mass starvation was just around the corner, as Ehrlich (1968) and others were predicting.

In a sense, the temporary abatement of the long-term trend toward cheaper food that occurred during the 1970s (Figure 4.3) was fortuitous. As commodity prices held steady, or even went up, incentives to adopt Green Revolution varieties of rice, wheat, and other crops were reinforced. This was true at the farm level, with individual producers willing to adopt more productive technology as long as output values were relatively high. It was also true for entire countries, which were more inclined to underwrite the research and extension needed to accelerate technological change if rising prices could be regarded as a signal of mounting food scarcity. In contrast, technological improvement seems less appealing for farmers and less urgent for society as a whole if commodity values are falling, as has happened since the early 1980s (Figure 4.3).

Spurred for a while by high output prices and having the fundamental advantage of enhancing agricultural TFP, the Green Revolution ultimately saved much of humankind from the cataclysm that many were forecasting in the late 1960s and early 1970s. As agriculture has intensified, increases in the supply of food consistently have exceeded those in demand – particularly in Asia, which is the world's most populous continent. Per-capita food supplies rose (Chapter 3) and, though little or no mention of it was made in the press, the long-term decline in commodity values that had stalled in the early 1970s resumed a decade later. As the century drew to a close, inflation-adjusted prices for the goods that nourish us had reached historical lows (Figure 4.3).

The easing of food scarcity that has taken place since 1950 is a remarkable achievement. Equally remarkable are the improvements in productivity that are primarily responsible for diminished scarcity. Thanks to advances such as the adoption of hybrid corn in the United States and semi-dwarf crops in Asia and other parts of the developing world, it has been possible for agricultural TFP to multiply precisely as real commodity values declined by 50 percent or more.

Productivity growth also has reduced the burden for farmers of falling prices. If improved productivity had not driven down costs, then agriculture's profitability – as measured by rates of return on land, capital, and other assets – would have declined over time. In fact, rates of return on the assets of commercial farmers have held remarkably steady as the years have gone by, averaging a little more than 10 percent in the United States (Hopkins and Morehart, 2002). This means that, in times past as well as today, $100 in productive capacity regularly has yielded output worth about $10 in one year, a comparable amount next year, and so forth.

This level of profitability is about what one expects in a competitive market that, in addition to consistently aligning consumption with production, reaches what economists call long-run equilibrium. In this equilibrium, rates of return are barely sufficient to maintain existing productive capacity without inducing any new investment (i.e., additions to capacity). A market is out of long-run equilibrium if the rate of return is higher. However, this triggers investment, which augments supply and causes a price decline that in turn diminishes profitability. Conversely, a rate of return that is too low causes disinvestment, which reduces supply and raises prices and profitability.

Staying in long-run equilibrium is more of a challenge than simply aligning consumption and output through price adjustments, as described in the Appendix. Since productive capacity is durable, economic decisions about replacing it, allowing it to depreciate, adding to it, and so on are based on expectations of market conditions in the future, which will affect the value of output that results from harnessing that capacity. If these expectations turn out to be erroneous, under- or over-investment results. This causes profitability to vary, the rate of return sometimes exceeding the "normal" level (as economists call it) consistent with long-run equilibrium and other times falling below that level. The finding by Hopkins and Morehart (2002) that normal profitability has been sustained in US agriculture is conclusive evidence of the capacity of competitive markets to adjust efficiently to major improvements in productivity that alleviate food scarcity.

4.4 Outlook for the Twenty-First Century

In light of agriculture's track record through the turn of the twenty-first century, the future direction of real commodity values will depend on changes to come in fundamental market forces. If demand outpaces supply, then prices will rise. But if the reverse is true, supply increases that exceed growth in demand will allow humankind to continue along the path of diminished food scarcity.

Our analysis of future trends focuses on cereal crops. Food grains are the source of many of the calories that people consume directly. When indirect consumption of feed grains through livestock products is factored in, cereals account for two-thirds of total caloric intake. Examining trends in the demand, supply, and prices of grain has another merit, which is that information about yields and other production variables are more reliable than data for other agricultural commodities.

The focus on cereals distorts projections of scarcity trends somewhat. As indicated in Chapter 3, yield improvements have been the primary reason why grain production has gone up in recent decades. In contrast, increases in per-hectare output of pulses, roots and tubers, fruit and vegetables have been slower (Table 3.4). Needless to say, reduced yield growth for cereals, perhaps matching rates for other crops, would make agricultural commodities scarcer in the future.

The fundamental indicator of scarcity, of course, is price. Our projections of this indicator during the first half of the twenty-first century rest on the assumption that, because of low elasticities, price goes up or down by 2 percent for any difference of 1 percent between demand and supply growth. Three of the demand-growth scenarios reported in Table 2.7 have been investigated:

- Slow increases would occur if the minimal demographic forecasts of the UNPD prove to be accurate and per-capita food consumption goes up by just 0.2 percent per annum.
- Medium growth would correspond to the UNPD's medium forecasts and an annual rise of 0.3 percent in per-capita consumption.
- Rapid increases would be observed if maximum demographic forecasts turn out to be true and per-capita consumption increases by 0.4 percent each year.

Given current information and assumptions, the second of these three scenarios represents the best guess one can make today about food demand in 2025 or 2050. Furthermore, the range within which demand will actually end up is almost certainly bounded by the other two scenarios. Consider the first of these. If human numbers increase slowly, this result will probably reflect marked improvement in living standards, in which case per-capita food consumption will go up each year by more than 0.2 percent. Thus, the first scenario serves very well as a lower bound for demand projections. Likewise, the third scenario is a good upper bound because, if the population is going up rapidly, increases in earnings will probably not be enough for per-capita consumption to rise at an annual rate of 0.4 percent.

Our projection of supply growth is quite simple. First of all, we suppose that there will be no overall increase in agricultural land use, which reflects what has taken place since the middle 1990s. Second, in light of the projection by Rosegrant *et al*. (2002) of negligible increases in the diversion of water to farmland (Chapter 3), no significant expansion of irrigation is assumed. Third, it is assumed that the linear increase in per-hectare output that has been observed since 1961 (Figure 3.2) will continue to hold. Yields may stagnate because irrigation does not increase. Alternatively, they may be higher in the future, if there is a bonanza of biotechnological innovation for instance. In any event, we assume linear growth in yields and, because there is no extensification, in output as well.[4]

Future increases in supply and demand are reported in Table 4.1, those between 2000 and 2025 in column 1 and those between 2000 and 2050 in column 2. Price changes under each supply–demand combination are reported in the same table. Only for the first demand scenario, in which the population is 7.47 billion in 2025 and 7.87 billion in 2050 (versus 6.06 billion in 2000), does food scarcity continue falling, as it has done since the middle 1900s. Under the medium scenario,

---

[4] There is a superficial resemblance between our supply forecast and Malthus's view of agricultural production. Both are obviously linear. But while we assume equal, additive growth in yields and production, Malthus did not consider changes in per-hectare output. Instead, he supposed that agricultural output would go up arithmetically due exclusively to extensification. Of course, Malthus contended that human numbers increase exponentially, which is different from the current trend (Chapter 2).

increases in demand exceed those in supply by a small margin. Accordingly, a modest rise in prices is to be expected, with average food values in 2025 one-tenth above turn-of-the-century levels. In contrast, a significant part of the progress that has been made toward diminished food scarcity since 1950 will be lost during the next 50 years if rapid growth in human numbers and per-capita consumption occur. In particular, prices in the middle of this century will be nearly double current levels under the third demand scenario (Table 4.1).

Our findings about scarcity trends are generally consistent with those of a number of other researchers. In one study, Rosegrant *et al.* (1995) assume a lower rate of yield growth – 1.24 percent per annum between 1990 and 2020 in their study as opposed to an average of 1.37 percent during the first part of the twenty-first century in ours. However, they also expect positive, though small, changes in agricultural land use – an average annual increase of 0.26 percent, to be precise. The combined supply growth that Rosegrant *et al.* (1995) estimate to result from extensification as well as intensification is a little above the supply increases we project based only on yield growth. However, their projections of demand are higher as well. As is the case with the medium demand scenario, they anticipate modest increases in prices during the years to come. It also bears mentioning that liberalized international trade for agricultural commodities – in particular, diminished incentives for over-production in wealthy nations – would put modest upward pressure on world food prices (Chapter 6).

The general lesson to be drawn from the price projections reported in Table 4.1 is that neither alarm nor complacency about the global food economy is in order. If recent trends in population, living standards, and agricultural production continue, then food prices will hold steady or rise slightly. Dramatic declines in market values, of the sort that occurred between 1950 and 2000, are not out of the question, although counting on such an outcome would not be prudent. Neither is a significant run-up in prices impossible. Increased scarcity could be self-containing, in the sense that sharply higher prices would spur capital formation and technological

Table 4.1 Growth in food supply, demand, and prices from 2000 to 2025 and 2000 to 2050

Supply growth	2000–2025	2000–2050
	36%	72%
	(1)	(2)
Low demand scenario		
Demand growth	30	44
Price change	(12)	(56)
Medium demand scenario		
Demand growth	41	79
Price change	10	14
High demand scenario		
Demand growth	53	120
Price change	34	96

innovation. However, mounting scarcity would also create a heavy burden for hundreds of millions of people in Sub-Saharan Africa, South Asia, and other parts of the world – people who are currently food-insecure even at today's low prices for agricultural commodities (Chapter 8).

Study Questions

1. Did central planning in the Soviet Union, China, and other communist states contain economic chaos or create it?
2. How does the achievement of competitive equilibrium affect the difference between what consumers are willing to pay for output and the variable cost of producing that output?
3. Who captures NEV and in what form?
4. What can governments do to lessen variation in the prices of agricultural commodities?
5. Compare and contrast supply management, support prices, and deficiency payments.
6. Why is decoupling between government payments and commodity output virtually impossible?
7. What has the general trend in inflation-adjusted grain prices been since 1950?
8. Explain the causes of high grain prices during the 1970s and describe the impacts of high prices on the Green Revolution, which was under way that same decade.
9. What combination of demand and supply trends would cause real prices to decline during the next few decades? What combination of trends would result in price increases?

Appendix: The Coordination of Decentralized Decision-Making

Consumption of any good or service is influenced by various factors. As indicated in Chapter 2, the amount of food people eat depends on their numbers and incomes. Choices among food items also reflect individual tastes. In addition, purchases of any single item depend on its price, the prices of a host of other goods, as well as tastes and preferences.

Food output is affected by a number of variables as well (Chapter 3). Productive capacity (i.e., land and capital) obviously has an effect. So do the prices of labor, materials, and other variable inputs. Another determining factor is the existing state of technology for transforming factors of production into agricultural output. Furthermore, production is responsive to the price of the good being produced, just as consumption is.

Underlying the alignment of consumption and production, then, are various adjustments by households and firms. To understand how all these adjustments end up being coordinated, some abstraction from complex reality is required. As explained in the appendices of the two preceding chapters, it is convenient to employ the assumption of *ceteris paribus* to create a demand curve – which relates consumption of a good to its price, alone – as well as a supply curve – which relates output of the same good only to its price. After doing this, one can analyze equilibrium in the market where consumers and producers do business, with primary focus on the role of prices in the achievement of that equilibrium.

The Alignment of Production and Consumption

There are two special sorts of market equilibrium. One of these occurs if production is economically infeasible, in which case nothing at all is exchanged between producers and consumers. Making caviar in Sierra Leone would be a good example. Infeasible production is signaled by the supply curve's intersection with the vertical axis (on which monetary values are measured) above the demand curve's intercept on the same axis, which means that the maximum amount that any consumer would pay for just one unit of output is less than the cost of producing the same unit. Non-scarcity is signaled if the supply curve rises out of the horizontal axis (on which quantity of output is measured) to the right of where the demand curve descends to that axis, in which case the good is free. Though vitally important, the air we breathe is an example of something that is not economically scarce.

Depicted in Figure 4.4 is the more typical case: a market for a good, like corn in the United States, that is both scarce and feasible to produce. With the demand curve sloping downward and the supply curve sloping upward, there is one and only one point where the two intersect. This point, at which production and consumption both amount to Q_E, is the market equilibrium. For US corn, Q_E is about equal to 9.20 billion bushels per annum.

Price adjustments drive the market inexorably toward equilibrium. As shown in Figure 4.4, the price, P_E, that causes annual consumption as well as yearly output of corn to equal 9.20 billion bushels is $2.00 per bushel. Consider what takes place if the price is higher, $2.50 per bushel for instance. At this higher value, production (9.41 billion bushels) exceeds consumption (8.80 billion bushels). In other words, there is a surplus of 0.61 billion bushels. With unsold inventories accumulating, producers cut prices. As the market value of the good being exchanged falls toward P_E, production goes down and consumption goes up, thereby causing the surplus to dwindle. Price adjustments cease entirely as the difference between output and consumption disappears – in other words, as equilibrium is achieved.

The tendency toward market equilibrium is similarly robust if price starts out below P_E. If a bushel of corn changes hands for $1.20, what people want to consume (10.18 billion bushels) exceeds what farmers want to produce (8.75 billion bushels). Responding to this shortage of 1.43 billion bushels, consumers bid the price up, which causes the gap between consumption and production to close. Once again, equilibrium is achieved once bushels of corn are changing hands for $2.00 apiece.

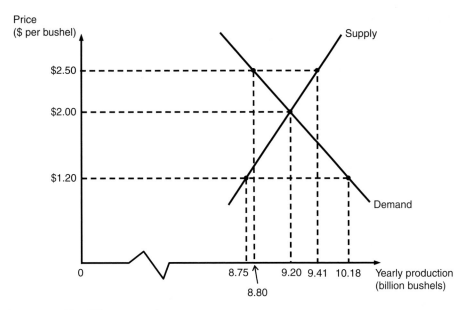

Figure 4.4 The US corn market.

Shifts in Demand and Supply

The equilibrating tendencies of markets are powerful, indeed – so powerful that markets adjust quickly to changes in demand or supply caused by things like population growth and technological innovation. That is, there is an expeditious adjustment from one equilibrium to another in response to a shift in demand, supply, or both. Almost always, the adjustment has an impact on output as well as price.

The easiest changes in equilibrium to analyze are those that result when either the demand curve or the supply curve shifts and the other curve remains in place. Consider a market in which demand has increased due to a rise in consumers' earnings. The price that formerly aligned consumption and production no longer does so. With no change on the supply side, production at the old equilibrium price is exactly the same. However, there is more consumption at that price because of the demand shift. The shortage induces a price rise, which ceases once consumption has fallen sufficiently and production has gone up enough for the two to equal one another. The end result of this adjustment is a higher price and more output.

Another example of a simple change in market equilibrium is a supply increase not accompanied by any alteration in demand. If farmers use more land, this increase in productive capacity leads them to produce more crops at any given price. At the old equilibrium price, there is a surplus. To be specific, consumption is unchanged while output is greater. As always, a surplus induces price-cutting, which causes consumption to grow and production to fall. Equilibrium is reestablished at a lower price and higher level of output.

The direction of changes in price and output is easy to determine when just one of the two curves, either demand or supply, shifts. With a change in demand, price and output always move in the same direction. If demand goes up, both are higher at the new equilibrium. If demand goes down, then these two variables decline. With supply changes, the two always move in opposite directions. An increase in supply causes output to rise and price to fall. The opposite happens if supply declines.

At least as interesting as the general direction of price and output adjustments in these simple cases are the relative magnitudes of these adjustments. As with so much else, the determining factor here is elasticity. Whether the relative change in price resulting from a shift in demand exceeds the relative change in output or vice versa depends on the elasticity of supply. Similarly, the sensitivity of consumption to price variations determines which changes more because of a supply shift, price or output.

The importance of elasticity is illustrated in Figure 4.5. Shown in Figure 4.5A is the shift in the spinach market's equilibrium caused by reduced prices for lettuce. The substitute good being cheaper, demand for spinach falls, which obviously causes price and output to decline. Since supply is elastic, relative change in the latter variable exceeds relative change in the former. A reasonable interpretation of all this is that farmers can move in or out of spinach production quickly and easily, which makes their output very sensitive to price changes. It is only to be expected, then, that the decline in demand results mainly in diminished production, with minimal price effects.

Represented in Figure 4.5B are the consequences of an improvement in agricultural technology, which increases food output at any given price. For a typical food item, own-price elasticity in the domestic market is fairly low, as indicated by the demand curve's steep slope. Given the limited responsiveness of consumption to price changes, technological change does not have much impact on output. Instead, the main consequence is a price reduction, which implies that consumers are the main beneficiaries of supply growth.

Of course, everything is more complicated if there are simultaneous shifts in demand and supply, as occurs all the time in the real economy. If these both decline, one cannot automatically say that equilibrium price, which is pulled down by decreased demand though pulled up by decreased supply, will be higher or lower. If demand rises while supply falls, equilibrium output can increase, decline, or stay exactly the same. Empirical investigation is needed to assess the net impacts of the changes taking place.

It is opportune at this point to repeat admonitions from Chapters 2 and 3 to be precise in the use of economic vocabulary, in particular not treating demand (a functional relationship between price and consumption) as a synonym for consumption or treating supply (a functional relationship between price and output) as a synonym for production. Failure to avoid this error can create enormous confusion, especially when simultaneous changes in demand and supply can result in variations in production and consumption that are either positive or negative.

In places like the United States, where countless markets reach competitive equilibrium routinely, bringing production and consumption into line with each

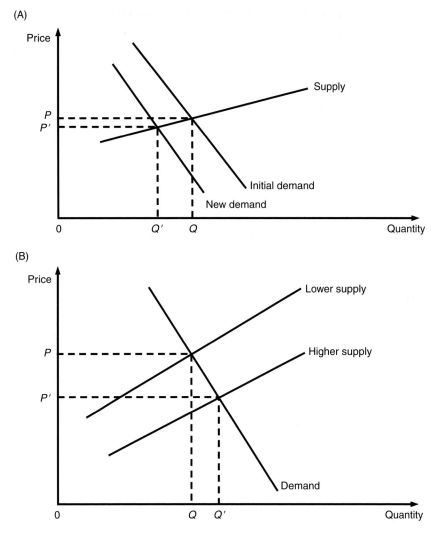

Figure 4.5 Elasticity and changes in market equilibrium. (A) A shift in spinach demand with elastic supply. (B) A change in food supply with inelastic demand.

other is taken for granted. Indeed, students in economics classes, who are tested on their understanding of demand-and-supply diagrams like Figure 4.4, sometimes seem to be the only people thinking about market equilibrium. But even if it is taken for granted, a system that coordinates the behavior of producers and consumers, and readily accommodates changes in demand and supply is precious. As emphasized at the beginning of this chapter, various totalitarian nations learned the hard way during the twentieth century that suppressing markets create economic chaos. People find this intolerable.

Net Economic Value and its Maximization

Compared to the chaos created by central planning, competitive equilibrium is eminently satisfying for just about everyone concerned. In particular, the fact that NEV is maximized when consumption falls into line with production in a market in which there are many buyers and many sellers can be demonstrated with reference to the hypothetical market for caviar in Astoria-by-the-Sea.

As indicated earlier in this chapter, the value that consumers place on the competitive level of output (Q_E) comprises the sum of MVs, which is represented as the area under the demand (or MV) curve between 0 and Q_E:

$$\text{WTP} = MV_1 + MV_2 + \cdots + MV_{125} = \$610 + \$606 + \cdots + \$110 = \$45{,}000.$$

Also, the variable cost of producing Q_E is represented by the area under the supply (or MC) curve between 0 and Q_E and comprises the sum of MCs:

$$\text{Cost} = MC_1 + MC_2 + \cdots + MC_{125} = \$30.00 + \$30.64 + \cdots + \$110 = \$8{,}750.$$

Subtracting variable cost from WTP yields NEV:

$$\text{WTP} - \text{Cost} = \$45{,}000 - \$8{,}750 = \$36{,}250,$$

which is obviously the triangular area between the demand and supply curves. As demonstrated above, CS and PS equal $31,250 and $5,000, respectively, under competitive equilibrium, so the two indeed add up to NEV.

To appreciate the efficiency of producing 125 ounces of caviar, consider the NEV of a lower production level, 120 ounces for instance. Now the gap between WTP,

$$MV_1 + MV_2 + \cdots + MV_{120} = \$610 + \$606 + \cdots + \$130 = \$44{,}400,$$

and variable cost,

$$MC_1 + MC_2 + \cdots + MC_{120} = \$30.00 + \$30.64 + \cdots + \$106.80 = \$8{,}208,$$

comes to $36,192. Also, over-production is every bit as inefficient as under-production, in the sense that NEV is not maximized. If, say, 130 ounces are produced, the difference between WTP,

$$MV_1 + MV_2 + \cdots + MV_{130} = \$610 + \$606 + \cdots + \$90 = \$45{,}500,$$

and variable cost,

$$MC_1 + MC_2 + \cdots + MC_{130} = \$30.00 + \$30.64 + \cdots + \$113.20 = \$9{,}308,$$

is $36,192, which is obviously lower than the net value ($36,250) of 125 ounces of sturgeon eggs.

Inefficient under-production always occurs if a market is monopolized. Consider, for example, how much caviar is supplied, Q_M, in Astoria-by-the-Sea if only one firm produces sturgeon eggs and that firm can raise or lower the price, P_M, as it sees fit. If different customers cannot be charged alternative prices (i.e., the monopolist cannot engage in "price discrimination"), then the difference between the monopolist's revenues (price times output, or $610Q - 4Q^2$) and its variable costs is maximized by producing 67 ounces and selling caviar for $342 per ounce (i.e., the MV of the 67th ounce).[5] The foregone NEV resulting from this under-production is represented in Figure 4.6. At Q_M, a positive margin exists between MV and MC: $342 versus $72.88. In other words, what someone bids for the 67th unit of output (i.e., the last unit supplied by the monopolist) exceeds the cost of producing it. Though somewhat smaller, the margin between MV and MC for the next unit (i.e., the 68th) is likewise positive. The same goes for every other unit up to Q_E. Adding up these margins yields the loss in NEV resulting from monopoly. As shown in Figure 4.6, this loss shows up as the shaded triangular area between the MV and MC curves and between Q_M and Q_E:

$$\tfrac{1}{2} \times (\$342 - 72.88) \times (125 - 67) = \$7{,}804.48.$$

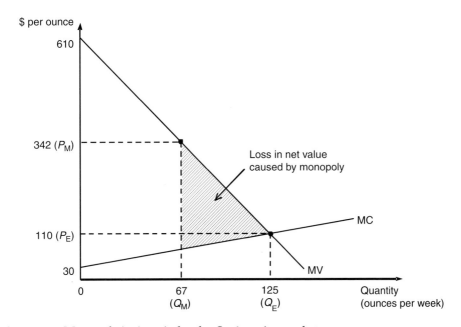

Figure 4.6 Monopoly in Astoria-by-the-Sea's caviar market.

---

[5] To find Q_M, one starts by differentiating revenues, which equal $610Q - 4Q^2$, with respect to Q in order to find the marginal revenue (MR) of output. Profit maximization requires that MR ($610 - 8Q$) equal MC ($30 + 0.64Q$). Solving for output, one finds that Q_M is approximately 67. Plugging this output level into the MV equation, one obtains P_M: $610 - 4 \times 67 = 342$.

The monopolist accepts this loss because raising output above Q_M would lead to a cut in price, in particular a reduction in the price paid for each of the first Q_M units.

Finally, we observe that, although NEV is always distributed between consumers and producers, the relative sizes of CS and PS depend entirely on elasticities of demand and supply. In the hypothetical market for caviar depicted in Figure 4.1, consumption is much less sensitive than production is to price changes. To be specific, the price of sturgeon eggs must rise by $4.00 to cause weekly consumption to decline by 1 ounce while a price change of $0.64 is enough to cause a 1-ounce variation in output. With supply much more elastic than demand, consumers capture the lion's share of NEV, in the form of CS. If consumption responded more than production to price changes, then the portion of NEV going to producers as PS would exceed the portion distributed as CS.

Since NEV and its distribution between consumers and producers depend on the location of demand and supply curves, displacement of either curve affects CS and PS. What about simultaneous increases in demand and supply, which have occurred and will continue to happen in the world food economy? One outcome is illustrated in Figure 4.7. Due to technological improvement, investment, or some other reason, displacement of the supply curve exceeds the shift in demand resulting from

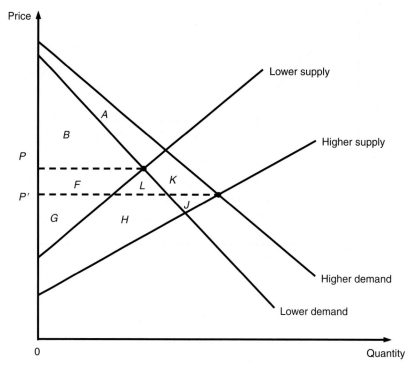

Figure 4.7 Supply growth that exceeds demand growth.

population growth and improved living standards. Food, then, is less scarce, as is indicated by the decline in its price – from P to P'. With demand increasing and price going down, CS is obviously on the rise, increasing from area B to the sum of areas B, F, L, K, and A. However, PS is going up as well, the triangular area formed by the original supply curve, the vertical axis, and the horizontal line extending out from that axis from P (i.e., areas F and G) obviously being smaller than the triangle bounded by the new supply curve, the vertical axis, and the horizontal line extending out from P' (i.e., areas G, H, and J).

5
Agriculture and the Environment

The advantages of allocating goods and services in markets with multiple firms and households are emphasized in the preceding chapter. Not only are individual decisions about production and consumption coordinated as a competitive market clears, but net economic value (NEV) is maximized. Where agricultural commodities are concerned, markets respond appropriately to changes in food scarcity. Since 1950, for example, real prices of farm products have fallen as supply growth has run ahead of increases in demand, in spite of distortions caused by government policies.

The advantages of competitive markets are comparable when decisions are being made about natural resources. For example, price signals are an excellent guide for the allocation of scarce farmland. Likewise, the marketplace does a good job of determining the production of copper and how this commodity is allocated among various uses.

Just as regulations and other forms of governmental intervention cause markets for food and other goods and services to perform inefficiently, interference with market forces often leads to the misallocation of natural resources, specifically including the resources that must be harnessed efficiently if everyone is to be adequately fed. Irrigation policy is a case in point. In country after country, water use by agriculture dwarfs that of any other economic sector and the prices paid by farmers benefiting from public irrigation projects amount to a tiny fraction of the cost of delivering water to their fields. In Ecuador, for example, these prices cover less than half the expense of operating and maintaining public systems. This means that farmers contribute nothing to capital costs, which are sizable. Paying low prices, they have little incentive to use water efficiently, which is a major reason why the returns on irrigation investment have been disappointing. As a rule, no more than 1 dollar's worth of benefits, in the form of additional agricultural output, has been generated for every 2 dollars spent on public projects. In addition, farmers lobby strenuously for new investment, largely because they are keen to capture subsidies. Most of this investment benefits people who are relatively well off, as opposed to the rural poor. Thus, income disparities in the countryside are aggravated. Furthermore, farmers care little about conserving water if they purchase it at low prices. Also, there is chronic lack of support for technical assistance on the proper use of irrigation water and the management of irrigated soils. Accordingly, water is wasted on a grand scale and extensive tracts of farmland are degraded (Southgate and Whitaker, 1994, pp. 62–67).

The misallocation caused by regulations, subsidies, and other sorts of governmental interference with market forces is often referred to as intervention failure. This is a curious term in a way. After all, the farmers lobbying for government favors like subsidized irrigation are perfectly aware of what they stand to gain. So are the public officials who deliver these favors, often in exchange for kickbacks on construction contracts or after having bought un-irrigated land at low prices before the irrigated project is announced publicly and then selling the same land at a premium afterward. The actual failure associated with policy-induced distortions in markets relates to losses suffered by society as a whole. These losses consist in part of higher taxes that the general population pays to cover the costs of subsidies and graft. They also take the form of environmental damage.

Intervention failure is pervasive and its environmental consequences reach far. However, it is not the primary focus of the literature on environmental economics. Most of that literature addresses the marketplace's tendency to treat many resources as valueless. The reasons for this market failure and the inefficiencies it creates have been examined in detail. Also, economists have shown how governmental intervention to correct market failure can enhance economic well-being.

The importance of natural resources in agricultural development are examined in this chapter, with attention paid to the inefficiencies caused by intervention failure and market failure. We begin by pointing out various linkages between the economy and the natural environment as well as trade-offs among different uses of scarce resources. We then turn to an analysis of market failure and policies for dealing with this problem. Next, two major environmental problems are analyzed. One of these is agriculture's geographic expansion at the expense of natural habitats and the other is degradation of the land resources that comprise the environmental base for crop and livestock production. This chapter concludes with commentary on the environmental consequences, both positive and negative, of agricultural development.

5.1 Environmental Trade-Offs

In any introductory economics class or textbook, the main focus is on exchanges in markets between households and firms. This trade is represented at the top of Figure 5.1. Firms use labor and other inputs provided by households to manufacture goods and services, which people consume. Corresponding to the exchange of inputs and outputs, of course, are compensating flows of monetary payments (e.g., what households spend on goods and services as well as wages, salaries, and other remuneration paid by firms), which have been left out of the figure for simplicity's sake.

To understand the economic significance of natural resources, a broader perspective is required. In particular, flows between the environment and the economy must

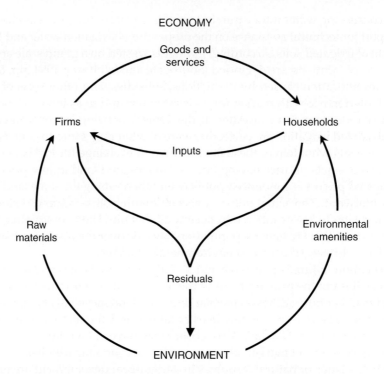

Figure 5.1 Linkages between the economy and the natural environment.
Source: Regional economics course notes, Emeritus Professor Daniel Chappelle, Michigan State University.

be identified and trade-offs have to be considered. Of the three flows depicted in Figure 5.1, perhaps the most obvious one has to do with the raw materials (timber, minerals, soil nutrients, etc.) that firms combine with inputs supplied by households to produce goods and services.

Allocation of raw materials has much in common with the allocation of other inputs. There is, for example, a market for iron. Allowing for the political agendas of state-owned companies in some places and taxes and regulations elsewhere, iron producers decide on output in response to price signals, pretty much as their counterparts do in other parts of the economy. The same is true of customers' purchases.

However, management of raw material flows also involves issues of timing that are not very important in other businesses. One intertemporal trade-off in the energy sector relates to differences in the accessibility and quality of different deposits of petroleum. For example, pumping oil from beneath deserts around the Persian Gulf happens to be cheap. In contrast, production expenses are much higher in recently developed offshore sites, near Newfoundland for instance. An important part of the opportunity costs of current extraction of Middle Eastern resources, then, comprises increased dependence in the future on more expensive sources of oil.

Resource heterogeneity is not required for current depletion to carry an intertemporal opportunity cost. All that is required is simple economic scarcity, in the sense that untapped stocks would not fully satisfy current and future demands if the resource in question were free. Analyzing this sort of intertemporal scarcity, McInerney (1976) and others show that the owners of scarce deposits decide on a time schedule of extraction by considering current and future prices of unextracted resources as well as the inflation-adjusted return on financial holdings. For example, an owner has no incentive to leave a deposit untapped if the rate of increase in its unextracted value (i.e., its delivered price less the cost of extracting it) is less than the real interest rate (i.e., the difference between the market interest rate and inflation). Under these circumstances, one is better off accelerating current extraction and buying bonds and stocks with the proceeds. But if the unextracted value is rising rapidly, there are incentives for conservation. Owners find that they make more money by holding on to the resource and letting its value appreciate rather than extracting it and investing the resulting income. Schedules of resource development are stable, not to mention efficient, if the pace at which unextracted values are rising exactly matches the real interest rate (McInerney, 1976).

Intervention failure can cause intertemporal allocation of natural resources to be inefficient. For example, Venezuela, Nigeria, and a number of other petroleum producers sell gasoline and diesel fuel in domestic markets at prices that are artificially low. In the short term, this policy benefits the owners of cars and trucks in those places, who are better off than most of their countrymen. But low prices also drive up current consumption, which reduces energy availability in the future. Another reason why raw materials obtained from nature are used inefficiently in many places is that property rights in resources are weak. At an extreme, private property was abolished in the Soviet Union, which meant that market signals had virtually no influence on the use and management of natural resources. The waste and misuse arising under these circumstances were enormous (Yergin, 1991, pp. 779–780). Likewise, marine fisheries are used and managed very inefficiently precisely because no one owns these resources and anyone with a boat and net is free to exploit them.

Weak or absent property rights are a major impediment to efficient management of the second flow depicted in Figure 5.1, which has to do with the physical residuals of production and consumption. Consider the manure that is a byproduct of dairy farming, which often finds its way into rivers, lakes, and streams. These waterways (and the aquatic organisms that inhabit them) have a certain capacity to break down manure, which is rich in nutrients, into something less noxious and harmful. However, this capacity of nature to neutralize waste flows is largely unowned, which means that waste disposal is free. Since this environmental service has no market price, it is inevitably overexploited by dairy farmers. Whenever discharges of manure exceed the assimilative capacity of aquatic ecosystems, as is bound to be the case with open access, water quality declines.

Various industrial uses of water are ruled out if rivers, lakes, and streams are polluted. To understand other trade-offs of residual flows that exceed the environment's assimilative capacity, one must consider the third linkage between the economy and the environment represented in Figure 5.1. Different from raw materials, which are combined with labor and other factors to produce goods and services used by households, environmental amenities are consumed directly by people. These include scenic views and clean ambient conditions. As a rule, demand for environmental amenities grows as income increases; in other words, these amenities are a normal good (Chapter 2). This is true, for instance, of people's desire to visit places that are pristine and exotic, like the Galápagos Islands in Ecuador and the Serengeti Plains of Tanzania. Also, polluting the environment with excessive discharges of residuals carries a high opportunity cost where affluence is widespread. For example, nobody is planning to build a new pulp-and-paper factory in Scandinavia these days. The disamenities that would result are reckoned to be excessive.

Of course, it is not just "undesirable" uses of the environment, like mining or waste disposal, that carry opportunity costs. Trade-offs also arise if a site is to be a source of environmental amenities. For example, a river with hydroelectric potential could be reserved instead for rafters and kayakers. All well and good, perhaps, although other people would pay a price, in the form of scarcer energy.

To summarize, nature is not merely a source of raw materials. It is not even just a place where inputs are extracted and wastes are dumped. It is also something that people enjoy directly, and this enjoyment grows as living standards rise. Something else to be recognized is that our diverse uses of the environment are not easily separated from one another. To the contrary, any given use – be it raw material extraction, the disposal of residuals, or the production of environmental amenities – routinely impinges on others, which creates trade-offs. Environmental management is further complicated by the persistence of many of these trade-offs. Current extraction of scarce petroleum obviously carries an intertemporal opportunity cost. Other trade-offs between different uses at varying times arise regularly. Mining, for example, can create contamination that lasts for decades. As far as many people are concerned, damming a river now to generate energy ruins its scenic and recreational appeal for a long time to come.

That the impacts of many uses of the natural environment are persistent can complicate valuation, which is needed to allocate resources wisely among competing ends. Direct use values (e.g., the worth of timber contained in things like pianos and baseball bats) are important. One must also consider option values – what people are willing to pay to leave, say, a forested wilderness untouched for the sake of possible future enjoyment. Even existence values matter. That is, we want pieces of nature to remain intact even though we are quite sure we will never use them. Direct use, option, and existence values all need to be taken into account if environmental trade-offs are to be resolved efficiently (Tietenberg, 2003, pp. 35–37).

5.2 Market Failure

Some of the pollution resulting from agriculture is long lasting. For example, there is a zone in the Gulf of Mexico just beyond the mouth of the Mississippi River where concentrations of oxygen reach very low levels every summer. The fault lies primarily with run-off of fertilizer and manure from farms and feedlots throughout the central United States – run-off that is carried by the river to the ocean. Even if this pollution came to an abrupt end, many years would be required for marine resources to recover fully (Rabalais *et al.*, 1999, pp. 6–33, 114–119).

However, much of the environmental damage that results when agriculturalists treat waste disposal as a free service does not persist all that long. This is often true, for example, of declines in water quality caused by the run-off of manure from dairy farms. To diagnose this sort of market failure, a suitable point of departure is a conventional model of market equilibrium, of the kind introduced in the preceding chapter.

Non-internalization and Inefficiency

Let us examine a hypothetical market for milk, one in which there is no market failure and, hence, all costs of production are internalized by dairy farmers. Obviously, each farmer takes into account the opportunity costs of labor, feed, and other commercial inputs. But full internalization also means that farmers consider the environmental damages associated with rising production. In practical terms, internalization could take the form of payments to neighbors who enjoy their backyards less on warm days because of odors and flies from nearby herds as well as monetary compensation to fishermen, boaters, and swimmers for the disutility they suffer because of water pollution. Alternatively, internalization might occur as farmers spend money on covered lagoons, manure injection equipment, and other measures. As discussed below, environmental costs might also be internalized because pollution is taxed or some other governmental action is taken.

Regardless of how internalization is accomplished, the resulting relationship between the price of milk and the production level, which reflects the marginal cost (MC) of output, is represented by supply curve S in Figure 5.2. Given the demand for milk, which is curve D in the same figure, market-clearing equilibrium occurs at P_E, with production equal to Q_E. As indicated in Chapter 4, price represents the marginal value (MV) of output. Since P_E also equals MC, Q_E is efficient.

Now consider what happens in the market if environmental damages are not internalized. In effect, dairy farms are receiving a subsidy – from those who bear the (uncompensated) costs of stinky air, flies, and manure run-off, to be specific. Any subsidy augments supply, and the implicit subsidy of non-internalization is no exception. As shown in Figure 5.2, farmers' neglect of environmental costs shifts the supply curve out to S', with more output at any given price. The short-run equilibrium price, P_I, is lower than the efficient price, P_E. It is also below the MC of Q_I, which is MC_I in Figure 5.2. Thus, output is inefficiently high: Q_I when it should be Q_E.

The inefficiency that results when dairy farms treat – or, better to say, are allowed to treat – the environmental damages of production as an external cost (or a negative externality) is precisely the opposite of the inefficiency associated with a lack of competition. A monopolist produces too little and charges too much (Figure 4.6). In contrast, non-internalization of costs causes output to be too high

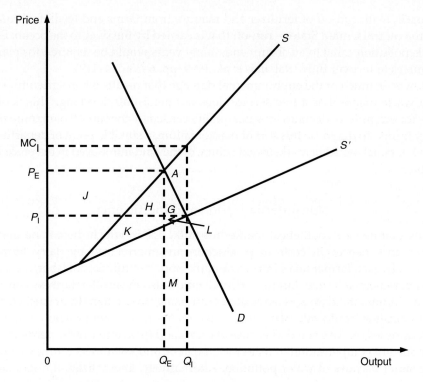

Figure 5.2 Market failure and its impacts on economic well-being.

and prices to be too low, as illustrated in Figure 5.2. The losses in NEV associated with the excessive production, Q_I minus Q_E, comprise the difference between the variable costs of this output (including environmental damages) and consumers' willingness-to-pay (WTP) for the same output. Variable costs amount to the area under the efficient supply (or MC) curve, S, between the two production levels: areas A, G, L, and M in Figure 5.2. WTP consists of the area under the demand curve between Q_E and Q_I: areas G, L, and M in the same figure. NEV losses, then, are represented by triangle A, which is bounded from above by supply curve S and from below by the demand curve, D.

Actual differences between efficient and inefficient prices and levels of output which result from negative externalities depend not just on the magnitude of non-internalized environmental costs but on demand elasticities as well. NEV losses do as well. Just as an outward shift in the supply curve in agricultural markets mainly benefits consumers – in the form of lower prices – if consumption varies little as prices go up or down (Figure 4.5B), a negative production externality is advantageous for consumers because of the wide gap that emerges between P_E and P_I if the demand curve is steeply sloped. However, their gains, along with the additional net benefits that producers may capture, do not fully compensate for the inefficiencies created because environmental costs are not being internalized. As shown in Figure 5.2, moving from the efficient combination of price and output (P_E and Q_E, respectively) to the inefficient combination, P_I and Q_I, creates additional consumers' surplus (CS): areas J, H, and G. Producers gain by imposing environmental costs on other people, although they lose because of the price decline. The former addition to producers' surplus (PS), comprising areas K and L in Figure 5.2, may or may not exceed the latter transfer to CS, which consists of area J. Significantly, the net increase in CS and PS – which does not include the transfer and is represented by areas K, H, L, and G – is less than the environmental costs (areas K, H, L, G, and A) that other people suffer if output reaches Q_I, which it certainly would if these costs were not internalized. The gap between the two, of course, consists of NEV losses (area A).

If consumption is very sensitive to price changes, NEV is still diminished by non-internalization. However, relative changes in price and quantity are very different. Just as an increase in commodity supply has little effect on price if demand is elastic, the difference between P_E and P_I is slight if price changes cause milk consumption to vary a lot. Instead, there is a substantial difference between Q_E and Q_I. NEV losses corresponding to this difference in production levels are sizable only if S is steeply sloped, as would be true for example if minor increases in milk production had substantial environmental impacts.

Of course, neither the price gap nor the difference in output levels is very large if environmental costs comprise a small part of production costs. This possibility is illustrated by the convergence of the two supply curves, S (which, to repeat, reflects all costs, including environmental damages) and S' (which does not reflect these damages), near the vertical axis in Figure 5.2. This convergence reflects the fact that, at low levels of output, emissions from the polluting industry are correspondingly low and, thus, the environment's capacity to receive and neutralize residual flows is not being overburdened. As output rises, residual flows go up as well and environmental quality begins to suffer. Furthermore, as pollution increases, the marginal utility (Chapter 2) of environmental quality, which obviously is becoming less available, is increasing. So at high levels of commercial output and residual flows, the environmental impacts of a marginal increase in production are sizable, as is the disutility that people associate with these impacts. Under these circumstances, inefficiencies may be large enough to induce a response from government.

Correcting Market Failure

As indicated in any textbook on environmental economics (Tietenberg, 2003), the governmental response to non-internalized damage to the environment can take different forms. One way to remedy market failure is to deal with its underlying cause, which is the lack of resource ownership. Take the case of a stretch of rangeland to which access is enjoyed by any and all. Degradation of the resource is guaranteed, as can be readily appreciated by thinking about the incentives facing a single livestock herder. Such an individual knows that any action he takes to conserve the resource, at a cost to himself, will only benefit others. For example, if he moves his herd off the range, thereby passing up an opportunity to fatten his livestock, other herders will rush their animals in, to consume the forage that he has so kindly left them. While the costs of conservation are internalized and the benefits are shared, the benefits of resource deterioration are internalized and its costs are an externality. That is, an individual herder gains by allowing his livestock to graze forage down to the roots. The resulting economic losses are shouldered by other herders, who must make do with degraded rangeland.

To curtail depletion of an open-access resource, property rights can be established where none previously existed. If rangeland is divided among private holdings, for example, the new owners will charge grazing fees that reflect resource values. This is one way that the spread of ownership alleviates non-internalization problems. Another way is that advantage is taken of a distinctive feature of property rights, which is that they can be bought and sold. In the market that emerges automatically once transferable ownership rights are created, resources end up in the hands of those who prize them most. Nobel laureate Ronald Coase was the first to show that some environmental problems can be solved by creating property rights, which are not just exclusively owned but freely traded (Coase, 1960).

Changing the legal status of resources is not free. Boundaries need to be surveyed and a registry established where a record of holdings is kept. Furthermore, a reasonably inexpensive way must be found to demarcate property lines, so that owners can ward off interlopers. For example, private ranches did not really come into being in the western Great Plains of the United States until after the invention of barbed wire, which proved to be the only economically practical fencing material for the region. Before then, private herds of livestock grazed on the open range, to which everyone had access (Anderson and Hill, 1975).

As economic historian and Nobel laureate Douglass North has documented, there has been a broad, long-term trend toward the conversion of open-access resources into properties. The spread of ownership is partly a consequence of increasing resource scarcity, which is an outcome of population growth and economic expansion. It is also the result of technological advance, exemplified by the invention of barbed wire, which lowers the cost of asserting, recording, and protecting ownership rights (North, 1973). This technological advance accelerated in the late twentieth century, thereby making possible the rapid expansion of private property in various parts of the developing world where this legal arrangement had been rare (de Soto, 2000).

This is not to say that open access will disappear entirely at some future date. Converting some resources into properties is prohibitively expensive. This is true, for example, of marine fisheries. Likewise, it is difficult to establish effective ownership of the environment's capacity to neutralize waste byproducts. If resources such as these are to be conserved, other approaches have to be used. More often than not, regulation has been favored, with limits set on pollution and other forms of resource degradation and fines charged for exceeding these limits. A more direct application

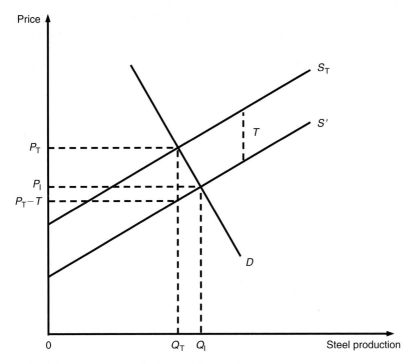

Figure 5.3 A tax on the output of a polluting industry.

of the "polluter pays principle" is to tax pollution, levying a fee on every discharged ton for example. As this principle is applied, the supply curve approaches S in Figure 5.2. If the pollution tax causes MC to be fully internalized, output will be efficient.

The market-level impacts of a pollution tax, one on the output of a polluting industry rather than on residual flows themselves, are easily represented for a special case – one in which residuals are directly proportional to marketable output and the marginal environmental damages of discharges are constant. Under these circumstances, the pollution tax is indistinguishable from a charge on the output of the polluting industry.[1] Consider how such a tax would work in a steel industry that, in the absence of governmental intervention, pays nothing for discharging contaminants into the atmosphere. Seeking to halt the resulting deterioration in air quality, government imposes a tax, T, on every ton of steel. This causes a parallel shift of the supply curve toward the vertical axis, signifying that less is produced at any given market price. As shown in Figure 5.3, the vertical distance between

---

[1] As a rule, taxing a polluting industry's output is a poor way to reduce environmental damage. For example, a tax on each bushel of wheat to reduce soil erosion hurts farmers who conserve their soil every bit as much as those allowing their land to degrade. Thus, such a tax distorts conservation incentives.

the supply curve with the tax, S_T, and the curve without the tax (and no other payments for environmental damages), S', equals T. This is because manufacturers, when deciding on production, pay attention to what they will actually receive per unit of output, which is the difference between the market price paid by consumers and T.

Any decrease in supply, caused by taxation or anything else, always causes the market-clearing price to go up and equilibrium output (and emissions) to fall. Demand in Figure 5.3 being fairly inelastic, consumers end up paying most of the tax. To be precise, the difference between the equilibrium price, P_T, they pay if manufacturers have to pay a pollution tax and P_I, which they pay if there is no tax or internalization of environmental costs, is just a little smaller than the tax, itself. Meanwhile, the net price received by producers, $P_T - T$, is only slightly lower than what they receive if nothing is done about environmental market failure, P_I. In other words, this group pays a small share of T.

As can be easily verified, distribution of the tax between consumers and producers is entirely different if demand is elastic. If consumption is highly sensitive to price changes, taxing the output of a polluting industry has very little impact on equilibrium price. With consumers paying a small portion of T, most of the burden falls on producers, in the form of a large difference between P_I and $P_T - T$. In contrast, there is a sizable reduction in output, and therefore pollution.

Aside from imposing regulations or levying taxes, a government seeking to reduce overuse of a natural resource that no one owns can create a market mechanism for this purpose. To deal with pollution, for example, an environmental agency can issue transferable (i.e., tradable) emissions permits, the specific number of which corresponds directly to the desired discharge level. These permits partially resemble the licenses that regulatory bodies give to firms to emit specific quantities of pollutants. However, licenses to pollute are no more marketable than licenses to drive are. When permits are bought and sold, companies that produce a lot of commercially valuable output with minimal damage to the environment bid pollution rights away from firms that are less adept at doing so. With this approach, often referred to as "cap and trade," any given level of environmental quality (i.e., the cap) can be achieved with maximum efficiency. In addition, polluting firms treat the market value of permits as a part of production costs. Hence, the goal of internalizing environmental damages is accomplished.

The efficacy of capping and trading is not just a matter of theoretical conjecture. In the 1990 Amendments to the Clean Air Act, the US Congress mandated cuts in emissions of sulfur dioxide (a precursor to acid rain) from the electric power industry. A regulatory approach, involving the issuance of non-marketable licenses for specific amounts of pollution to individual generating plants, was not adopted. Instead, the industry as a whole was given an overall emissions target, with a corresponding quantity of permits issued. Firms then proceeded to buy and sell transferable permits among themselves. Sulfur dioxide emissions have been reduced throughout the country by 50 percent and the cost, $1 billion, is below even the most optimistic projections made when the program got under way (Kerr, 1998).

Yet another way to deal with environmental market failure is to invest in the development of technology that reduces the environmental impacts of economic activity. This sort of development narrows the gap between market supply with and without internalization – S and S', respectively, Figure 5.2 – and therefore the differences between efficient and inefficient prices and production. Improvements in agricultural technology often have precisely this impact.

It needs to be stressed that governmental intervention to correct market failure is never free. The administrative resources of the state are scarce, so regulating a polluting industry, taxing emissions, or implementing a transferable permits scheme always creates trade-offs. Also, raising taxes to pay for public programs diminishes private economic activity. Clearly, the benefits of an intervention need to be compared to its costs.

Furthermore, the benefits of a single corrective intervention may not be as great as one might think. In the real world, there are multiple inefficiencies, related to market failure, lack of competition, regulations, and subsidies. Some of these inefficiencies are mutually reinforcing, as with the classic intervention failure of subsidizing a polluting industry. But it is also possible for one inefficiency to cancel out another one. This is the case if monopolization, which results in prices that are too high and output that is too low, coincides with negative externalities, which pull prices down and cause production to be excessive. Under this circumstance, imposing a tax to deal with externalities might actually detract from economic well-being, not enhance it (Davis and Whinston, 1965).

Safe to say, dealing appropriately with environmental market failure can be a tricky business.

5.3 Environmental Deterioration in the Absence of Agricultural Intensification

Non-internalized damage to natural resources is ubiquitous in the food economy. Along with the manure running off dairy farms and feedlots, fertilizer and pesticides find their way into rivers, lakes, and streams. Among the impacts of the resulting pollution are fish mortality, the impairment of recreation, and higher costs of treating the water people drink. Also, fields, pasture, and rangeland are often placed under stress because of crop and livestock production. Aside from the productivity declines associated with land degradation, downstream impacts are created as eroded soil is deposited in reservoirs, irrigation canals, and navigable waterways.

In recent decades, agriculture has intensified and the externalities of crop and livestock production have multiplied. However, the environmental pitfalls of intensification need to be put into perspective. As emphasized in Chapter 3, the Green Revolution has allowed humankind to respond to unprecedented growth in the demand for food without resorting to widespread agricultural extensification. As a result, many natural habitats have been left intact. Furthermore, land degradation is often severe where agricultural productivity is depressed, which is not a coincidence. Unproductive farmers tend to mine soil nutrients. This mining may yield high dividends for a while if the initial nutrient content of the soil is high. But as the process continues, a point is reached at which per-hectare output trails off, which can trap farmers who engage in this practice in a downward spiral of mounting poverty and environmental deterioration.

Land Degradation

Taking various forms and reducing agricultural production in different ways, land degradation has been examined carefully in a number of specific locales. Likewise, estimates are available of the global area affected by erosion and chemical and physical deterioration. According to the Global Land Assessment of Degradation (GLASOD) sponsored by the United Nations, 22 percent of the Earth's agricultural land, including 38 percent of the entire area planted to crops, shows signs of one sort of degradation or another. Of the 1.9 billion hectares affected, more than four-fifths have suffered primarily from erosion. Another eighth has been affected by salinization (i.e., the build-up of minerals harmful to plant growth in the rooting zone), acidification, and other adverse changes in soil chemistry. The other 5 percent shows signs of physical problems, such as soil compaction and waterlogging (Oldeman et al., 1991). It has been estimated that another 5–10 million hectares slip into the degraded category every year (Scherr and Yadav, 1997).

The toll this degradation takes on agricultural output varies from place to place. In temperate settings, where soils with favorable characteristics comprise a large portion of agriculture's land base (Wood et al., 2001, pp. 45–48), the negative impacts of resource deterioration on crop output have been small relative to the yield gains resulting from technological improvement. Having surveyed available research, Mitchell et al., (1997) conclude that 100 years of erosion in the United States have lowered yields by 3 or 4 percent (p. 54). This penalty is tiny relative to the yield increases that have resulted from advances like hybridization (Chapter 3). In contrast, the impacts of land degradation are more worrying in the tropics and subtropics, where the erosive impact of rainfall concentrated in heavy storms is high and where the vast majority of soils have one sort of deficiency or another. Applying the GLASOD methodology for resource assessment, Van Lynden and Oldeman (1997) find "moderate" production impacts on 24 percent of the agricultural land categorized as degraded in South and Southeast Asia and "strong" impacts on an additional 13 percent of that area. Scherr and Yadav (1997) suggest that erosion and chemical and physical deterioration have lowered global agricultural output by 5–15 percent.

Sometimes, land degradation is a consequence of intensified production. For example, soil compaction can occur if heavy farm equipment lacking flotation tires is driven frequently across a field. Also, excessive fertilization changes soil chemistry for the worse. For example, applying ammonium nitrogen drives down soil pH, perhaps enough to hinder crop growth. Irrigation can create problems as well. If an irrigated field is not properly drained, soils are apt to become waterlogged.

Also, excessive irrigation combined with poor drainage can cause the subterranean water table to rise, thereby bringing soluble minerals up into the rooting zone or even to the surface. The resulting soil salinization may grow so extreme that crop production is preempted.

Where agricultural intensification is coinciding with resource deterioration, intervention failure is often the culprit. Fertilizer subsidies in India are illustrative in this regard. Aside from a general policy of providing fertilizer to farmers at below-market prices, the national government has made nitrogen (from urea and other sources) cheap relative to phosphorus and potassium. In response, Indian farmers apply too much of the first nutrient and too little of the second and third. Accordingly, crop growth is held back by the limited availability of phosphorus and potassium and the returns to nitrogen application are diminished. It is estimated that a switch to more balanced fertilization, as agronomists recommend and as would occur if price distortions were eliminated, would cause annual production of rice and wheat to increase by 160 million metric tons and 25 million metric tons, respectively (Roy, 2003).

In Asia, where governments spurred the Green Revolution by subsidizing inputs such as fertilizer and irrigation water, land degradation resulting from the misuse of these inputs is a major concern. Damage to agriculture's resource base has happened for entirely different reasons in places bypassed by the Green Revolution. Irrigation development has not proceeded very far in Sub-Saharan Africa, for example. Accordingly, problems like waterlogging are not very common in the region. Furthermore, African farmers, rather than applying fertilizer at a high rate, use that input very sparingly. This is a major reason why erosion and chemical and physical deterioration of soil resources are widespread.

Differences in fertilization in various parts of the world are striking, as are the changes that have occurred in recent years. A quarter century ago, fertilizer applications were particularly heavy, averaging 145 kilograms per cultivated hectare, in the old Soviet Union and its satellites (Table 5.1, column 1). More than anything else, this reflected startling inefficiencies in the use of agricultural resources (and just about everything else, of course) under socialism. Between the collapse of communism and a fall in human numbers, which has lowered food demand, application rates have plummeted, reaching a mere 34 kilograms per hectare of cropland as the twentieth century drew to a close (column 2). During the same period, there has been a modest decline in fertilization in Western Europe, the United States, and other affluent places, mainly because agricultural resources are being used more efficiently.

In most of the developing world, fertilizer application rates have been increasing. Already high 25 years ago in East Asia, fertilization has risen to elevated levels (235 kilograms per hectare) not observed in any other part of the world, rich or poor. Formerly low, application rates have tripled in South Asia and are now approaching levels in wealthy nations, although it bears repeating that policy-induced price distortions lead to inefficiently unbalanced fertilization in India. Significant increases also have occurred in Latin America and the Caribbean and in the Middle East and North Africa. Sub-Saharan Africa has been the lone exception to the upward trend. No part of the world had lower levels of fertilization a quarter century ago. Since

Table 5.1 Average fertilizer application rates, 1979–1981 and 1998–2000

Region	Kilograms per hectare of cropland in 1979–1981 (1)	Kilograms per hectare of cropland in 1998–2000 (2)
East Asia	112	235
Eastern Europe and Former Soviet Union	145	34
Latin America and Caribbean	59	90
Middle East and North Africa	42	79
South Asia	36	107
Sub-Saharan Africa	16	13
Western Europe, United States and Canada, Australia and New Zealand, Japan and South Korea, etc.	133	125

Source: World Bank (2003b), p. 126.

then, per-hectare application rates have actually gone down, from 16 kilograms in 1979–1981 to a mere 13 kilograms in 1998–2000.

Henao and Baanante (1999) have assessed the extent of nutrient mining throughout Africa, excepting Egypt (where fertilizer use is heavy) though including small island nations near the continent. Part of their study included estimation of the nutrients that are lost annually because of crop uptake, leaching, soil erosion, and other processes. Also considered was nutrient replacement, by fertilization and other means. In three places – Mauritius and Reunion in the Indian Ocean as well as Libya – replacement was found to exceed losses. Everywhere else, the reverse was true. Annual depletion of nitrogen, phosphorus, and potassium was modest in South Africa, amounting to just 14 kilograms per hectare. Elsewhere, accelerated depletion was happening. Net per-hectare losses exceeded 60 kilograms per annum throughout the eastern part of the continent, from Ethiopia and Somalia south to Tanzania and over to Madagascar. The same was true of the old Zaire, which is now the Democratic Republic of the Congo (DRC), as well as Ghana, the Ivory Coast, Nigeria, and other parts of West Africa.

The same investigators attempted to put a price tag on nutrient mining. Their approach was simple, involving the multiplication of net annual losses – 385,800 metric tons in North Africa, 110,900 metric tons in South Africa, and 7,629,900 metric tons in the rest of the continent – by the market values of nitrogen, phosphorus, and potassium in the form of commercial fertilizer. The estimate of annual cost they arrived at was $1.5 billion (Henao and Baanante, 1999). While this finding reinforced the point that nutrient mining creates sizable economic losses, Henao and Baanante (1999) conceded that their analysis is not at all definitive. Not examined, for example, is the crop production that is passed up because of poor nutrient management.

The cumulative effects of nutrient mining were on display in the Midwestern United States before the Second World War. By that time, crops had been produced without significant fertilization for several decades. As a result, the organic content of soils was seriously depleted. This took a toll on yields since soils without much organic content have little capacity to hold water that plants must draw on in dry periods, which occur frequently during the growing season. Also, the loss of organic content made the land more erodible while simultaneously reducing the growth of plants that furnish protection from the elements. Thus, high winds during the 1930s created a Dust Bowl (Dennis Avery, Personal Communication, August 4, 2003).[2]

There are very few places in Africa where farmers can exploit a generous endowment of fertile land. Most of the continent's soils are of ancient geological origin and thus highly weathered, which means that nutrient mining takes an immediate toll on plant growth. The report from Henao and Baanante (1999) that soil fertility is being depleted rapidly, and has reached an advanced cumulative stage in some places, obliges us to reconsider conventional measures of the economic productivity of African agriculture. As indicated in Chapter 3, value added in the sector is found by subtracting the costs of fertilizer, capital goods, and other inputs aside from labor from the value of crops and livestock output. Not incorporated in this calculation are the economic losses suffered because farming and ranching damage natural resources. Obviously, this omission leads one to exaggerate agricultural value added at any particular date, not to mention growth in this productivity indicator over time. For example, the conventional measure of value added per agricultural worker in Rwanda, a small nation in the highlands of East Africa, went from $271 in 1979–1981 to $251 in 1999–2001 (World Bank, 2003b, p. 130). Since no other country in the region has had a higher annual rate of nutrient loss, 136 kilograms per hectare, part of this decline undoubtedly reflects land degradation. It is equally certain that, if estimation of agricultural value added took this deterioration into account, an even greater decline in the productivity of Rwandan agriculture would be revealed.

Mainly because of data limitations, attempts to incorporate resource values in economic assessments of African agriculture have been few and far between. Of the small number of available studies, one undertaken in Mali, where nutrient depletion is moderate by continental standards (Henao and Baanante, 1999), has yielded solid evidence that the costs of resource depletion are, indeed, significant. In this study, the Universal Soil Loss Equation (USLE), which was developed originally to estimate erosion on fields in the United States (Wischmeier, 1976), was adapted to African conditions. In addition, data from the International Institute for Tropical Agriculture (IITA), which is located in Nigeria and is part of the CGIAR (Chapter 3), were used to analyze the linkage between erosion and crop yields. It was found that

---

[2] Learning well the lessons of the Dust Bowl, US farmers applied much more fertilizer after the Second World War. Sometimes they did so excessively. For example, phosphorus concentrations had become so elevated by the 1970s in the Midwest that many of them were able to stop applying the nutrient for at least a few years without sacrificing yields (Richards *et al.*, 2002).

losses in agricultural output resulting from erosion were equivalent to 1.5 percent of Mali's GDP and well above the cost of conserving soil (Bishop and Allen, 1989).

Findings like this have given considerable impetus to initiatives to reverse land degradation in Sub-Saharan Africa and other settings. Some of the people and agencies involved in these initiatives have specific ideas about nutrient replenishment. Advocating manure use and the planting of trees and other vegetation that transfer nitrogen from the atmosphere to the soil, they have undertaken pilot projects to make the case that alternatives to commercial fertilizer are viable. Although some of these projects yield interesting results, the general case for alternative fertilization (i.e., organic farming) is not very convincing. In many areas undergoing nutrient mining, manure is in very short supply. Livestock herding is impeded by tsetse flies and other pests in many parts of Africa. In India, manure from livestock is available, although a lot of it is gathered for cooking fuel. In either setting, it is difficult to envision manure being spread in sufficient quantities to arrest land degradation.[3]

Regardless of the advantages that commercial fertilizer has over the alternatives, there are many places where this input is not being used to halt deterioration of the environmental base for agriculture. Replenishing nutrients is not free, of course, and fertilizer prices are higher now than they were in the early 1980s in a number of African countries because national governments, obliged to reduce public spending and fiscal deficits, have cut subsidies. However, there is a more fundamental reason for the lack of fertilization, namely insufficient investment in the public goods required for economic progress in the countryside. Inadequate transportation infrastructure, for example, drives up the expense of fertilizer and other purchased inputs. Since deficient infrastructure also diminishes the prices farmers receive for their commercial output, input demand is reduced as well. Where public goods are lacking entirely, the bulk of the rural population has no choice other than to engage in subsistence farming – using no purchased inputs and producing barely enough to feed itself, if that. The end result is a combination of grinding poverty and extreme land degradation.

Agricultural Extensification

If a rural family cannot escape poverty either by raising agricultural productivity or by abandoning subsistence farming in favor of some other line of work, it may still forestall a deterioration in living standards by relocating to a place where land degradation has not proceeded very far. This response underlies much of the agricultural encroachment on tropical forests and other natural habitats that is happening in Africa, Asia, and Latin America.

---

[3] There has been relatively little analysis of the relative merits and costs of different nutrient sources, especially in the developing world. In a study focused on Denmark, it was found that switching from chemical to organic fertilizers would drastically reduce crop output and that many of the environmental impacts would actually be adverse (Bichel Committee, 1999, pp. 20–27, 51–70).

Deforestation sometimes has nothing to do with agriculture. Hydroelectric development, for example, results in the permanent flooding of some tree-covered land. Also, logging may do so much damage to the vegetation remaining after timber has been harvested that, for all intents and purposes, deforestation has occurred. Moreover, there are various reasons why habitats are displaced by cropland and pasture, and rural people desperate to escape poverty are not the only agents of deforestation and related land-use change. In particular, commercial farmers and ranchers are responsible for some of agriculture's geographic expansion. Rather than being concerned about mere survival, they are motivated by the search for profit.

Brazil is one place where encroachment on natural habitats has at least as much to do with commercial gain as with the escape from rural poverty. No tropical or subtropical nation has more tree-covered land; in addition, no country, either near or far from the equator, loses more hectares of forest annually (FAO, 2003c, Data Table 2). However, agricultural land use is expanding mainly in the *Cerrado*, a vast savannah of grasslands southeast of the Amazon Basin. In 1990, the Empresa Brasileira de Pesquisa Agrícola (EMBRAPA), which is the state agency responsible for agricultural research and development in Brazil, estimated that 136 million hectares in the region were suitable for large-scale farming, of which 47 million hectares were being cultivated at the time. To expand agriculture, land preparation is needed to overcome soil acidity, aluminum toxicity, and limited availability of nitrogen and phosphorus. However, commercial farmers are seeing fit to deal with these environmental limitations. Much of the recent expansion in the area planted to soybeans in Brazil – from less than 4 million hectares in the early 1970s to more than 12 million hectares in the late 1990s – has taken place in the *Cerrado* (Schepf *et al.*, 2001).

During the 1980s, intervention failure was singled out as a cause of excessive deforestation in the Brazilian Amazon. Subsidized loans were criticized, as were income tax credits given to companies investing in livestock operations in the region (Mahar, 1989). But by the middle 1990s, it had become clear that the land-use changes resulting from these policies had not been as great as many had thought. Having examined data both on deforestation in different places and on the geographic incidence of cheap loans and tax credits, Schneider (1995) reported that agricultural land clearing had been especially rapid in the southern reaches of the Amazon Basin, which had not received many subsidies. In contrast, subsidies had been directed mainly to the eastern part of the watershed, which had experienced less deforestation.

The same investigator has pinpointed another intervention failure of greater importance for Brazil's forests. Thanks in no small part to generous tax treatment given to agriculture, prices for rural real estate skyrocketed in southern Brazil after the early 1970s. Farmers with smallholdings, whose incomes were not high, were not much interested in competing for land in order to take advantage of things like accelerated depreciation and the deduction of farming losses from non-agricultural earnings. Many of them found it more rewarding to sell their farms in the south and move to agricultural frontiers in the Amazon, where real estate remained relatively cheap. As Schneider (1992) has argued, it was no coincidence that, as land prices in southern Brazil rose, immigration and agricultural land clearing in the Amazon increased.

Similar policy regimes have had the same impacts in other countries. In Colombia, for example, competition for prime farmland is biased by the favorable tax treatment of agriculture as well as the fact that owners of large holdings are normally first in line for loans from the government that carry a low interest rate, are apt to be forgiven, or both. Each of these benefits matters most to the wealthy, which causes them to bid more for rural real estate. In contrast, disadvantaged small farmers are consigned to inferior parcels, along Andean hillsides and in tropical forests for example (Heath and Binswanger, 1996).

Driven up by government policies in a number of countries, the net returns captured by people who colonize agricultural frontiers are also enhanced because the environmental consequences of this activity are not internalized. Some of these consequences are local. For example, removing trees and other vegetation from the upper reaches of a watershed can lead to increased flooding and sedimentation downstream. Also, deforestation can cause the local climate to grow hotter and drier. Other consequences of agricultural land clearing are of global importance. One of these is diminished biological diversity. Increased atmospheric concentrations of greenhouse gases are another.

None of these impacts is easy to evaluate. Consider the economic damages associated with reduced biodiversity. There is little doubt that tree-covered ecosystems close to the equator harbor a large share of the world's flora and fauna and that, as these ecosystems are encroached on, some species are driven to extinction. All agree that a cost of this encroachment relates to the fact that pharmaceutical advances are impeded because of a reduction in the supply of biological specimens collected in natural habitats. However, this cost may not be very large. One group of economists has estimated the one-time payments that the pharmaceutical industry would make for species-rich forests in the tropics. Assuming a high probability of finding something unique in the wild that yields a valuable drug, they have found that payments of this sort may exceed $10 per hectare in a handful of places – in particular, where there is rapid destruction of ecosystems with many endemic species (i.e., plants and animals found nowhere else). But as a rule, the value of tropical forests as a site for collecting biological specimens amounts to just a few dollars per hectare (Simpson *et al.*, 1996).

Aside from the brake that is placed on pharmaceutical research, there are other consequences of diminished biodiversity, although none of these others appears to have greater economic significance. In contrast, the economic costs associated with global warming, which some contend is accelerated by deforestation, could be much larger. A simple measure of these costs is found by multiplying the damages resulting from the emission of 1 ton of carbon (in the form of carbon dioxide) by the 100–200 tons that are sequestered in a forested hectare. Having evaluated rising sea levels and other adverse impacts of a warmer climate, Nordhaus and Boyer (2000) have estimated that the discounted value of current and future damages resulting from the emission of 1 ton was about $7 in 1995 (p. 133). Using this figure, one concludes that deforestation creates global warming costs ranging from $700 to $1400 per hectare.

Table 5.2 Forested area in 2000 and deforestation during the 1990s

Region	Forested area in 2000, '000 hectares (1)	Annual change, 1990–2000, '000 hectares (%) (2)
East and South Asia	507,403	−801 (−0.2)
Eastern Europe and Former Soviet Union	937,169	+814 (+0.1)
Latin America and Caribbean	964,358	−4,669 (−0.5)
Middle East and North Africa	27,748	+65 (+0.2)
Sub-Saharan Africa	643,604	−5,296 (−0.8)
Western Europe, United States and Canada, Australia and New Zealand, Japan and Korea, etc.	789,168	+496 (+0.1)

Source: FAO (2003c), Data Table 2.

To deal with the market failure that results because farmers and other agents of land-use change do not internalize global warming damages and other environmental costs, these agents could be charged a tax for every hectare they clear. Another option would be to pay landowners to sequester carbon in forests. Economic expansion and population growth may cause the present value of damages to rise as high as $62 per emitted ton by the end of the century (Nordhaus and Boyer, 2000, p. 133). If carbon-sequestration payments reflected this growth in damages, landowners would respond by planting trees. The extent of the world's forests could grow from 3.5 billion hectares today to 4.0–4.5 billion in 100 years (Sohngen and Mendelsohn, 2003).

For the time being, mechanisms to pay for carbon sequestration, biodiversity conservation, and other environmental services provided by tree-covered land remain poorly developed. Nevertheless, as indicated in Table 5.2, deforestation is not a serious problem in many parts of the world. Forests are spreading in Western Europe, the United States, and other wealthy places, where agricultural development driven by extensification has ended. Similar changes in land use are occurring in Eastern Europe and the Former Soviet Union, where demand for agricultural commodities is not growing and where inefficient modes of production characteristic of socialism are being abandoned. Forested area is even increasing in the Middle East and North Africa, albeit from a very small base.

There have been interesting changes in Asia. True, Indonesia (which has more tropical forests than any other country in the region), the Philippines, and a number of other nations lost more than 1 percent of their forests every year during the last decade of the twentieth century. However, Myanmar (formerly Burma) is the only place where the number of hectares cleared annually went up between the early 1980s and the 1990s. Everywhere else, this number has declined. Furthermore, forested area has been increasing, not going down, in India and China. Mainly because annual expansion of forests in the latter country has been averaging 1.2 percent, losses of tree cover throughout the region have come down to 0.2 percent per annum (WRI, 1992, pp. 287; FAO, 2003c, Data Table 2).

Elsewhere in the developing world, forests remain under threat. The situation in the American tropics is similar to that of many nations in South and Southeast Asia. That is, hectares cleared annually have gone down, although encroachment on forests continues. In Brazil, yearly deforestation during the 1990s was 0.4 percent, or 2.3 million hectares. While no other country loses more forests, this rate is in fact lower than the 2.5 million hectares cleared each year during the early 1980s. Another interesting case is Colombia, where 890,000 hectares were lost annually two decades ago. This was the second highest rate in the entire world, not just in Latin America and the Caribbean. Since then, land-use change has decelerated markedly, averaging 190,000 hectares per annum during the 1990s. The only large nation in the Western Hemisphere with annual deforestation above 1 percent is Mexico. All other countries exceeding this rate are small and densely populated. Among these are Ecuador, Haiti (where land clearing accelerated as the twentieth century drew to a close), and a number of Central American nations. In contrast, deforestation in the temperate Southern Cone of South America was modest in the early 1980s and remains so today. This reality and the fact that land-use change in Brazil is small relative to the country's remaining forests explain why the annual rate for all of Latin America and the Caribbean is 0.5 percent, not something higher (WRI, 1992, pp. 286–287; FAO, 2003c, Data Table 2).

Nowhere are forests in greater danger than in Sub-Saharan Africa. Different from what is taking place in Asia or the Western Hemisphere, deforestation is accelerating in many parts of the region, with the number of hectares cleared (Table 5.2, column 1) as well as land-use change as a portion of remaining forests (column 2) both going up in recent years. Among countries following this trend are several in West Africa – Benin, Cameroon, Ghana, Liberia, Sierra Leone, and Togo – and quite a few in the eastern and southern parts of the continent – Botswana, Kenya, Namibia, Uganda, Zambia, and Zimbabwe. Deforestation is also accelerating in Mali and Sudan in the semi-arid Sahel, just south of the Sahara Desert, as well as in Burundi, the DRC, and Rwanda, in Central Africa. In the DRC, which has more tropical forests than any other nation aside from Brazil, land-use change as a portion of remaining tree-covered land is only 0.4 percent per annum. All other countries where habitat destruction has accelerated in the late twentieth century, not to mention several where the number of hectares cleared annually fell after the early 1980s, have yearly deforestation rates at or above 1.0 percent. Land-use change is especially dramatic in Burundi and Rwanda – two small nations located to the northeast of the DRC and wracked by genocidal conflict in recent years. In Rwanda, annual clearing increased from 5,000 hectares in the early 1980s to 15,000 hectares during the 1990s. Even worse is the case of Burundi, where annual clearing went from 1,000 to 15,000 hectares during the same period. The more recent rate is equivalent to 9 percent of the forests that remain in the country (WRI, 1992, p. 286; FAO, 2003c, Data Table 2).

The driving forces of deforestation vary considerably from one part of the developing world to the next, and even within any particular region. As emphasized in Chapter 3, low and stagnating yields cause farmers to convert natural habitats into cropland and pasture. This linkage is apparent in some Asian nations. But elsewhere, rapid intensification coincides with modest extensification. Also, there is great interest in the region in conserving forests that are the source of environmental services. For example, tree planting is being undertaken in China to prevent land degradation, both in upper watersheds of great hydrological importance and where deserts threaten to spread (Richardson, 1990, pp. 227–251). Mechanisms for dealing with global externalities, such as biodiversity loss or warming of the planet, are poorly developed. This is one reason why deforestation is happening in Latin America. However, it is also true that the benefits of agricultural extensification are considerable in some parts of the region, that is in places that are not currently being farmed but are well suited to agriculture.

In other parts of the developing world, the productive potential of settings undergoing deforestation is more limited and the driving forces of this environmental change are different. Land in these settings is often cleared by small farmers who hope to escape poverty. This poverty traces in part to depletion of soil and other natural resources in places from which these farmers have migrated. Sub-Saharan Africa is not the only region where this sort of deforestation is occurring. However, it is no coincidence that peak deforestation and peak land degradation (brought about because of accelerated nutrient mining) are both happening in the same part of the world.

5.4 Agricultural Development and the Environment

In impoverished, rural settings, the natural resources required to produce a given amount of food are considerable. Machinery, agricultural chemicals, and other commercial inputs are not used. Instead, edible output depends on whatever natural fertility the soil may hold, the amount of rain that happens to fall, seeds saved from last year's crop, and of course work done by farm households. Yields are modest.

Although not a basis for widespread wealth, this sort of farming can go on for a very long time, and indeed did so throughout the world for millennia. If soil fertility, which is depleted by a few years of cropping, is replenished during a fallow period of adequate length, then the entire cycle of shifting cultivation (or swidden farming) is sustainable. People can live as well as their grandparents did, and their grandparents before them.

However, shifting cultivation becomes unsustainable as demand for food grows, due to an increase in population or economic activity. Land does not lie fallow for as many years between one cropping period and the next. As a result, soil fertility diminishes as time passes. As long as crops are produced in traditional ways, about the only way to compensate for fertility decline in one place is for farmers to colonize other settings. Thus, a cycle of land degradation and deforestation is frequently set in motion.

As recent experience in Sub-Saharan Africa demonstrates, any gains in living standards accruing to people engaged in this cycle – gains, it must be remembered, that carry a substantial though largely non-internalized environmental penalty – are meager. For accelerated growth in GDP per capita, a break with farming traditions is required. Input mixes must be altered so as to raise yields and enhance productivity. The general pattern of agricultural development during the twentieth century has been to increase fertilization, irrigation, and mechanization, with a corresponding decline in the relative importance of labor and land inputs (Chapter 3).

At least during the initial stages of this development, environmental deterioration continues to mount. The application of commercial fertilizer, for example, is often accompanied by increased run-off of nutrients into rivers, lakes, and streams, which creates water pollution. Likewise, the introduction of irrigation where farmers are accustomed only to rain-fed modes of production can lead to waterlogging, salinization, and other problems. As indicated in this chapter, market failure is partly to blame for this damage, since agriculturalists do not internalize all environmental costs of crop and livestock production. But as we also stress, deterioration of natural resources often results from governmental interventions, such as fertilizer and irrigation subsidies, intended to accelerate farming's transformation.[4] Regardless, environmental quality is not moving in the same direction as GDP per capita, which goes up as agricultural development makes food more abundant.

A point is eventually reached at which increased affluence ceases to carry an environmental cost and beyond which growth in GDP per capita actually coincides with improvements in natural resources. This change is easy to understand in the case of agriculture. Income growth allows for expanded investment in research and development, which reduces the environmental impacts of crop and livestock production in various ways. For example, US farmers' adoption of *Bacillus thuringiensis* (Bt) cotton, which resists insect attacks, has allowed them to reduce annual pesticide applications by 2 million pounds since 1996. Agricultural biotechnology promises to yield related environmental dividends in the near future. In 2003, regulatory approval was given to a genetically modified variety of corn that resists rootworm, which does more damage to US corn harvests than any other pest. As this variety is adopted, annual pesticide applications will fall, by as much as 14 million pounds (Rauch, 2003).

Genetic modification can also help to conserve soil, which is the single greatest environmental challenge facing agriculture if GLASOD and other resource assessments are to be believed. Much of the challenge relates to weed control in fields

---

[4] India, where intervention failure contributes substantially to unbalanced fertilization (see above), is also a place where the environmental damage associated with irrigation is exacerbated by bad public policy. In the Punjab and neighboring regions, where precipitation is sparse and yet the Green Revolution has had a major impact, there is heavy reliance on groundwater. Access to this resource is open to any and all farmers. In addition to paying nothing for water, they are sometimes provided free electricity to run their pumps. Under this policy regime, subterranean resources are being depleted rapidly, as reflected by declines in the water table of as much as 0.2 meters per annum (Kanwar, 2003).

that are not plowed or otherwise tilled. However, biotechnology has produced crop varieties that are not harmed by herbicides, which break down into harmless residuals soon after application. Using these varieties, farmers can maintain yields while avoiding the damage done to fields by plowing (Rauch, 2003).

Accelerated technological progress is probably the main explanation of why economic growth in affluent places has beneficial consequences for natural resources. But there is another reason for the positive linkage between living standards and environmental quality in such settings, which is that a shift in demand occurs. As noted early in this chapter, the income elasticity of demand for a wide variety of environmental amenities is high, quite high in many cases. Moreover, this income elasticity goes up as living standards rise. Environmental amenities, then, are very different from food, for which Engel's Law that income elasticity is a diminishing function of earnings (Chapter 2) applies. Suffice it to say that, as GDP per capita rises high enough that concerns about hunger are left far behind, the value of extra commodity output falls relative to the MV of environmental quality.

This change in demand is bound to affect public policy. For example, support for things like farm subsidies, which stimulate agricultural output, weakens relative to the demand for governmental action to limit environmental market failure. This change may be under way in the United States, where agriculture is a major source of water pollution. Of 3,692,830 miles of streams and rivers in the country, environmental assessments have been carried out for 699,946 miles, or 19 percent of the total. Of the assessed portion, 269,258 miles, or two-fifths, have been found to have impaired water quality. Moreover, agriculture is the leading source of pollution for 128,859 (48 percent) of these miles (USEPA, 2000, p. 14). In light of these facts, containing the run-off of fertilizer, manure, and other pollutants has become a central focus not just of environmental policy, but also of agricultural policy – which is little concerned these days with raising production.

The tendency of increasing living standards to coincide initially with mounting pressure on natural resources and later with improved environmental quality is expressed by the Environmental Kuznets Curve (EKC), which is illustrated in Figure 5.4. Shaped like an inverted U, this function is inspired by the proposition that income inequality rises in societies emerging from penury but declines after GDP per capita passes a certain threshold (Kuznets, 1955). Of course, an EKC's specific dimensions matter a great deal. For example, a high peak implies that resource degradation grows severe as economic expansion takes place in a setting where living standards are modest. Another possibility is for the peak to be low, though occurring at a high level of average income. Under this circumstance, mild trade-offs between environmental quality and economic growth are a prolonged feature of the development process.

The existence and dimensions of EKCs for various sorts of damage to natural resources have been investigated empirically. For example, EKCs for tropical deforestation have been estimated in different parts of the world (Barbier, 2001). In one study, Cropper and Griffiths (1994) found that the threshold at which income growth reduces deforestation is well above GDP per capita in the vast majority of

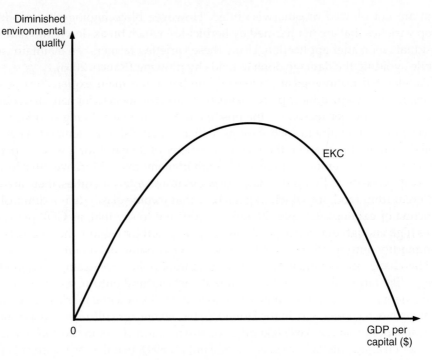

Figure 5.4 The Environmental Kuznets Curve.

African and Latin American countries with extensive tracts of tree-covered land. This finding implies that, all else remaining the same, economic growth in those two continents is likely to coincide with the loss of natural habitats.

Estimation of EKCs for tropical deforestation yields results that are in line with other research addressing linkages between income growth and other kinds of environmental damage. Surveying the available literature, Dasgupta *et al.* (2002) conclude that various EKCs peak when GDP per capita is between $5,000 and $8,000. This is worrying in the sense that living standards are below the level of environmental turnaround in most of the developing world. However, the same investigators identify a number of reasons, all related to globalization, for being optimistic. One of these is the freer flow of information, which makes people more aware of pollution and more willing to do something about it. Another is trade liberalization, which has discouraged governments from subsidizing and protecting industries that use a lot of energy and pollute heavily. Yet another reason for optimism is the spread of cost-effective approaches to environmental regulation.

There is no basis for supposing that the environmental benefits of globalization apply any less to agriculture than to the rest of the economy. Aside from promoting greater awareness of resource deterioration and its consequences, the free flow of information facilitates the development and spread of technology for raising crops and livestock in environmentally friendly ways. In addition, technological change is facilitated by the income growth that results from the expansion of commerce. Ultimately, resource conservation in much of the developing world depends not on maintaining the self-reliance of farmers in Africa, Asia, and Latin America, but instead on ending their economic isolation. As long as this isolation continues, a large share of the world's rural population will continue to engage in nutrient mining, thereby degrading land resources. Their chances of rising out of poverty will be slim to none.

Study Questions

1. Discuss the impacts of irrigation subsidies – which are a classic intervention failure – in terms of economic efficiency, the distribution of income, and the natural environment.
2. Describe three linkages between the economy and the natural environment as well as the trade-offs associated with selected uses of the natural environment.
3. How does non-internalization of the environmental costs of production affect equilibrium price and quantity in a competitive market?
4. What is the relationship between the own-price elasticity of demand and the price changes resulting from improved internalization of the environmental costs of production?
5. In terms of the area affected, which sort of land degradation is the most globally prevalent?
6. Analyze the causes of inadequate fertilization in Africa.
7. What kinds of intervention failure and market failure contribute to tropical deforestation?
8. Explain the EKC, both generally and with reference to the US food economy.

6
Globalization and Agriculture

A recurring theme of this book is that specialization and trade create significant benefits, not least the improvements in living standards that drive up the demand for environmental quality and enable its creation. As emphasized at the beginning of Chapter 3, development of the US food economy has much to do with farmers' specialization in production agriculture and their trade with agribusinesses that are efficient suppliers of inputs, marketing services, and the like. In *The Wealth of Nations*, Adam Smith suggests that exchange among individual family members is mutually advantageous (Smith, 1964, p. 401). What he had in mind, no doubt, is that parents provide their young children with the necessities of life. In return, offspring can be counted on for sustenance later on, once the parents have grown too old to work.

In the global context, any single country finds it rewarding to offer a limited list of goods and services in international markets and import a much wider array of items from places where these are produced more efficiently. Accordingly, trade across boundaries and frontiers has been happening for centuries, even millennia. In modern times, this commerce has flourished, as technological improvement has driven down transportation and communication costs. During the nineteenth century, for example, sailing craft made of wood started to be replaced with metal-hulled vessels powered by steam, which drastically reduced the expense of transoceanic shipping. Trade was stimulated and, not coincidentally, economies grew.

If anything, there has been an acceleration of technological change facilitating the movement of goods and services, not to mention capital and other inputs, over great distances. Indeed, the pace of change is now so rapid that a new term, globalization, has been coined to describe the resulting expansion of trade, investment, and related interchange. As the twentieth century drew to a close, a revolution occurred in information technology, which has made data transmission nearly free. Money can now be sent from one part of the world to another with a few keystrokes on a computer terminal. So can knowledge. Likewise, it has become much easier for buyers and sellers, including participants in markets for farm products, to deal with one another.

Many are uncomfortable about proliferating flows of goods, money, and ideas around the world. Globalization's opponents are not limited to boisterous protesters, who have tried to disrupt every economic summit since the November 1999 meeting of world leaders in Seattle. International trade often imposes a sacrifice on one part of the economy or another. Even though this sacrifice is exceeded by the benefits of commercial exchange, and usually by a wide margin, those bearing costs often clamor for trade restrictions. Of course, protectionism, which is accomplished with tariffs and other sorts of interference with market forces, drives down living standards.

Agriculture is at the forefront of the debate over globalization, which this chapter addresses. Rarely having spoken with a firm and united voice in international trade talks, developing countries did exactly that in September 2003 in Cancún, Mexico. The "Group of Twenty," which originally included China, India, and Brazil and has since recruited a number of other nations, took a firm stand against agricultural protectionism and subsidies in rich nations. Thanks in part to this stand, the World Trade Organization (WTO) agreed in July 2004 to phase out these policies.

This step by the WTO reflects a conviction that the food economy needs less protectionism and more globalization. This conviction is rooted in the economic case for free trade that is nearly as old as the discipline of economics, itself.

6.1 The Theory of Comparative Advantage

Several years ago, Stanislaw Ulam, a mathematician, challenged economist Paul Samuelson to name "one proposition in all of the social sciences which is both true and non-trivial." A Nobel laureate, Samuelson (1969, p. 1) confesses that a suitable response did not occur to him immediately. But he eventually thought of one – David Ricardo's (1965, pp. 77–93) theory of comparative advantage.

Now nearly two centuries old, the theory goes beyond the argument for free trade offered in *The Wealth of Nations*, which stressed the virtues of trade but was vague about how individual agents (be these people, firms, or countries) choose to specialize. Part of Ricardo's contribution was to distinguish between absolute advantage, which reflects productive capacity, and comparative advantage, which depends on different agents' opportunity costs. He also showed that competitive markets foster specialization according to comparative, not absolute, advantage. When all agents allow themselves to be guided by market forces, trade among them creates greater prosperity. Samuelson has no doubt that Ricardo's case for free trade is the most significant idea ever to come out of the social sciences: "That it is logically true need not be argued before a mathematician; that it is not trivial is attested by the thousands of important and intelligent men who have never been able to grasp the doctrine themselves or to believe it after it was explained to them" (Samuelson, 1969, p. 1).

The essence of the theory of comparative advantage is captured by a vivid illustration for which we can thank Gregory Mankiw, a leading US economist. The illustration has to do with Tiger Woods, who has a number of skills and capabilities – including though not limited to his being one of the best golfers of all time. Mankiw supposes that Woods can earn $10,000 in 2 hours by filming a television commercial, which is probably a very conservative estimate of the compensation he would receive. Another option for Woods is to spend the same 2 hours mowing his lawn. In light of his athleticism and intelligence, nobody else can do the job as quickly. For example, Forrest Gump needs 4 hours to accomplish the task. During those same 4 hours, Gump's best option is to work at a fast-food restaurant, for a $20 paycheck.

The inefficiency of specializing according to absolute advantage is undeniable. If Woods, mindful of his superior capacity for mowing lawns, actually does the job, he incurs a substantial opportunity cost – to be specific, the $10,000 that would come his way for filming a commercial. In contrast, Gump, who has no absolute advantage in lawn-mowing, holds a comparative advantage, as indicated by the fact that his opportunity cost is just $20. As long as Gump charges less than Woods's opportunity cost (to repeat, $10,000), Woods gains by hiring Gump. Gump gains as well, provided Woods offers more than $20 (i.e., Gump's opportunity cost). Clearly, the incentive for the two men to strike a bargain is compelling (Mankiw, 2001, pp. 56–57).

Specializing according to comparative advantage and trading amount to what economists call a "positive-sum game"; that is, interchange that creates gains for all participants. The outcome is a positive sum when individuals, like Woods and Gump, are involved. As indicated in this chapter's appendix, the same holds for entire countries that trade with one another. Some of globalization's opponents deny that small, poor countries can benefit by trading with larger, more productive nations. This is tantamount to saying that Forrest Gump stands to gain nothing from being hired to take care of Tiger Woods's lawn. Others do not see how a rich country ever profits from commercial exchange with less productive places. This is equivalent to doubting that Woods is better off if he hires Gump. The two groups of opponents cannot be simultaneously correct. More than that, the theory of comparative advantage reveals that both camps can be simultaneously wrong, and indeed almost always are.

Distribution of the Net Benefits of Trade

Within a trading nation, commerce with foreigners benefits producers with a comparative advantage – producers that become exporters as access is gained to other countries' markets. However, the biggest winners are consumers. They are able to purchase imported goods at prices reflecting marginal opportunity costs in exporting nations, which by definition are lower. Considerable though the gains enjoyed by exporters and consumers are, there are still trade-offs. For instance, costs are incurred by producers lacking a comparative advantage. This group faces heightened competition from low-priced imports. Furthermore, consumers must take into account the higher prices they pay for goods their countries export. The conceptual framework for analyzing net economic value (NEV) and its components presented in Chapter 4 can be used to assess all these economic impacts.

Two national markets for the same commodity, wheat, are shown in Figure 6.1. In one country, the exporter, the good would not be very scarce in the absence of trade. This is reflected by the relatively low price, P_{EA}, that would be observed there under autarky. In contrast, the price in the importing country would be relatively

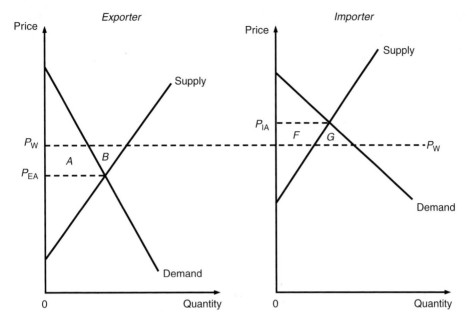

Figure 6.1 Trade and net economic value.

high, P_{IA}, if it were economically isolated. NEV in each country would consist of the area between the domestic demand and supply curves if there were no international commerce.

Suppose now that economic isolation ends. Merchants do what they do so well, buying wheat where it is cheap (in the exporting country) and selling the commodity where it is dear (in the importing nation). This business, called arbitrage, causes prices in the two markets to converge to a single price, P_W, which is above P_{EA} and below P_{IA}. At P_W, the excess of production over consumption in the exporting country is the same as the excess of consumption over production in the importing country. The comparative advantage of the exporting country is indicated by the fact that the international price exceeds what the marginal cost (MC) of wheat would be in the country if there were no trade. The latter would equal P_{EA} if the market were competitive. Also, the importer's lack of comparative advantage is reflected by the international price's being lower than P_{IA}, and therefore MC without trade.[1]

The increments in NEV created in both countries by commercial exchange show up in Figure 6.1, as do changes in consumers' surplus (CS) and producers' surplus (PS) in each trading partner. With the price of wheat going up in the exporting nation, there is an unmistakable gain in PS, represented by areas A and B. Part of this growth comes at the expense of consumers, the decline in CS being represented

---

[1] As this example makes clear, comparative advantage depends not just on the supply side of the economy, but on the demand side as well. With enough demand growth in an exporting nation, the price observed there without trade rises above the world price, thereby causing that nation to become an importer instead.

by area *A*. However, the gap between the former gain and latter loss is positive, obviously comprising area *B*. The changes in NEV in the country with a comparative advantage in wheat production are the mirror image of the impacts on CS and PS in the trading partner. In the latter, a declining price creates additional CS (areas *F* and *G*) while simultaneously reducing PS (area *F*). But since the former impact outstrips the latter, there is a net gain: area *G*.

To summarize, trade creates two costs, one resulting because consumers must pay more for the goods their country sells internationally and the other comprising the losses of producers obliged to compete with cheap imports. But these two costs are less than the benefits of trade, which accrue to consumers given access to low-priced imports as well as producers given access to foreign markets in which prices are high. Each trading partner experiences growth in NEV, represented by areas *B* and *G* in Figure 6.1.

6.2 The Net Costs of Trade Distortions

In a way, the preceding analysis, which focuses on a pair of national markets for a single good, exaggerates the negative impacts of international trade. Take area *A* in Figure 6.1, which represents the decline in CS resulting from higher prices of exported goods. This loss for consumers is far outweighed by the benefits coming their way because imports exert downward pressure on other prices. For example, Guatemala, like many other countries, has a short list of exports. To be sure, its citizens pay a little more for coffee, bananas, and a few other things than they would in autarky. But they would also pay much more for automobiles, computers, and many other goods and services for which no comparative advantage exists in the local economy. Only an economic masochist would take to the streets to protest international trade because of its effects on consumer prices.

Focusing on a single commodity can likewise distort one's view of the impacts of trade on PS throughout an entire economy. As a rule, diminished prices, production, and earnings in those industries and sectors that lack comparative advantage are more than offset by expansion in other markets. As long as labor can move from one domestic site of employment to another, machinery can be retooled, and related adjustments are possible, then the overall impact of trade on PS is bound to be positive.

Regardless of all the well-being created by commercial exchange across national frontiers, costs are still incurred. Transferring factors of production from one part of the economy to another is not free. If the expense of moving assets out of an industry or sector lacking comparative advantage is prohibitively high, then earnings for the owners of these assets decline. For example, a worker might have acquired skills that apply only to a particular job. If demand for these skills falls, or if the job is eliminated, then his or her wages suffer. For someone going through this because of expanding trade, knowing that the overall national interest is being served is cold comfort. He or she eagerly seeks protection from imports.

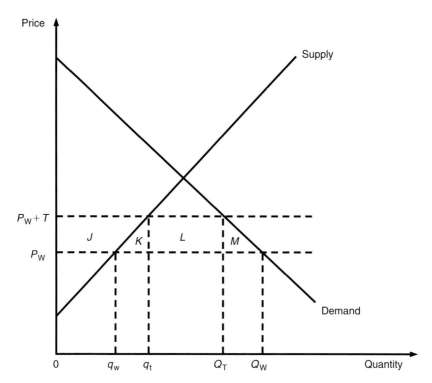

Figure 6.2 The impacts of protectionism.

To be sure, this protection comes at a price. Depicted in Figure 6.2 is one possibility, which applies to the common case in which the importer's purchases comprise a tiny portion of international supplies. Imposing a tariff (T) causes the price within the importing country to rise from the world value (P_W), which remains the same, to $P_W + T$. This causes domestic production to go up, from q_w to q_t, and national consumption to fall, from Q_W to Q_T. Government collects revenues (area L) on imports, which obviously are lower ($Q_T - q_t$) with the tariff than without ($Q_W - q_w$). Also, additional PS is created (area J). However, the sum of these revenues and the incremental PS falls short of the reduction in CS (areas J, K, L, and M). The bottom line is that NEV falls, the net reduction being represented by areas K and M.[2]

Protection of the kind shown in Figure 6.2 is ubiquitous in the global food economy. For example, the European Union and the United States keep high-cost sugar producers in business by limiting imports from the tropics and subtropics, where the commodity can be produced more cheaply. Similarly, Japan is nearly self-sufficient

---

[2]From the standpoint of domestic producers and consumers, a quantitative restriction on imports has exactly the same impacts as a tariff. For example, limiting purchases of foreign goods to $Q_T - q_t$ would cause the domestic price to go up to $P_W + T$. The increment in PS would be area J and CS losses would comprise areas J, K, L, and M. However, area L would not end up in government coffers, as happens with a tariff. Instead, it is captured by people allowed to import the restricted good, and perhaps by officials suborned by these people to acquire their licenses.

in rice thanks to quotas and other measures that keep out imports from places where the grain can be raised more efficiently, such as Thailand and Vietnam. The resulting loss of CS is far greater than the gain in PS, much of which is expressed as an increase in real estate values (Chapter 4).[3]

Distortion of international agricultural trade is by no means confined to the protection of producers that lack comparative advantage. In the United States and elsewhere, price supports have been used to enhance the incomes of farmers who have no trouble competing internationally. This policy stimulates output and exports, which in turn depresses the prices that other countries' producers receive in global markets. Evaluating the overall impacts of subsidized exports on NEV can be a little complex. In importing countries, there is a gain, with additions to CS exceeding losses in PS. In nations with subsidized exports, PS grows more than CS declines. However, there is another negative impact, associated with using government funds to distort commodity markets. By and large, this last cost, which includes the losses in economic well-being suffered by households that must pay higher taxes, exceeds the net gains in PS and CS in both exporting and importing nations. In other words, the world does not benefit when efficient farmers are subsidized.

International trade is not distorted solely for the sake of producers, either those who face competition from imports or those desiring a boost in exports. Consumers are sometimes accommodated, as in the case of food aid that is dumped on the commercial market of a recipient country rather than being donated to those who cannot afford an adequate diet. More often, consumers benefit from an over-valued currency, which makes imports artificially cheap (Perkins *et al.*, 2001, pp. 695–698).

Currency over-valuation often comes about because a country's inflation is not matched by a devaluation of its currency. This sort of over-valuation and the import bias it creates can be illustrated with a hypothetical example, one having to do with an imaginary trading partner of the United States called Macondo. Suppose that US inflation is zero and Macondo's rate is 50 percent. In other words, prices of all goods and services in Macondo rise by one-half during the calendar year while, on average, US prices do not vary at all. If the exchange rate between Macondan pesos and US dollars does not fully adjust in response to this difference in inflation, then US goods will grow relatively cheaper and Macondan output will become relatively more expensive in both national markets. This change in relative prices, or terms of trade, causes Macondo's imports from its trading partner to rise and its exports to fall.

---

[3]It is not coincidental that Ricardo, who gave us the theory of comparative advantage, also sharpened our understanding of land values. In his day, Britain's corn laws – agricultural protectionism put in place to prevent commodity price declines after the Napoleonic Wars came to an end – were hotly debated. The policy's advocates contended that prices needed shoring up because costs of inputs, including real estate expenses, were high in the country. Ricardo countered that this view of cause-and-effect was erroneous, and that land rents comprise a residual return left over after the costs of labor and other non-land inputs are deducted from the market value of output. His position, and that of other economists, was that prices and specialization are best left to free markets and that, as these are determined, rents will adjust upward or downward as appropriate, with no effect on output (de Vivo, 1987). While caveats can be made to this argument, its fundamental logical is irrefutable, and remains compelling today.

Consider cross-border trade of wheat, which Macondo sells abroad, and corn, which the United States exports. At the beginning of the year, the price of corn in its country of origin is $2.00/bushel. The exchange rate is 10 pesos per dollar, which means that a bushel of corn costs 20 pesos in Macondo. Meanwhile, the price of a bushel of wheat is 40 pesos or $4.00. Macondo's international terms of trade (i.e., the price of its export divided by the price of its import) are 2/1, expressed either in US currency ($4.00 divided by $2.00) or pesos (40 divided by 20).

If trade is not to be affected by Macondo's inflation, the peso must devalue. In particular, the exchange rate has to rise by the end of the year to 15-to-1, implying that the peso's value falls from $0.10 to $0.06⅔. With this adjustment, the price of Macondan wheat in the United States remains unchanged at the end of the year: 60 pesos/bushel ÷ 15 pesos/dollar = $4.00/bushel. Accordingly, US consumers have no reason to buy any more or any less of the imported commodity. By the same token, Macondo's imports of corn do not vary. Since devaluation causes foreign currency (i.e., the dollar) to become 50 percent more expensive, prices of everything purchased with that currency likewise go up by one-half. For example, the price of a bushel of US corn in the importing nation rises from 20 pesos ($2.00/bushel × 10 pesos/dollar) to 30 pesos ($2.00/bushel × 15 pesos/dollar). Since this price rise equals inflation (i.e., average price increases), Macondo's consumers have no reason to alter their purchases of imported corn.

But if the peso does not fall all the way from $0.10 to $0.06⅔, then by definition it will be over-valued and trade will be affected. This over-valuation may be caused by governmental intervention in currency markets. It can also result if Macondans working in the United States are sending dollars back home to relatives. As these dollars are converted into local currency, which the relatives need to buy goods and services locally, there is a bidding up of the peso's value. Regardless of whether over-valuation occurs for this reason or some other, however, it is certain to create an import bias. Suppose, for example, that the exchange rate only goes up to 12-to-1, rather than all the way to 15-to-1, as Macondo experiences 50 percent inflation. After 12 months, the price of the country's wheat in the US market will have risen from $4.00 to $5.00/bushel (60 pesos/bushel ÷ 12 pesos/dollar), with the price of US corn remaining unchanged. With this change in relative prices, US consumers will buy less wheat and more corn. Their counterparts in Macondo will respond exactly the same way as the terms of trade shift from 2/1 at the beginning of the year to 5/2 (60 pesos/bushel of wheat ÷ 24 pesos/bushel of corn) at the end of the year. Not only do Macondan wheat growers lose ground in the US market, but sales in their own country become harder to achieve.

Currency over-valuation creates substantial benefits for consumers with a large appetite for imported goods. Since consumers like these tend to be politically influential, governments – especially in the developing world – are often reluctant to devalue. However, an excessively strong currency always makes things difficult for the producers of tradable commodities, including farm products. Competing against imports is tough. So is exporting from a country with a distorted exchange rate. Of course, simultaneously subsidizing consumers and implicitly taxing producers is unsustainable. Countries with over-valued currencies lose foreign exchange as production for foreign markets is discouraged and as people rush to purchase imports that are artificially cheap. Once reserves of foreign exchange are depleted, major changes, not least devaluation of the national currency, are imperative to restore the trade balance.

6.3 The Debate over Globalization

As technological change has made communication and transportation less expensive, international trade has expanded, including in the food economy. While the net benefits of trade are clearly positive, selected industries and sectors have been obliged to retool, relocate, and otherwise adjust, always at a cost. Since greater adjustments are required as the pace of globalization accelerates, complaints about globalization have multiplied as well.

Multinational Firms

Some of the unease about globalization relates to large companies operating in more than one country. There are vague accusations that these companies control culture and access to knowledge, often at the expense of free expression and local advancement. Multinationals are also charged with exploiting workers and despoiling the environment.

To be sure, businesses are more than willing to shift employment from high- to low-wage settings, provided differences in worker productivity are not too great. However, it is difficult to make the case that, within any single country, multinationals' employees are treated unfairly. On average, wages paid by companies with operations in multiple countries are 50 percent above the compensation offered by enterprises that operate in just one place. The discrepancy is especially pronounced in low-income settings, where multinationals' wages are approximately double average compensation (Edward Graham of the Institute for International Economics, cited in Crook, 2001).

A considerable amount of environmental damage can be traced to multinationals. A spectacular example – one with direct relevance to the food economy – is the Bhopal disaster of 1984, when an explosion at a pesticide plant in India owned by Union Carbide killed thousands of people and left many more permanently injured. Thanks to subsequent legal action, the company paid fines and restitution amounting to millions of dollars and agreed to a range of safety measures.

Incidents such as Bhopal, which are by no means excused by pointing out that a host of state-owned firms and public agencies in various parts of the world have done comparable environmental damage or worse, demonstrate the need for effective public intervention to contain market failure. Absent this intervention, which can take the form of pollution taxes for example, private businesses harm natural resources excessively (Chapter 5). The real question where multinationals are concerned relates to their discouraging governments from dealing with market failure, by threatening to relocate to places where environmental controls are more lax. At worst, governments can engage in a "race to the bottom," hoping to attract mobile enterprises, and the investment, jobs, and exports they bring, by turning a blind eye to pollution or offering other inducements.

The race to the bottom is not as relentless as one might think. Indeed, it is implausible if governments are reasonably representative and if environmental impacts are mainly confined within national frontiers. Why, one wonders, would a representative government waive environmental controls for a polluting industry if the ultimate effect is to diminish national well-being – in particular, because the benefits of industrial expansion are outweighed by environmental damages suffered by the government's constituents? If a large share of environmental damage spills across national frontiers, then a government that fairly represents its constituents while neglecting other countries' interests might indeed engage in a race to the bottom, creating substantial trans-boundary externalities along the way. Of course, a government that is not representative, because it is corrupt or dictatorial, for example, can also be induced to compete by beggaring the natural environment. The point is that the race to the bottom is not ineluctable. Rather, it happens because mechanisms for dealing with trans-boundary spillovers are faulty, governments are inordinately swayed by special interests, or both.

Multinational firms certainly have been known to exploit deficient arrangements for the internalization of environmental impacts (be these local or transboundary) and to strike deals with special interests. However, these companies are not always enthusiastic participants in a race to the bottom. For one thing, many of them are susceptible to consumer boycotts organized to protest natural resource degradation, coziness with despots, and so forth. Moreover, with information flowing as freely as it does in a globalized world, the risks and expenses of doing business in places where the rule of law is weak are becoming clearer every day, including to multinationals. Insofar as these firms stay away from such places, pressure grows to strengthen the rule of law. Multinationals' behavior, then, sometimes promotes a race to the top, not the bottom.

Food Self-Sufficiency and Export Cropping

Aside from looking askance at multinational corporations, many critics of globalization are generally suspicious of international trade. In particular, these critics propose that poor countries, regardless of any comparative advantage they may have in agriculture, foreswear export cropping for the sake of self-sufficiency in food production.

Agricultural self-reliance makes no economic sense. In every part of the world, people above the international poverty line (approximately $2.00 in daily earnings) are able to buy enough food, which aside from times of armed conflict is readily available in local markets. It is the poor, then, who are chronically or occasionally food-insecure. Just as trade results in prosperity, autarky creates poverty, and therefore hunger.

The sacrifice in living standards that happens if a poor country's comparative advantage in export cropping is not exploited has been amply demonstrated by economic research. A study in Liberia, for example, found that a typical family could have three times more rice and also manage natural resources better by producing commercial, tree-crop output for foreign markets and purchasing rice, rather than growing its own grain (Epplin and Musah, 1987). Surveying long-term investigation in the Gambia, Guatemala, Kenya, the Philippines, and Rwanda, the director of the International Food Policy Research Institute (IFPRI) has concluded that "agricultural commercialization raises the income of the rural poor, thus improving their food security" (von Braun, 1989, p. 4). Cash cropping is not a cause of hunger, but rather an important part of the cure.

There is another detriment of agricultural self-reliance, which is the risk it creates of output shortfalls. In many settings, the index of agricultural production failure, which is defined as the standard deviation of production divided by mean output, is 20 times the world level, which happens to be very low. Accordingly, supply reductions occur frequently in agricultural autarkies and, since demand for staple commodities tends to be inelastic (Chapter 2) in these settings, these shortfalls provoke price spikes, which are very burdensome for food-insecure people. With more open markets, price variability is modest. A crop failure in one part of the world causes only minor changes in food prices and consumption because world production and reserve stocks are shared. Higher prices at the global level induce every supplier to put a little more food on the market from stocks or production. Higher prices also tell every food consumer to eat a little less. Through international trade, food from regions with plentiful supplies is made available to regions where there is acute scarcity.

International agricultural trade is opposed categorically on other grounds. Some critics worry about the energy required to move commodities from place to place, although this is not a problem if fuel is not priced too inefficiently. Pointing out that women often specialize in the production of food crops, others worry that they will be disadvantaged by increased emphasis on cash cropping, even though there is no evidence to support this claim. At the end of the day, there is no escaping the conclusion that self-reliance is a recipe for deprivation. Food security requires trade.

Free Trade versus Fair Trade

Far less extreme than categorical opposition to export cropping is the stance that trade, though generally desirable, should be "fair" rather than "free."

In principle, the difference between the two is clear enough. As a practical matter, "free trade" can be defined as international commerce that is unencumbered by barriers, other than those allowed by the WTO. Spoken of approvingly by many, "fair trade" is supposed to be consistent with the protection of workers, women, minorities, children, and the environment. In general, free trade leaves social and environmental issues to be resolved by the country or countries affected. In contrast, fair trade would impose internationally established social and environmental rules (or chapters) on any nation, rich or poor, before it would have access to the markets of other fair-trading countries. For example, Runge *et al.* (2003, pp. 172–176, 205) call for a global environmental organization that would address trans-boundary pollution and discourage relaxation of national ambient standards in a race to the bottom, much as the WTO deals with multilateral trade issues.

In the final analysis, fair trade's appeal is superficial. The best chances a poor country has of improving living standards, generally, and investing more in things like environmental quality, specifically, lie with its exploitation of comparative advantage, almost always in industries and sectors that make intensive use of cheap, unskilled labor. What bureaucrat in an international agency is competent to set a "fair" or "just" wage for this input? If the wage is set too low, living standards will be unduly depressed. More likely, it will be set too high, in which case investors will be discouraged from creating jobs and household earnings will be hurt by unemployment. It must also be remembered that international trade rules that would enforce wage standards by curtailing imports from non-complying poor countries would hurt consumers elsewhere, by raising prices.

While many objectives of fair trade are commendable, its pursuit, if unsuccessful, can be damaging if movement toward freer trade is impeded. It is notable that efforts to begin a new round of multilateral trade negotiations in Seattle in late 1999 broke up when the European Union and the United States failed to convince low-income nations that environmental and social (labor) chapters should be included in the negotiating agenda. The ultimate result of this failed attempt to implement fair trade was to postpone the beginning of multilateral talks until a meeting two years later in Doha, Qatar.

6.4 Agricultural Trade: Recent Trends and the Current Debate

Whatever the merits or shortcomings of the arguments against globalization, international trade has increased steadily since the middle of the twentieth century, just as it did during most of the preceding 150 years. Exactly as predicted by the theory of comparative advantage, living standards have improved as well, especially in the world's most open economies.

The fact that trade has been allowed to grow since the Second World War reflects important lessons learned painfully during the 1930s. At the beginning of that decade, leading industrial nations had veered sharply toward autarky, with protectionism in one country begetting protectionism elsewhere – all in vain hopes of maintaining output and jobs in industries facing foreign competition. The ensuing collapse in international commerce largely explains the length and severity of the Great Depression, which did not end in the United States until global conflict induced an industrial recovery in the early 1940s. Keenly aware of the linkage between protectionism and economic collapse, which in turn threatens world peace, 44 countries, including the victorious non-communist powers, took steps toward freer trade (or liberalization) at a meeting in Bretton Woods, New Hampshire in July 1944 – many months before the final defeat of Germany and Japan. The WTO would not come into being until the middle 1990s.[4] But its precursor, the General Agreement on Tariffs and Trade (GATT), was accepted.

Thanks to the GATT, the cause of unencumbered commerce among nations was advanced. During the first round of negotiations, which got under way in Geneva in 1947, the 23 participating countries, all of which were developed, agreed to cut average industrial tariffs by 20 percent. Subsequent talks halved these average duties and then halved them again – to around 4 percent in 2000. In the Uruguay round, which ended in 1994, agreement was reached to establish the WTO, with mechanisms for dispute settlement and sanctions for the violation of trade accords. These mechanisms and sanctions were never a part of the GATT. Also accomplished during the Uruguay round was a commitment by participating governments to convert non-tariff barriers into tariffs, which as a rule negotiators find easier to ratchet downward.

As protectionism has been rolled back, international commerce has expanded. From 1961 through 2001, for example, growth in trade outstripped increases in global output. In keeping with this broad pattern, world agriculture became more trade intensive, with commodity exports going up at a 4.0 percent annual rate during the last four decades of the twentieth century as output growth was averaging 2.3 percent per annum (World Bank, 2003a, p. 110; FAO, 2004). However, agriculture has not been at the forefront of trade intensification. Amounting to 27 percent of global merchandise trade in 1961, exports and imports of farm products had fallen below 10 percent of the world total by the turn of the twenty-first century (FAO, 2004). As indicated in Table 6.1, agricultural trade consistently has grown at a slower rate than imports and exports of manufactured goods since 1980.

Table 6.1   Growth in agricultural and manufactured exports, 1980–2001

	Annual export growth, 1980–1990 (%) (1)	Annual export growth, 1990–2001 (%) (2)
Entire world		
Agriculture	4.3	3.6
Manufacturing	5.9	4.8
Developing countries		
Agriculture to	3.4	4.8
Developing countries	3.6	7.8
Developed nations	3.4	3.3
Manufacturing to	7.6	8.9
Developing countries	7.3	10.0
Developed nations	7.8	8.3

Source: World Bank (2003a), p. 110.

---

[4] Negotiators at Bretton Woods did agree to establish the International Monetary Fund, which provides short-term financial assistance and guidance to countries experiencing chronic outflows of foreign exchange (because of persisting trade deficits and other reasons), as well as the World Bank, which underwrites development projects.

Agriculture and International Trade Negotiations

A fundamental reason for agriculture's relatively weak performance is that developing countries kept their distance from the GATT. For many years after the Second World War, Latin American nations, which had been very open to trade prior to the 1930s and therefore had suffered a lot during the Great Depression, tried to industrialize by keeping out imports (Chapters 7 and 12). Participation in the GATT obviously would have interfered with this strategy. In addition, as European colonialism came to an end, newly independent nations in Asia, the Middle East, and Sub-Saharan Africa tended to follow Latin America's lead, with few joining negotiations aimed at trade liberalization. With the developing world largely outside the GATT during that body's early years, virtually nothing was done about the trade barriers and export subsidies that undermine the comparative advantage of poor countries. Perhaps protection of farmers in Western Europe, Japan, and other places would not have been overcome; having suffered food shortages during the 1940s, those parts of the world were committed to agricultural self-sufficiency. But the debate was never joined and agriculture stayed off the GATT agenda for decades. Not coincidentally, commodity exports from Africa, Asia, and Latin America suffered. In 1961, the developing world was a net exporter of farm products, with sales to other regions 89 percent above agricultural imports. Subsequently, the farm subsidies that Western Europe had adopted to achieve self-sufficiency caused that region to become a food exporter. As a result, the developing world became a net agricultural importer, with exports 7 percent below imports in 2001 (FAO, 2004). Poor countries' share of total agricultural exports fell from 41 to 30 percent during the last four decades of the twentieth century (Figure 6.3).

To this day, international exchange of edible commodities continues to be more restricted than other categories of trade. The average global tariff on agricultural commodities was 62 percent at the turn of the twenty-first century, which was far above the average tariff of 12 percent that the United States levied on agricultural imports (Gibson *et al.*, 2001). Growth in agricultural exports from developing countries to affluent nations did not accelerate in the 1990s (Table 6.1, column 2), mainly because the latter nations' farmers were still protected from foreign competition. Meanwhile, annual increases in agricultural trade within the developing world were

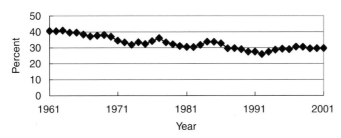

Figure 6.3 Agricultural exports from developing countries relative to world totals. *Source*: FAO (2004).

picking up, as were manufactured exports from developing countries – to rich and poor countries alike.

Since the WTO's inception in 1994, farm products have been a primary concern in international trade talks. Getting under way in November 2001, the Doha round of negotiations is focused largely on sectors in which the developing world holds comparative advantage. Among these are textiles and other labor-intensive industries. Another such sector is agriculture. This emphasis reflects poor countries' involvement in trade talks. Of the more than 140 nations that belong to the WTO, about 100 can be categorized as developing. Moreover, these countries have coalesced around a negotiating position reflecting their shared interests – a position expressed in the Dhaka Declaration of 2003. Signed by 39 African, Asian, and Latin American governments, the document calls for free and secure access by developing countries to rich nations' markets for food and other products. It also demands an end to export subsidies, especially for agricultural commodities. In addition, environmental and labor chapters are rejected, as occurred at the Seattle summit in November 1999.

The Dhaka Declaration is not a manifesto for free trade. For example, it endorses the protection of poor countries' "infant industries" from global competition. Also, some fear that aggressive bargaining by leading developing nations could lead to a collapse of multilateral trade talks. Ominous in this regard was the break-up of the Cancún meeting of September 2003 – after the Group of Twenty vociferously rejected a vague proposal by the European Union and United States to curb agricultural trade distortions.

When multilateral talks stall, bilateral or regional negotiations tend to be stimulated, as indicated by the fact that the United States was discussing trade accords with 14 different nations as of late 2003. The WTO's predecessor generally took a benign view of bilateral agreements that help poor countries. In 1971, for example, the GATT authorized a departure from its general rule that, if a GATT member extends trade preferences to another member, the same preferences must be given to all members. To be specific, wealthy states were allowed to liberalize trade for up to a decade with one or more poor countries, alone.

Notwithstanding waivers of this sort, replacement of the WTO and its multilateral framework with a series of bilateral and regional accords would not be good for poor countries. For one thing, rich nations seldom choose to strike trade deals with low-income places, which stand to gain much from access to markets in Europe, North America, and other settings. Moreover, even when a deal is reached, preferences often do not extend to agricultural commodities, the output of labor-intensive industries, and other things in which developing countries hold comparative advantage. This is an important omission because the tariffs that rich countries apply on these goods are much higher than average duties. Hertel *et al.* (2002) have found, for example, that tariffs in the developed world on the output of labor-intensive industry in less affluent places are typically double or triple the average duty of 4 percent. An additional shortcoming of bilateral or regional dealing is that poor countries often are pitted against each other. For example, the European Union gave tariff preferences on bananas imported from former British and French colonies in the West Indies. Due to these preferences, which the United States challenged successfully in the WTO, banana exports to Europe by more efficient producers in Central and South America fell sharply. Finally, the *ad hoc* liberalization of trade with poor countries has retarded their growth by overlooking their own major barriers to trade.

Clearly, globally free trade is best, and the surest way to achieve it is through multilateral negotiations under the auspices of the WTO.

The Gains from Freer Trade

There are some ironies in the negotiating stance of poor countries belonging to the Group of Twenty. As emphasized in the next chapter and other parts of this book, much of the harm suffered by agriculture in individual countries is inflicted by national governments. Too frequently, public officials peg commodity prices at artificially low levels, thereby discouraging production. In addition, farmers' access to new technology is impeded because, aside from Brazil, China, and India, too little is invested in agricultural research and development. Moreover, the trade barriers that developing nations impose on each other's output are substantial, typically exceeding the barriers to agricultural imports in places like Europe. In 1995, low-income countries faced tariffs averaging 15.1 percent on their exports to rich nations, but 18.3 percent on exports to other poor places (Hertel *et al.*, 2002). Duties on farm imports generally exceed 60 percent in the Caribbean, Sub-Saharan Africa, and North Africa; in South Asia, the average agricultural tariff exceeds 110 percent (Gibson *et al.*, 2001).

Another reason why focusing on agricultural protectionism in rich nations makes little sense for the developing world is that it has a greater stake in other sorts of trade. Fixating on agriculture would have been understandable three, four, or five decades ago, when Africa, Asia, and Latin America had limited comparative advantage in industry – and, it must be reiterated, most governments in these regions chose to be mute spectators in the GATT. In 1965, agriculture accounted for 45 percent of poor countries' exports, with manufacturing the source of another 24 percent. But by 2000, the former share had fallen to one-tenth and the latter had risen to three-quarters (Hertel *et al.*, 2002). One supposes that developing nations are training their sights on agricultural protectionism in other places these days because the same nations impose some of the world's highest industrial tariffs, with duties averaging 11.5 percent in poor countries as opposed to 1.5 percent in affluent settings in 1995. The scale of self-inflicted damage in the developing world is sizable. In particular, reducing tariffs on industrial goods by 40 percent would expand the volume of global trade by about $380 billion, with a large portion of this increase captured by East Asia and other developing regions where manufacturing is on the rise (Hertel *et al.*, 2002).

There was a further irony concerning the break-up of talks at Cancún, which is that rich countries stand to gain more from abandoning their own agricultural protectionism than poor countries do. Throughout the Organization for Economic Cooperation and Development (OECD), which is a grouping of affluent nations, annual spending on farm subsidies exceeds $300 billion. This expenditure, made possible by taxpayers of course, would fall drastically if agricultural commodities were traded freely. In addition, some food prices would fall, thereby benefiting the non-farming majority of the OECD population.

Diao *et al.* (2001) have examined the improvement in well-being that would result in the near term from the elimination of tariffs, subsidies, and other distortions in agricultural markets. This improvement, called static resource allocation gains, would amount to $31 billion for the world as a whole. Most of this amount would accrue to taxpayers and consumers in rich nations (Table 6.2, column 1). Outside of Japan, Norway, and the European Union, free trade would cause commodity values to rise. According to this study, average prices of farm goods would go up by 11.6 percent, with approximately half the increase due to the scrapping of tariffs (including in the developing world), another third resulting from elimination of price supports in the United States and Europe, and the other sixth associated with ending export subsidies of the sort favored by the European Union. Higher prices would, of course, benefit farmers in poor countries, especially because the developing world would return to being a net exporter of agricultural commodities. On the other hand, non-agricultural households in these places would lose purchasing power. But since this latter impact would be exceeded by the benefits of specializing according to comparative advantage, static resource allocation gains would be positive outside the OECD (Diao *et al.*, 2001).

The same three investigators have also estimated the dynamic benefits of liberalized trade in farm products. Benefits of this sort arise because, without trade

Table 6.2 Annual welfare gains from the elimination of agricultural trade distortions (1997 dollars per annum)

	Static resource allocation gains (billion) (1)	Static gains plus dynamic benefits (billion) (2)
Entire world	31.1	56.4
Affluent nations	28.5	35.2
United States	6.6	13.3
Canada	0.8	1.4
European Union	9.3	10.6
European free trade area	1.7	0.2
Japan and South Korea	8.6	6.2
Australia and New Zealand	1.6	3.5
Developing countries	2.6	21.3
China	0.4	2.2
Rest of Asia	1.5	5.1
Southern Africa	0.3	0.8
Mexico	−0.2	1.6
Rest of Latin America	3.7	6.1
Other	−3.1	5.4

Source: Diao *et al.* (2001), pp. 37–39.

barriers and other distortions, savings, investment, and productivity change follow a more efficient course over time. Impacts of this sort were modeled for a 15-year period, with the resulting gains converted into constant, annualized equivalents. Interestingly, annualized dynamic benefits turned out to be nearly as large as static resource allocation gains. In addition, the combined sum of these benefits and gains would be more evenly divided between OECD members and the rest of the world: $35.2 billion versus $21.3 billion, for an annual total of $56.4 billion (Table 6.2, column 2).

International commerce in agricultural products, not to mention many other goods and services, which is entirely free of tariffs and other barriers is an ideal that will not be realized any time soon. Recognizing this, economists have investigated the consequences of partial, as opposed to complete, liberalization. In one simulation exercise, which was undertaken to identify the stakes involved in the Doha round of trade negotiations, it was assumed that agricultural tariffs would be reduced to an average of 5 percent in affluent countries, with none of their duties remaining above 10 percent on any item. For developing nations, higher trade barriers on farm products were stipulated: an average duty of 10 percent and no individual tariff above 15 percent. On the manufacturing side, an average tariff of 1 percent and a maximum duty of 5 percent were assumed in OECD members. For less affluent countries, the rates were 5 and 10 percent, respectively. No export subsidies of any kind were allowed in the simulation (World Bank, 2003a, p. 48).

Under these conditions, which would represent a considerable (though not unrealistic) step toward freer trade, $291 billion would be added to world income in 2015; this actually comprises 75 percent of the potential economic gains from full liberalization. Of the total impact, $159 billion would accrue to developing countries and the remaining $132 billion to OECD members. In addition, the annual benefit of freer trade would grow with time, surpassing $300 billion soon after 2015 and continuing to increase in subsequent years (World Bank, 2003a, p. 50).

Two-thirds of the additional global income that this study suggests freer trade can create would be due to changes in the food economy. To be sure, some of the increase would be captured by OECD consumers. But most would go to developing countries, as liberalization allowed them to exploit their comparative advantage in agriculture. Indeed, 80 percent of the $200 billion or so that freer trade in farm commodities could add to world GDP in 2015 would result from agricultural policy reform in developing countries, themselves (World Bank, 2003a, p. 50).

The vast majority of the world's farmers have a positive interest in agricultural trade liberalization. In the United States, raising international commodity prices by 11.6 percent, as Diao *et al.* (2001) predict would happen, would cause annual farming receipts to go up by $22.4 billion, after averaging $193.2 billion from 1998 through 2000. This potential gain exceeds the $16 billion per annum in deficiency and other direct payments that farmers received from the federal government during the same period. It also exceeds yearly outlays by the Commodity Credit Corporation for farmers, which averaged $20.5 billion from 1998 through 2000 (USDA, 2001). In the

European Union, the dismantling of export subsidies and barriers to imports would have a negative impact on farm incomes. Allowing rice imports into Japan would have a similar effect. Far outweighing these losses, however, would be the growth in earnings experienced by farmers in other parts of the world as they gained access to European and Japanese markets. Of course, the spending of these additional earnings would lead to GDP expansion in exporting nations.

Finally, while higher food prices would be a burden for many non-farming households in the developing world, agricultural trade liberalization would create advantages for them as well. During the 1980s, Rodney Tyers and Kym Anderson estimated that commodity prices would become more stable with free trade, with the coefficient of variation declining by 50 percent for beef and lamb and by 75 percent or more for dairy products, rice, and wheat (World Bank, 1986, p. 131). Reduced price variability would be a boon for impoverished Africans, Asians, and Latin Americans, many of whom can barely afford an adequate diet under normal market conditions but go hungry when the cost of food spikes.

6.5 Why Not More Trade?

Whether individual persons or entire nations are involved, voluntary exchange of goods and services creates prosperity. Trade has been going on for as long as people have had things to barter. During the past two centuries or so, international commerce has expanded, sometimes because of improved technology for transportation and communications and sometimes because governments have been persuaded to dismantle trade barriers. All economic research points to major improvements in living standards still to be gained from freer trade. In particular, consumers and others benefiting from trade would be able to compensate firms and workers – ideally with job retraining and relocation assistance – that lack comparative advantage, and therefore suffer costs as international commerce is liberalized.

Economic studies of the gains from trade do not always convince governments and their constituents. In spite of the economic sacrifice, protectionism is often chosen in response to the complaints of workers or industries that lack comparative advantage and are reluctant to bear the cost of adjusting to another line of work. Barriers to trade are also erected to create profits for the favored few. The US sugar quota – a perennial target of criticism for advocates of free trade – is an excellent case in point.

There are no doubts at all about the gross inefficiency of this policy. Its costs, which comprise higher prices for beverages and food, far outweigh its benefits. The problem is that the former are spread very thin, amounting to under $100 per capita for the entire national population. In contrast, the benefits are highly concentrated, comprising tens of millions of dollars for a handful of US producers. Under these circumstances, the latter group has strong incentives to lobby for continued protection, while most consumers do not find it worth their while to agitate for free trade in sugar. Foreign producers, who of course have no vote, have practically no influence. The ultimate result is that the quota remains in place.

Although the advocates of protectionism can be relied on for a stout defense of their special interests, at the expense of society as a whole, it would be unduly pessimistic to despair about liberalization, of the sort shepherded originally by the GATT and now by the WTO. The agreement that trade ministers reached in July 2004 for WTO members to begin phasing out export subsidies and other policies that distort international commerce in agricultural commodities suggests that the food economy is no less susceptible than manufacturing and other sectors to globalization, which continues to gather force.

Study Questions

1. Why would Tiger Woods and other people suffer economic losses if he mowed his own lawn?
2. How are consumers, producers with comparative advantage, and producers without comparative advantage affected by freer trade?
3. How do the benefits of an import tariff, which comprise increased PS as well as tariff revenues, compare to the lost CS associated with a tariff?
4. How are imports and exports affected by currency over-valuation? What about the effects of currency under-valuation?
5. Under what circumstances do governments engage in a race to the bottom, so as to attract multinational firms?
6. How might an emphasis on food self-sufficiency at the expense of export cropping aggravate hunger problems?
7. Why is fair trade, as opposed to free trade, difficult to implement?
8. Why were the United States and its western allies wary of protectionism at the end of the Second World War and what steps were taken on behalf of freer trade?
9. Why was there less progress toward free agricultural trade than toward freer trade in manufactured products during the second half of the twentieth century?
10. Explain and assess the negotiating stance of the Group of Twenty in current WTO negotiations.
11. Compare and contrast the static and dynamic gains from freer worldwide trade.
12. What do you think are the prospects for freer worldwide trade?

Appendix: A Two-Country Illustration of Comparative Advantage

The theory of comparative advantage is often explained not with reference to a pair of individuals, like Tiger Woods and Forrest Gump, but is instead illustrated with a couple of countries. For example, one might consider Argentina and Mozambique, which we suppose for simplicity's sake produce the same two goods: beef and sugar.

Let us say that Argentina has twice the agricultural resources that Mozambique possesses. The former nation's absolute advantage in the two commodities is indicated by the location of its production possibilities frontier (PPF$_A$), which depicts how much beef and sugar can be produced with the resources available in the country (Figure 6.4). Beef output is 1,000,000 head (i.e., the vertical-axis intercept of PPF$_A$) if all inputs are employed in ranching. Likewise, sugar production amounts to 500,000 tons (PPF$_A$'s intersection with the horizontal axis) if Argentina's agricultural capacity is dedicated entirely to raising this single crop. Linear combinations of the two commodities (e.g., 500,000 head of cattle and 250,000 tons of sugar) are also feasible. Lacking any sort of absolute advantage, Mozambique can raise 100,000 head of beef and no sugar or 200,000 tons of sugar and no beef – represented, of course, by the vertical-axis and horizontal-axis intercepts, respectively, of the country's production possibilities frontier (PPF$_M$) – or a linear combination of these two extremes (i.e., PPF$_M$ itself).

Even though monetary values do not appear in Figure 6.4, opportunity costs are readily identified. In particular, MC, which consists of the units of one commodity that must be foregone for the sake of a small increase in the other good, is expressed by the PPF's slope. In Argentina, increasing sugar production by 1 ton requires a two-head reduction in beef output. Conversely, the MC of beef is half a ton of sugar. PPF$_A$ being steeper than PPF$_M$, MCs are different in Mozambique. There, a 1-ton rise in sugar production requires that beef output be cut back by half

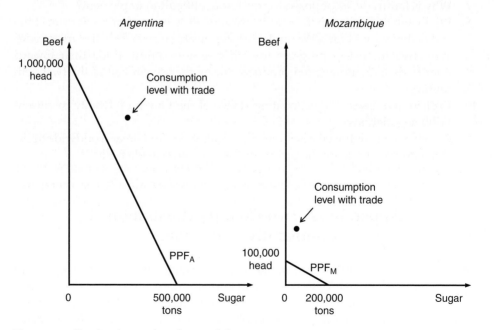

Figure 6.4 Production and trade possibilities.

a head, just as increasing beef production by one head involves a 2-ton sacrifice of sugar.[5]

In light of these MCs, raising sugar in Argentina would be every bit as inefficient as a decision by Tiger Woods to mow his own lawn. This is proved by demonstrating that both countries can reach superior levels of beef and sugar consumption if each produces the commodity in which it holds a comparative advantage and imports the commodity in which it is not so favored. To be specific, let Argentina specialize in ranching, because the MC there of beef production (i.e., half a ton of sugar) is lower than the MC in Mozambique (i.e., 2 tons). Likewise, sugar's MC is lower in the African nation, so it should produce nothing but that crop. Aggregate production, then, consists of 1,000,000 head of beef and 200,000 tons of sugar. One possible allocation of this output would result if Mozambique exchanged 150,000 tons of its sugar for 250,000 head of Argentine beef. Significantly, this would allow both countries simultaneously to achieve consumption levels beyond their respective PPFs, which represent limits on what people eat – not just what they produce – in the case of autarky (i.e., economic isolation). Clearly, 250,000 head of beef and 50,000 tons of sugar cannot be consumed in Mozambique if the country does not trade. By the same token, 750,000 head and 150,000 tons are outside PPF_A, which means that this consumption level is out of reach if Argentina is an autarky. Obviously, trade is mutually beneficial – that is, a positive-sum game.

Finally, the mutual benefits of trade are bound to materialize given the prices that will be observed if international commerce is unimpeded. Specific prices for beef and sugar cannot be identified in the hypothetical case of Argentina and Mozambique. However, something can be said about the price ratio (or terms of trade) – the price of sugar (P_S) divided by the price of beef (P_B).

Without trade, P_S/P_B in each country would reflect local opportunity costs. This represents equilibrium for a closed economy since no sugar grower can gain by switching to beef production and no cattle rancher has an incentive to sell his herd and start planting sugarcane; both activities are equally profitable. If Argentina were an autarky, for example, P_S/P_B would equal 2/1. Likewise, the terms of trade in Mozambique would be 1/2 if that country were economically isolated. Each of these two price ratios corresponds exactly to the slope of the national PPF, of course.

If trade begins, merchants will buy each good where it is relatively cheap and ship it where the good is relatively expensive. With beef initially worth 1/2 ton of sugar in Argentina, that country, which as we have shown already has a comparative advantage in beef production, will be an international source of that commodity. For identical reasons, Mozambique will export sugar, in which it has a comparative advantage.

---

[5]Most introductory texts in economics feature a PPF or two in the first few pages. Limits on productive capacity are depicted not by straight lines, as in Figure 6.4, but instead by curves bowed away from the origin. Such a curve indicates that, as production of a commodity goes up, its MC rises; for example, the beef given up for a unitary increase in sugar output would grow as sugar production rose. With linear PPFs, of course, the MC of sugar (i.e., sacrifices in beef production) does not vary at all as sugar output increases or decreases.

As beef moves out of Argentina, P_S/P_B falls in that country. Similarly, the price ratio goes up in Mozambique. In addition, these changes in the terms of trade are reinforced by the Latin American nation's imports of sugar and imports of beef by the African trading partner. In each case, consumers end up paying less for a good in which their respective countries lack comparative advantage. One cannot say what exactly the terms of trade will be with international commerce, but it is certain that P_S/P_B will be lower than 2/1 (relative prices in Argentina if the country is economically isolated) and above 1/2 (i.e., the price ratio in Mozambique if it is an autarky).

Once trade occurs and prices adjust, specializing according to comparative advantage is irresistible, in the sense that this is the only way to cover opportunity costs. Suppose that, with trade, P_S and P_B equal $750 and $450, respectively;[6] terms of trade, then, would be 5/3, which is obviously between 2/1 and 1/2. Producing one head of beef 4in Argentina carries a sacrifice of 1/2 ton of sugar (see above), which is worth $375 in the marketplace. Since beef sells for $450 per head, the rancher finds the trade-off profitable. By the same token, the market value of a ton of sugar produced in Mozambique exceeds the opportunity cost of this output (1/2 head of beef multiplied by $450). In contrast, growing sugar in the South American country is unprofitable, as is raising cattle in the African state. In each case, the market value of output is exceeded by opportunity costs.

So market forces oblige everyone to specialize according to comparative advantage. The only way for producers lacking comparative advantage to survive is to petition government for protection. But whenever petitioning of this sort is successful, the gains for society of trade, which are sizable, are lost.

---

[6] Note that, with these prices, the value of each country's exports equals the value of its imports: $750/ton × 150,000 tons of sugar = $450 × 250,000 head of beef. Thus, the mutually beneficial pattern of specialization and trade depicted in Figure 6.4 is financially feasible for both Argentina and Mozambique.

7
Agriculture and Economic Development

When trade occurs among people and countries that specialize according to comparative advantage, prosperity spreads. Where such trade is suppressed, as happens routinely in the global food economy, economic well-being is diminished.

Considerable though the rewards of rational specialization and trade are, free commercial exchange does not amount to a comprehensive program for broad-based development. Impoverished households may trade frequently with others of comparably limited means. Indeed, careful examination of poverty-stricken communities invariably reveals a lot of sensible bargaining and bartering among community members. This should surprise no one. Since survival itself often hangs in the balance for the parties involved, how can their bargaining and bartering be anything but sensible? If people remain poor or nearly so in spite of specializing and trading rationally, their inability to offer much of value in the marketplace is to blame.

In remedying this deficiency of limited productive capacity, development inevitably must deal with the foundation on which economic performance ultimately rests – a foundation comprising three pillars. One of these pillars consists of human, natural, and other resources. Another is institutional – the rules, informal and formal alike, that govern one individual's dealings with others. Without a framework of rules that define contracts, property rights, and so on, the marketplace, and hence mutually beneficial exchange, simply cannot exist. An economy's institutions reflect its third foundational pillar, which is the prevailing culture. By the same token, cultural attitudes are influenced by institutions. For example, consistent enforcement of contracts and property rights, which is what economists and others have in mind when they talk or write about the rule of law, creates a culture of trust, which facilitates trade as well as productive investment. But deficiencies in the rule of law create distrustful attitudes; as a result, commercial exchange and the formation of productive assets are hindered and the prospects are dim for an institutional framework conducive to markets (Tweeten and Brinkman, 1976, p. 60).

If resources are available and an economy's institutional and cultural underpinnings are sound, then the stage is set for capital formation, or investment, which enables people to produce more and therefore achieve better standards of living.

The whole process begins with savings, which are left over as consumption expenditures are subtracted from income. An economy need not supply all its own savings. Poor nations, for example, receive foreign aid. Another example of using savings from other countries is the United States, which attracts tens of billions of dollars from foreign investors annually. However, most of the savings available in most countries are locally generated.

If savings are simply stuffed in mattresses, then capital does not accumulate. To avoid this outcome, financial intermediation, which involves channeling the savings provided by households to enterprises that invest in productive capacity, is needed. Banks and other businesses providing this service need to be efficient, of course. If they are not (e.g., because of undue political interference), then capital formation fails to have the desired impact on productive capacity. But if financial intermediation is reliably efficient, then high-payoff investment opportunities are consistently exploited, capital builds, and GDP per capita grows.

A sizable economic literature addresses the gains in living standards that are created as savings are generated and efficient investment takes place. Agriculture is a major focus of this literature since that sector dominates many poor economies. Where markets are allowed to allocate resources efficiently, so that investment projects featuring the highest returns are undertaken, there is a tendency for economic structure, which describes the portions of GDP produced in various sectors, to diversify. In particular, agriculture's GDP share tends to decline as an economy develops. This structural transformation has to do with Engel's Law (Chapter 2), which holds that the income elasticity of food demand declines as GDP per capita increases. In addition, as crop and livestock production grows more efficient, labor and other inputs are released from farming and are employed instead in manufacturing, services, and other sectors.

Structural transformation of this kind is only symptomatic of the fundamental development process, which it bears repeating is driven by savings and efficient investment. Nevertheless, ratcheting down agriculture's GDP share has been a major goal of development strategies pursued in various parts of the world during the twentieth century. Accelerating economic diversification and growth by distorting markets in favor of manufacturing and other non-agricultural sectors and at the expense of farmers and ranchers has never been successful. Worse than that, this approach sometimes has created famine. Furthermore, sight has been lost entirely of one of the essential challenges of development, which is to establish the institutional and cultural underpinnings that an economy needs if maximum well-being is to be derived from available resources.

7.1 Growth and Economic Structure

First investigated systematically during the 1950s, the economic diversification that always accompanies development can be detected by examining a cross-section of rich, poor, and in-between nations, by investigating structural change over time

Table 7.1 Living standards and economic structure

	GDP per capita		Agriculture's value added as % of GDP		Industry's value added as % of GDP		Services' value added as % of GDP	
	1990	2001	1990	2001	1990	2001	1990	2001
	(1)		(2)		(3)		(4)	
Tanzania	$174	$275	46	45	18	16	36	39
Laos	211	352	61	51	15	23	24	26
Bangladesh	282	351	29	23	21	25	50	52
Bolivia	676	885	17	16	35	29	48	56
Ecuador	1,037	1,383	13	11	38	33	49	56
Algeria	2,472	1,764	11	10	48	55	40	36
Thailand	1,529	1,880	12	10	37	40	50	49
Chile	2,297	4,430	9	9	41	34	50	57
Poland	1,544	4,519	8	4	50	37	42	59
Sweden	27,712	23,313	3	2	32	27	64	71

Source: World Bank (2003B) for GDP and GDP shares (pp. 190–192) and population in 2001 (pp. 14–16); 1990 population from World Bank (1992, pp. 218–219).

in specific settings, or by drawing on both these approaches. Reported in Table 7.1 are data of the sort used in the combined approach.

Listed in that table are the same 10 countries for which demographic data are reported in Chapter 2. Poverty is rife in Tanzania, as it is in most of Sub-Saharan Africa. Laos, which has been ruled by communists since the 1970s, is one of the most destitute places in Asia, as is Bangladesh. Bolivia is one of the poorest countries in the Western Hemisphere, and living standards are not much better in Ecuador. People are a little better off in Algeria, which is a major exporter of oil and natural gas. Impoverished during the 1970s, Thailand has experienced rapid economic growth since then. Having made a successful transition away from communism, Poland is now more prosperous than Chile, which has had the fastest economic expansion in Latin America since the early 1980s. Sweden has long been one of the world's richest nations, although GDP per capita slipped there during the 1990s.

Tanzania and Laos are typical of the world's least developed nations, in the sense that agriculture dominates each country's economy. There are no data on how many Tanzanians raise crops and livestock, but a sizable majority undoubtedly does so. Stable during the 1990s, the ratio of value added[1] in agriculture to GDP is high. No recent data on the structure of employment are available for Laos, where four-fifths of the labor force worked in agriculture in 1980. Farming still accounts for half the country's GDP.

---

[1] As explained in Chapter 3, a sector's value added comprises the difference between the value of its output and its input purchases from other sectors. Value added is distributed between wages and other payments to labor and dividends and other payments to capital.

In the middle of the twentieth century, when investigation of economic diversification began in earnest, two structural changes received primary attention, one being agriculture's relative decline and the other being the relative rise of manufacturing and other industrial parts of the economy. To be sure, such a transformation is explained in part by shifts in consumption related to Engel's Law. But it also results because improved inputs for agriculture from science and industry have created the means to produce more food with fewer resources. For example, Johnson (2000) stresses that farmers' use of mechanical reapers, tractors, and other manufactured inputs in Europe and the United States during the 1800s and early 1900s released labor from rural areas precisely as industry's demand for this factor was expanding. In light of historical experience of this sort, development was viewed during the 1950s and afterward as going hand in hand with the expansion of manufacturing and other industrial enterprises (Kuznets, 1965; Chenery and Syrquin, 1975). Likewise, Johnson (2000) contends that the industrial revolution would have proceeded much more slowly without the change in farming methods.

Reading down the list of countries in Table 7.1, one can see that agriculture's GDP share (column 2) decreases as the standard of living rises. The relationship between industrialization and average income is a little more complex, although industry's share (column 3) clearly goes up as median earnings rise from a few hundred to a couple thousand dollars per annum. Consider Laos, where GDP per capita went from $211 per capita in 1990 to $352 in 2001 and the ratio of industrial value added to GDP increased from 15 to 23 percent. This ratio is higher in Bolivia and Ecuador, where average incomes are well above the Laotian level.

During the past 50 years or so, services have grown faster than either agriculture or industry in many parts of the world. As indicated in Table 7.1, the ratio of industrial value added to GDP tends to stabilize as average income surpasses $2,000. Note that the ratio for Sweden in 2001 was comparable to those for Bangladesh and Bolivia. Moreover, industry's GDP share has been going down in Chile and Poland, where GDP per capita is approaching $5,000. In these two countries as well as in Sweden, the service sector is the largest and fastest-growing part of the economy.

Services are, if anything, more diverse than either agriculture or industry. Obviously, the sector's output comes in myriad forms. Some services require no skill and, quite often, back-breaking effort. Carrying goods to or from market, cleaning houses, or standing guard are all good examples, and activities like these employ a large segment of the labor force in impoverished settings. Demand for the same services exists in wealthy places; as Gregory Mankiw points out, there are strong incentives for Forrest Gump to be hired to mow Tiger Woods's lawn (Chapter 6). However, the rising preeminence of services in more developed parts of the world – not just affluent countries, like Sweden, but nations with prospects of becoming so, like Chile and Poland – relates to the growth of activities requiring a lot of training and education. This is true of financial intermediation, which is an important part of the service economy. It is also true of the work done by schools and universities, which equip people for employment in a sophisticated, affluent economy. Likewise, the health care system, which keeps people healthy and productive, is another part of the service economy that grows in importance as living standards rise.

Just as industrialization in decades past provided the means to change agriculture and reallocate labor from farms to factories, the expansion of services has had profound and beneficial impacts on the rest of the economy. If medical attention and pharmaceuticals are readily available, for example, disease does not impair the productivity of agricultural labor very much.[2] As emphasized in Chapter 3, the efficiency of farms in places like the United States can be traced directly to marketing and other services supplied by agribusinesses. Also, hybridization and the Green Revolution were made possible by human capital formation and agricultural research and development underwritten by governments, private foundations, and international organizations. The rising importance of services in the food economy – services provided by private firms as well as the public sector – is reflective of general trends in the economy, including productivity growth.

By the late twentieth century, the world's most successful economies had become entirely dominated by the service sector. In Sweden, services' value added exceeds seven-tenths of GDP (Table 7.1, column 4). Three-fifths of the country's male workforce and nearly nine out of every ten Swedish women are employed in banks, educational institutions, hospitals and clinics, and other enterprises that are neither agricultural nor industrial. In the United States, where about 2 percent of the labor force farms and less than 20 percent is industrial, the ratio of services' value added to GDP is 73 percent (World Bank, 2003b, pp. 48 and 192).

The Diversity of Structural Transformation

Conditioned as it is by geography, history, and a host of other factors, development never follows the same path in any pair of countries. In any particular setting, growth in living standards and structural transformation have been influenced greatly by how the service sector happened to evolve.

Since many service jobs in impoverished settings require few skills and pay poorly, service employment represents an alternative for people who otherwise would engage in subsistence farming, which is even less remunerative. This is the predicament of the majority of Tanzanians, who can neither read nor write and who must choose between seeking out a living on the land and seeking menial work in Dar es Salaam or some other city. Since poverty limits the size of the domestic market, growth prospects for industry are limited. As already observed, most of the working population is employed in agriculture, which accounts for more than one-third of Tanzania's GDP.

The situation in Laos is comparably bleak, although there is a twist. An article of faith for the country's communist leaders has been that the service economy is inherently exploitative. Professing to save producers and consumers alike from

---

[2] Disease hampers agricultural development noticeably in Sub-Saharan Africa (Chapter 15).

"intermediaries" – "middlemen" who buy output from the former group and, after transporting this output and transforming it in ways that consumers desire, sell it to the latter group at higher prices – these leaders actively suppressed the private supply of services. In spite of some relaxation of doctrinaire communism in recent years, Laos still has a poorly developed service sector. It is not coincidental that, to this day, most Laotians are needy farmers.

Although most Bangladeshis are rural (Table 2.2), services now account for half the country's GDP. The value of the sector's output, not to mention the value of agricultural and industrial output, would be much higher if there were less illiteracy. The same can be said of Bolivia, where most of the rural population has decamped to cities and towns but lacks the skills needed for all but the most menial employment. In Ecuador, the service sector currently employs about the same share of the labor force as in Sweden: 63 percent of male workers and 84 percent of the females (World Bank, 2003b, p. 46). However, low skills prevent most of these people from producing anything of great value, which means that living standards are not very high.

In a way, the economy of Algeria, where GDP per capita exceeds $1,500, is under-performing in exactly the same way as the Laotian economy. Fossil fuel production accounts for a large share of industrial activity, which in turn makes up half the economy. Oil and gas earnings go mainly to government and state-owned enterprises, which in one way or another have crowded out private service providers. As a share of GDP, value added in the service sector has declined in recent years, to 36 percent. The only country in our sample of 10 with a lower share is Laos. Algeria's problems are representative of those arising in other developing countries rich in fossil fuels – especially nations in the Middle East and North Africa (Chapter 13).

In Thailand, GDP expansion of 3.8 percent per annum from 1990 to 2001 (World Bank, 2003b, p. 188) has coincided with modest structural change. Industry's share of the national economy grew, but only from 37 to 40 percent, and the service sector's share stayed about the same. However, much non-agricultural expansion has taken place in rural areas, where three out of every four Thais live (Table 2.2). This has provided an impulse for agricultural growth, which has nearly kept pace with increases in GDP.

Under communism, manufacturing was the mainstay of the Polish economy. A sure sign that a transition truly has happened is that services have emerged quickly as the dominant sector. As in so many other places – affluent, impoverished, and otherwise – the service workforce is mainly female. Three-fifths of the country's women are employed outside of agriculture and industry. For men, that portion is only two-fifths. One out of every five Polish men and an equal share of Polish women labor on farms (World Bank, 2003b, p. 47). Another sign that the Poles have put communism decisively behind them is that GDP has been going up, with annual growth averaging 4.5 percent from 1990 to 2001 (World Bank, 2003b, p. 187).

Aside from Poland, none of the 10 countries listed in Table 7.1 experienced faster growth in average living standards than Chile, where GDP per capita in 1990 was little more than half its 2001 level. The service sector has become progressively more important with each passing year. This has helped the country to exploit its comparative advantage in the production of fruit, wine, and other such goods. As a result, expansion of agriculture has exactly matched that of GDP. Rather than representing a drag on the national economy, agriculture has experienced robust development and, in so doing, has been an important part of Chile's economic success.

As already indicated, Sweden is a wealthy place where the service sector accounts for most employment and output. It is representative of what one finds in Western Europe, North America, Japan, and other affluent parts of the world.

Living Standards and Income Distribution

Fifty years ago, Simon Kuznets, who along with Hollis Chenery pioneered the study of structural transformation in developing economies, hypothesized that income inequality tends to be modest in very poor settings, rises as an emergence from penury occurs, and then declines after the average living standard passes a certain threshold (Kuznets, 1955). This relationship is depicted in Figure 7.1. Like its environmental variant (Figure 5.4), the Kuznets Curve is shaped like an inverted U. The only difference between it and the Environmental Kuznets Curve (EKC) has to do with their respective vertical dimensions. While environmental damage is plotted up and down in Figure 5.4, income inequality is measured vertically in Figure 7.1.

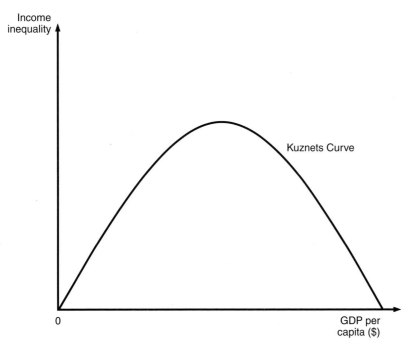

Figure 7.1 Living standards and income distribution.

The standard index of income inequality is the Gini coefficient, or Gini concentration ratio (Perkins *et al.*, 2001, pp. 118–123). Its minimum and maximum values are, respectively, 0 (reflecting complete equality) and 100 (for the case of all income being captured by a single individual, with absolutely nothing left for anyone else). These extreme values are never observed in the real world, of course. As a rule, a Gini coefficient below 40 indicates modest inequality. In contrast, inequality is considered high if the concentration ratio is over 50.

Although our sample of 10 countries is small, the pattern of living standards and Gini coefficients reported in Table 7.2 is broadly consistent with Kuznets's hypothesis. Of the first three countries, Bangladesh is an especially good example of an impoverished country with little income concentration. Per capita Gross National Income (GNI), which is generally quite close to GDP per capita if variations in purchasing power parity (PPP) are allowed for, is higher in the next four countries, although three of these have markedly higher concentration ratios, Algeria being the single exception. The country with the highest reading is Chile, which shares with most of Latin America a tradition of acute inequality. In contrast, Poland's Gini coefficient is on a par with Bangladesh's. Income is less concentrated in Sweden than in any of the other nine countries.

Some of the economic realities underlying the Kuznets Curve are alluded to in the preceding discussion of income growth and structural change. In a place like Bangladesh, half or more of the labor force farms and is poorly compensated for doing so. Another third is in the service sector, with most of this segment receiving a low wage for unskilled work. The same is true of many industrial laborers. The vast majority of Bangladeshis being poor (Table 7.2, columns 1 and 2), their country's Gini coefficient is just 31.8 (column 3).

Table 7.2 Average income and income inequality

	2001 GDP per capita (1)	2001 GNI per capita (PPP measure) (2)	Gini coefficient (survey year) (3)
Tanzania	$275	$520	38.2 (1993)
Laos	352	1,540	37.0 (1997)
Bangladesh	351	1,600	31.8 (2000)
Bolivia	885	2,240	44.7 (1999)
Ecuador	1,383	2,960	43.7 (1998)
Algeria	1,764	5,910	35.3 (1995)
Thailand	1,880	6,230	43.2 (2000)
Chile	4,430	8,840	57.5 (1998)
Poland	4,519	9,370	31.6 (1998)
Sweden	23,313	25,400	25.0 (1995)

Source: GDP per capita from Table 7.1; GNI per capita from World Bank (2003B), pp. 14–16; Gini coefficients from World Bank (2003B), 64–66.

If development starts to take place under conditions such as these, the impoverished and unskilled majority often does not benefit very much. The poor may continue to farm. Alternatively, they may opt for unskilled service employment, as many Ecuadorians have done. Either way, their earnings rise modestly, at most. In contrast, the minority of people with a good high school education – or, better yet, a university degree – see demands for their skills go up substantially, and their incomes rise accordingly. This causes concentration ratios to climb into the middle 40s and beyond.

One place (which happens not to be listed in Table 7.2) where rapid economic expansion is undoubtedly coinciding with rising income inequality today is India. Global communications allow credit-card companies, airlines, and other businesses to transfer customer-service operations outside the United States.[3] This outsourcing is creating hundreds of thousands of well-paying jobs for educated Indians with good English (Chapter 11). Since there is no comparable improvement in earnings for the impoverished majority of their countrymen, Gini coefficients are bound to rise.

The situation is different in Eastern Europe, mainly because schooling long has been widely available there. Indeed, this is the only part of the non-affluent world where small elites do not parlay superior education and other advantages into earnings that greatly exceed those of their fellow citizens. Instead, a broad segment of the population is able to benefit from the demands for skilled workers and professionals which development creates. Thus, the distribution of income is not highly skewed in places like Poland.

Few nations can match Sweden's success in extending educational and other opportunities to the entire national population, which means that most Swedes are able to exploit the opportunities arising in an affluent, service-based economy. Also, heavy taxation, particularly of wealthy citizens, pays for one of the world's best social safety nets. As a result, the Gini coefficient, 25.0 (Table 7.2, column 3), is unusually low.

7.2 Agriculture's Role in Economic Development

As a rule, agricultural value added as a share of GDP diminishes as incomes increase. This decline is by no means undesirable, since it reflects increased specialization and trade as well as a general enhancement of productive capacity – or, to be more precise, enhanced productive capacity for each member of the workforce. Put another way, agricultural development ought to be pursued in ways that cause the GDP shares of non-agricultural sectors to go up, to the benefit of overall living standards.

Johnston and Mellor (1961) highlight five specific contributions that agriculture makes to economic growth and diversification.

1 *Increase food supplies for the domestic, non-agricultural economy.* If food supplies are unreliable, or can only be obtained at a high price, then wages are bid up and

---

[3] For example, this book was copyedited and indexed – very ably! – in India.

employment falls in the industrial and service sectors. This hampers economic diversification, which is also held back because households have less to spend on non-food items. By the same token, savings are reduced, which diminishes the build-up of productive capacity required for GDP growth. A sustained increase in food supplies for the domestic economy has precisely the opposite effects. As food grows less scarce, upward pressure on wages abates and non-agricultural employment rises. Also, cheaper food allows households not only to eat better, but to diversify consumption expenditures and increase savings.

2 *Augment foreign exchange earnings*. As indicated in the preceding chapter, most developing countries have a comparative advantage in one or more agricultural commodities. As this advantage is exploited, foreign exchange is earned. Among other things, this allows for the importation of capital goods, which are an important component of productive capacity. But if comparative advantage is overwhelmed by currency over-valuation and other policy-induced distortions (Chapter 6), then exports decline and less foreign exchange is earned. To avoid a contraction of imports, a country may seek foreign aid, which is supposed to facilitate development. More often, imports are maintained in the face of diminished exports by international borrowing, which of course adds to indebtedness.

3 *Reallocate labor*. It bears repeating that a major contribution of agricultural development to overall economic progress has been to release labor from the countryside, and to enable people to seek employment instead in industry or (more often than not these days) the service sector. Obviously, this transfer is impeded if a large segment of the workforce remains on farms because this is the only way for the entire population to be fed.

4 *Increase tax revenues and transfers of savings*. As agriculture develops, it becomes a source of tax revenues that can pay for public goods – public goods not just for rural areas but for the entire economy. In many parts of the United States, for example, property taxes on farm real estate have been spent on education. Many of the young people benefiting from this investment in human capital have subsequently found productive work in manufacturing and the service sector. Another impact of increased agricultural productivity is that savings by farm households increase. These savings help to finance the improvement of capital inputs, as required for productivity growth. If other parts of the economy are growing more rapidly than agriculture and if financial intermediation is good, then some of the savings of rural households can be reallocated to non-agricultural sectors, thereby accelerating growth and diversification. Of course, an agricultural economy consisting largely of subsistence farms has few financial resources to share.

5 *Enhance demand in rural areas for non-food products*. Likewise, an impoverished farm economy is a poor market for goods and services produced by non-agricultural sectors. But as farmers grow more prosperous, their increasing demand for these products becomes an important force for structural transformation and GDP growth.

Although it might seem tempting to do so, the observations that Johnston and Mellor (1961) offer about how agriculture can, and often does, contribute to economic progress should not be regarded as a precise blueprint for development, one that is universally applicable. In places that already enjoy a high level of affluence, living standards among farmers are comparably high and mainly reflect remunerative employment opportunities in the economy as a whole, as opposed to creating those opportunities. This is true in the United States, for example, where increases in non-agricultural wages since the Second World War have had a major impact on farmers' earnings (Gardner, 2002, pp. 271–277). In poorer countries as well, prescriptions for development must be tailored to local circumstances. For example, Johnston and Mellor (1961) generally favor increasing the domestic supply of food, which diminishes its price, in order to accelerate economic growth and structural transformation. However, some nations have prospered by relying on imports instead. Lacking the natural resources needed for food self-sufficiency, Singapore and Hong Kong have specialized in non-agricultural activities, in which each country has a strong comparative advantage, and bought farm goods from foreigners.

As Johnston and Mellor (1961) readily concede, complementarities and trade-offs exist among the contributions that agriculture makes to development. One complementarity arises because an increase in commodity exports, which enhances foreign exchange earnings, also causes rural incomes to rise. This allows for higher savings and increased purchases of non-food items in rural areas. One example of a trade-off has to do with increased exports, which can cause prices paid by local consumers to rise as commodities are directed away from the domestic market or as resources that would otherwise yield output for that market are instead used to produce output for international buyers. Likewise, various trade-offs are associated with a government's attempts to extract tax revenues from the agricultural sector. If the tax burden on farmers grows too onerous, production will suffer, for domestic as well as international markets. Savings of rural households will also fall, as will their spending on non-food items. Excessive taxation of agriculture may accelerate the exit of labor from farming. Otherwise, it diminishes the full range of contributions the sector can make to a larger and more diversified economy.[4]

No pair of countries is likely to strike precisely the same balance among these and other trade-offs and complementarities. Something important to recognize, however, is that a good way to ameliorate trade-offs and to enhance complementarities is to make agriculture more productive. As farmers grow more efficient, domestic food supplies and exports can increase simultaneously and this expansion can coincide with the inter-sectoral reallocation of labor. Rural households are more prosperous, which enables them to pay more taxes, increase savings, and

---

[4] Of course, penalization of non-agricultural sectors is no less disadvantageous than suppression of the farm economy, which obviously depends on the former sector for markets, improved inputs, and jobs.

spend more on the output of non-agricultural sectors – all at the same time. Rising agricultural productivity, then, is a major driver of overall economic progress, especially in the many poor nations where farming remains an important part of the economy in terms of GDP and employment shares.

7.3 Trying to Develop at Agriculture's Expense

Lessons about the various ways that agriculture contributes to development have not been learned easily. To the contrary, development strategies predicated on neglect – or, worse yet, penalization – of that sector were pursued in various parts of the world during the twentieth century. The record of this experimentation has turned out to be dreadful, not just economically but all too frequently in terms of lives lost as well.

Experiments in Communist Nations

In late 1917, Vladimir Lenin brought off a coup d'etat that allowed for the application of scientific socialism, as he preferred to call it, in the nation immediately rechristened the Union of Soviet Socialist Republics (USSR). He and other communist leaders were united in their enthusiasm for industrializing the country, which at the time was overwhelmingly agricultural. For obvious ideological reasons, there was little enthusiasm for private sector participation in new or expanded manufacturing enterprises. Some in the new regime contemplated exploiting the Soviet Union's comparative advantage in agriculture, using the foreign exchange earned from crop exports to pay for imported technology and inputs so that industries owned by the state could expand. However, this option was ruled out once foreign trade was choked off due to the communists' repudiation of debts from the czarist era. The only way to finance socialist industrialization, as called for in the New Economic Policy (NEP), was to tax peasants, who had ended up with most of the country's farmland after the break-up of large estates and who comprised more than 80 percent of the population (Skidelsky, 1996, p. 50).[5]

By and large, taxation of the rural economy was accomplished indirectly, which was possible because of complete socialization of the service sector. In particular, all marketing was taken over by the government, which proceeded to pay low prices for farm products and to charge excessively for manufactured goods purchased by peasants. This manipulation of the domestic terms of trade (i.e., relative prices within the national economy) resulted in the extraction of wealth from the

---

[5] State seizure of agricultural land right after the October 1917 "revolution" was not an option since "all land to the peasants" was one of the three major promises that Lenin made when he seized power. Of course, what the communist state gave peasants, in the form of parcels carved out of old estates, it took from them in the form of coercive food procurement at near-confiscatory prices.

countryside, obviously not as voluntary savings but rather in the form of compulsory payments to state-owned monopolies (Skidelsky, 1996, p. 51).

The peasants were not entirely without recourse. An obvious response to low official prices was to cut back on marketable output. Another was to feed more grain to livestock, which was privately owned. Yet another was to produce vodka and other liquor, which were fairly easy to sell through illegal channels. As a result of responses such as these, revenues collected by the government through the indirect taxation of farmers proved disappointing. This put a brake on the expansion of state-owned industry.

Josef Stalin, who prevailed in the succession struggle that followed Lenin's death in 1924, had a brutally simple remedy for the shortcomings of the NEP. Soon after he consolidated power, in the late 1920s, virtually all agriculture was collectivized, with individual holdings – aside from parcels smaller than one hectare retained by individual families – incorporated into large-scale farms organized by the state. This nationalization of the rural economy was accomplished with unspeakable cruelty. Between 1929, when Stalin became dictator, and 1941, when the Soviet Union was invaded by Nazi Germany, nearly 15 million people died, the majority rural folk. They either starved in famines engineered by communist authorities to drive recalcitrant farmers, denigrated at the time as *kulaks*,[6] off the land or were worked to death as slave laborers in mines, factories, and infrastructure projects (Conquest, 1986, pp. 299–307).

Collectivization of agriculture was pursued elsewhere behind the Iron Curtain after the Second World War, again with enormous loss of life. At least 25 million Chinese, mainly peasants, perished after Mao Tse-tung opted for the nationalization of farming in the late 1950s, particularly after he decided on a crash program of industrialization (including in rural areas) that left agriculture starved of state support (Short, 1999, pp. 486–505). The consequences of imposing communism on the countryside have been no less appalling in smaller countries, most notably Cambodia under the Khmer Rouge and North Korea under the Kims, *père et fils*.

After the Second World War, some communist governments refrained from the Stalinist approach to agriculture. Polish farms, for example, were not taken over by the state. This proved fortuitous after communism fell in the country; the simple fact that eggs, milk, and other food items were readily available from private producers allowed the market economy to gain a decisive foothold quickly (Yergin and Stanislaw, 1998, pp. 267–268). Elsewhere, collectivization was eventually abandoned. Begun experimentally in 1978, soon after Mao's death, China's Household Responsibility System (HRS) allows farm families to decide what to produce and how. Likewise, marketing decisions are a matter of individual prerogative. The HRS, which is a capitalist arrangement, has been received with enthusiasm in the countryside, where it has had a beneficial impact on living standards.

---

[6] The literal meaning of *kulak* in Russian is fist. Communists applied the term to individual farmers so as to suggest that they, the farmers, were inordinately acquisitive, to the detriment of the wider good, and therefore deserving of extermination.

Moreover, its launch changed China's economic trajectory, ushering in a period of accelerated growth that continues to this day (Naughton, 1995, pp. 138–142).

Governmental Intervention in the Developing World

Except for communist states, most African, Asian, and Latin American nations have avoided the disaster of socialist agriculture, collectivized farming only having been dabbled with here and there. However, scores of governments in the developing world have succumbed to the temptation to manipulate market forces so as to accelerate structural transformation – exactly as was tried in the Soviet Union under Lenin. Especially during the first three or four decades after the Second World War, domestic terms of trade were manipulated to the detriment of agriculture so that manufacturing could grow faster. That is, government policies drove down prices of farm products relative to the prices of manufactured goods. This approach, called import-substituting industrialization (ISI), has proven no less disappointing than the NEP was.[7]

As implied by the term, import-substituting, special emphasis is placed in ISI on national production of domestically consumed manufactured items. Various rationales have been offered for this emphasis over the years. Besides saving foreign exchange that otherwise would be spent on imported products and enhancing manufacturing employment, an oft-repeated justification is that "infant" industries should be given an opportunity to acquire comparative advantage in international markets through "learning by doing" in national markets, from which cheaper imports are kept out for a while (Perkins *et al.*, 2001, pp. 680–683). A structural shift from agriculture to industry results as domestic prices of manufactured goods are sustained above international levels. The same shift is reinforced by holding down the domestic prices of farm products. One way to accomplish the latter distortion is to impose controls on food prices, which is obviously welcomed by non-agricultural workers. Such controls also benefit firms employing these workers, since pressure to raise wages is reduced. To diminish the prices received by commodity producers serving foreign markets, exports can be taxed. In addition, ISI is promoted by currency over-valuation (Chapter 6) and its resulting import bias. This bias is counteracted by trade barriers in the case of domestically produced manufactured goods, but not in the case of farm products and other primary commodities.

Sizable economic losses result as domestic terms of trade are distorted to the benefit of import-substituting industries. Consumers are denied access to manufactured imports that are cheaper and, quite frequently, of higher quality. At the same time, agriculture and other parts of the economy enjoying a comparative advantage in the international marketplace are penalized since prices for these sectors' output are

---

[7] The resemblance between ISI and the NEP is more than passing. In Africa, agricultural marketing boards established during the colonial era were harnessed enthusiastically by the region's leaders after independence to tax farmers indirectly, by paying them below-market prices for their output, to be specific.

depressed and higher prices are paid for inputs purchased from protected sectors. For a few years after ISI is adopted, the sacrifices made by consumers and competitive sectors seem worthwhile because domestic industry expands rapidly as imported products are driven from national markets. But unless manufacturers enjoying trade protection truly learn by doing, which rarely happens, this period of rapid expansion comes to an end once they dominate domestic markets. With no great incentive to operate efficiently, protected industries never acquire comparative advantage, and hence are unable to compete internationally. Once they have taken over the domestic market, further industrial growth is constrained by expansion of the national economy, which is often sluggish (Perkins et al., 2001, pp. 704–705).

Many countries, mainly the larger Latin American nations, experienced rapid expansion in manufacturing immediately following the adoption of ISI, but later found economic growth and industrialization difficult to sustain. A general consequence of this stagnation was mounting trade deficits. To repeat, exports rarely are generated by industries accustomed to protection. Moreover, this part of the economy tends to rely on imported capital goods and raw materials, the prices of which are kept low by currency over-valuation. Meanwhile, exports from agriculture and other sectors with a comparative advantage are reduced by currency over-valuation and other distortions. Thus, the overall impact of ISI is a deteriorating balance of trade as imports run ahead of exports.

During the 1970s, trade deficits were sustained throughout Latin America and other parts of the world where ISI had been adopted thanks to substantial foreign borrowing.[8] International debts also mounted because government spending consistently exceeded tax revenues. Rising indebtedness could not continue indefinitely and, immediately after Mexico declared a moratorium on interest payments in August 1982, Latin America was plunged into a severe and prolonged recession. At the end of the day, the countries that have put this crisis behind them have largely abandoned ISI and its costly distortions.

7.4 Agricultural Development for the Sake of Economic Growth and Diversification

In light of the economic dislocation suffered in the late twentieth century by countries that earlier had pursued ISI, it is fortunate that there is an alternative path to structural transformation and economic growth. The other approach, sometimes called the Outward-Looking Strategy (Perkins et al., 2001, pp. 705–723), was first pursued by a handful of places in East Asia, which opted to produce manufactured items for foreign markets precisely as Latin America and other parts of the

---

[8] As indicated in Chapter 4, grain prices were high during the 1970s in part because of international purchases by Latin America and other developing regions that were paid for with loans.

developing world were shying away from international competition and instead providing strong incentives for domestic manufacturers to concentrate on national markets. The four East Asian "tigers" – Hong Kong, Singapore, South Korea, and Taiwan – chose export-oriented industrialization over ISI mainly for a simple, practical reason. Particularly during the 1950s and 1960s, their respective domestic markets were quite small, comprising in each case a few million consumers with low average earnings.

Regardless of the motivating forces, choosing the Outward-Looking Strategy has proven fortuitous. As a rule, domestic prices are distorted less than in countries practicing ISI. At times, the four tigers and other practitioners have undervalued their currencies, which has created an export bias (i.e., precisely the opposite of the import bias resulting from currency over-valuation). However, chronic trade surpluses cannot be sustained indefinitely any more than chronic deficits can, in part because the accumulation of foreign money in a country with an under-valued currency eventually leads to a strengthening of that currency, which in turn diminishes exports and stimulates imports.

Aside from distortions and imbalances of this sort, governments not absorbed entirely with the task of market manipulation have tended to concentrate more on the build-up of productive capacity, broadly defined. Education has received ample support. In addition, the legal guarantees that commerce requires – reliable and uniform enforcement of contracts, property rights, and other elements of the rule of law – are better established. As a result, the structural transformation and economic growth that have taken place in Hong Kong,[9] Singapore, South Korea, and Taiwan are the envy of the world.

Positive elements of the Outward-Looking Strategy have been adopted elsewhere. Having cut taxes and played on the advantage of an educated workforce, Ireland – the "Celtic tiger" – has enjoyed an economic boom in recent years. As highlighted earlier in this chapter, income inequality remains high in Chile, which suggests that improvements in living standards have not been very broad based. Nevertheless, it comes closer than any other country in Latin America to being the region's tiger. In addition, interference with market forces is fairly restrained in Malaysia and Thailand, which have invested substantially in human capital and other elements of productive capacity.

In settings such as Thailand, which are well endowed with arable land and where agriculture continues to be an important sector, economic progress in the countryside has made important contributions to overall development. Much more than in Latin America and Sub-Saharan Africa, farmers in Asia have intensified their operations, making use of fertilizer and other purchased inputs to raise yields,

---

[9] For example, Hong Kong's industrial development focused initially on lines of manufacturing, like textiles, that made heavy use of low-paid, unskilled labor, which was abundant during the 1950s and 1960s. Subsequently, a switch was made to electronic goods and other products requiring more skilled labor and capital. These days, the former British colony, like other rich places, specializes in banking and other services. Most industrial operations – the majority of which are owned by Hong Kong businessmen – have moved to China and other countries.

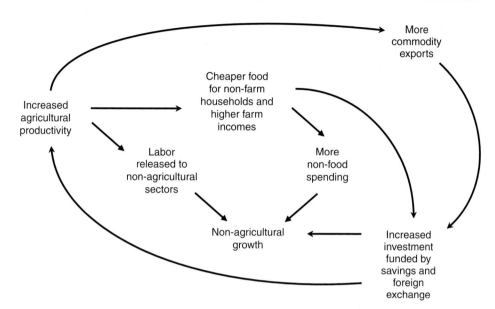

Figure 7.2 Productivity growth in agriculture and economic development.

especially during the Green Revolution (Chapter 3). Also, product composition has changed in response to evolving export opportunities as well as shifts in domestic demand resulting from higher incomes for consumers. Since total factor productivity (TFP) has gone up, the various benefits of agricultural development identified by Johnston and Mellor (1961) and surveyed in the second section of this chapter have actually materialized. Aside from increasing food supplies for domestic and international markets and releasing labor to non-agricultural sectors of the economy, expenditures on non-food items have been stimulated, as have savings (Mellor, 1995, pp. 10–16).

Given that the "expenditure multiplier" of increased farm earnings is very high in countries with modest living standards, careful attention needs to be paid to the virtuous cycle set in motion by improvements in agricultural productivity. As emphasized in this chapter and throughout this book, these improvements depend on undistorted prices, non-agricultural development (which supplies machinery and other inputs needed to raise productivity), as well as agricultural research and development, marketing infrastructure, and other public goods. As illustrated in Figure 7.2, one link in the cycle driven by productivity growth has to do with its direct effects – to be specific, lower food prices and more exports. Also, spending on non-food items increases as food becomes less scarce. The resulting economic diversification is further accelerated by the increased availability of workers formerly employed on farms. Diminished food scarcity also allows savings to increase, as does growth in farm earnings. As these savings and the foreign exchange generated by exports are invested in agriculture and other sectors, productivity is given a boost throughout the economy, which allows the entire virtuous cycle to continue.

7.5 Summary and Conclusions

The suppression of private farming by communist authorities not only took an enormous economic toll in the Soviet Union, Maoist China, and other nations, but cost tens of millions of lives during the twentieth century. In Latin America and Africa, manipulation of market forces to agriculture's disadvantage is less egregious today than it was when ISI was being pursued vigorously. However, investment in public goods for agriculture in these two regions still compares poorly with Asia's record (Mellor, 1995, pp. 307–329). Consequently, major opportunities for development, within the agricultural sector and beyond it, continue to be missed.

In a fundamental sense, the greatest error committed in nations where communism or ISI was implemented has been to focus too much on the fates of particular sectors and to ignore the true underpinnings of economic progress. Excessive attention to sectors has even been a problem at times in the East Asian tigers. For example, the South Korean government has lavished cheap credit and other support on industries that seemed to have good export potential; however, its success at "picking winners" has been mixed (Perkins et al., 2001, pp. 164–165). As Tweeten and Brinkman (1976, p. 60) point out, an economy's prospects are based on three things that may have very little to do with different sectors' GDP shares: the endowment of human and natural resources, institutions, and the prevailing culture. Up to a point, substitution among these three factors is possible. For example, natural resource scarcity can be compensated for by the formation of human capital. It can also be made up for by strengthening the rule of law, which is an institutional reform of critical importance in many places. However, cultural deficiency – for example, distrust that is acute and pervasive enough to discourage commerce and choke off investment – is something that must be addressed directly through education and by reinforcing the rule of law rather than being compensated for by the acquisition of additional resources.

In successful economies, resources (be these human, natural, or something else) are matched up, first, with institutions that are functional in the sense that the rule of law is secure and, second, a culture of trust made possible by functional institutions. Rarely if ever are price distortions debilitating in these economies, neither are public goods severely under-supplied. Agriculture makes an appropriate contribution to economic progress and few people are food-insecure.

Unsuccessful economies, on the other hand, often possess abundant environmental wealth, and may even have an impressive endowment of human capital. The limiting factors, then, are institutional and cultural, with true rule of law only to be imagined and with everyone so distrustful that trade and investment are greatly discouraged. Under these circumstances, governments frequently try to stimulate development by meddling with prices, typically ignoring public goods as a result. The outcome, in terms of agricultural development, the general trajectory of the economy, as well as food security, is invariably disappointing.

Study Questions

1. Aside from resources (human, natural, and otherwise), what does development require?
2. How does an economy's structure normally change as it develops?
3. Are the richest countries of the world the most industrialized, in terms of GDP shares? Why or why not?
4. Explain and illustrate the Kuznets Curve.
5. Identify various roles of agriculture in economic development as well as complementarities and trade-offs among these roles.
6. Compare and contrast the NEP of the early Soviet Union and ISI.
7. Why do trade deficits mount and growth flag the longer a country pursues ISI?
8. Compare and contrast the Outward-Looking Strategy for economic development, as pursued by the four East Asian tigers for example, and ISI.
9. How has agricultural development in Asian nations pursuing the Outward-Looking Strategy compared with agricultural development in Latin America and Sub-Saharan Africa?

8
Striving for Food Security

The world's food economy has accomplished wonders. Population growth between 1950 and 2000 was historically unprecedented. Also, food consumption per capita rose in various places because of income growth. Incredibly, global supply consistently went up faster than worldwide demand, and thanks to dramatic yield growth during and since the Green Revolution, declining food scarcity was achieved in many parts of the world without geographic expansion of farming and ranching on a huge scale. In addition, countries where agricultural productivity improved have experienced impressive economy-wide gains.

By no means is the task of feeding each and every human being fully achieved. To the contrary, hundreds of millions still go hungry chronically, and many more do so frequently. No one can dispute the shared judgment of the directors of International Food Policy Research Institute (IFPRI) and Bread for the World that "the global community stands indicted for knowing much about how to reduce hunger, but not doing so" (Runge *et al.*, 2003, p. xvii).

This chapter addresses the challenge of achieving food security. We start by defining the term and reporting how many people are not eating enough in various parts of the world. Also, food aid is examined. A fundamental conclusion is that hunger and poverty are inextricably tied to each other and the steps taken to enhance food security are all but indistinguishable from those required for broad-based economic progress. These steps are described in the latter part of the chapter.

8.1 What is Food Security?

Hunger being a recurring concern for humankind, the precise meaning of food security has been argued over at length. The current consensus is expressed by the FAO's declaration that the goal of ending hunger around the world will have been achieved "when all people at all times have physical, social, and economic *access* [emphasis added] to sufficient, safe, and nutritious food that meets their dietary needs and food preferences for an active and healthy life" (FAO, 2002b, p. 49).

The key word in this standard, to which organizations like the World Bank and US Agency for International Development (USAID) subscribe, is access. Not addressed is an important dimension of proper nourishment, which is utilization of food once it has been consumed. Genetic factors – a person's predisposition to certain allergies, for instance – can hinder absorption of particular vitamins or nutrients, or even interfere with the adequate intake of calories. So can improper food preparation, eating disorders, pathogens, and simple bad habits. A problem for countless Africans, Asians, and Latin Americas is limited access to clean water; these people often have intestinal parasites, which can cause malnutrition even among those who are actually eating enough.

The case is strong for paying attention to utilization in food security programs. To help households make better use of the resources available to them, nutrition and health education, vitamin supplements, oral rehydration, treatment of parasites, and immunization all need to be offered. Increased funding of these measures could eliminate a lot of malnutrition, especially among children. But while the advantages of an integrated approach are widely recognized, improved utilization is often not subsumed in food security initiatives.

Something else to understand is that these initiatives are normally predicated on food being available in sufficient quantities. True, availability is a critical issue if prolonged civil strife causes food-distribution networks to break down and farms to be abandoned. But as emphasized by Amartya Sen, a Nobel laureate who has investigated a number of famines, the food that hungry people need is often at hand, as was the case when starvation took an enormous toll in Bengal in 1943 (Sen, 1981, pp. 57–63). As indicated in Chapter 3, global production in recent decades has consistently been enough to feed the entire human race.

Since aggregate availability is not the fundamental problem, the focus is on access – access of a distinctly economic nature, to be precise. The skeletal inmates of refugee camps that have sprung up in places like the Horn of Africa as local strongmen and gangs have prosecuted their sordid, sanguinary rumbles are not really emblematic of global food insecurity. The fundamental problem is more mundane and widespread, though no less painful. Quite simply, the meager earnings of hungry people do not cover the expense of an adequate diet.

8.2   Who and Where Are the Food Insecure?

As much as one-third of the human race suffers from one or more forms of undernourishment. Approximately 2 billion people are anemic, because of iron-poor diets and other reasons. At least 90 percent of this group is in the developing world, where iodine and other micronutrient deficiencies are also common problems (Babinard and Pinstrup-Andersen, 2001).

The cost of public health and educational campaigns needed to deal with problems like these is modest. For instance, iodine deficiency, which in extreme cases impairs motor development, can be corrected simply by ingesting a little iodized

Table 8.1 Undernourishment in the developing world (% in brackets)

Region	1969–1971 (1)	1979–1981 (2)	1990–1992 (3)	1997–1999 (4)	2010 (5)
East and Southeast Asia	475 million (41)	378 million (27)	275 million (16)	193 million (10)	123 million (6)
South Asia	238 (33)	303 (34)	289 (26)	303 (24)	200 (12)
Latin America and the Caribbean	53 (19)	48 (14)	59 (13)	54 (10)	40 (7)
Middle East and North Africa	48 (27)	27 (12)	25 (8)	32 (9)	53 (10)
Sub-Saharan Africa	103 (38)	148 (41)	168 (35)	194 (34)	264 (30)
All developing countries	917 (35)	904 (28)	816 (20)	777 (17)	680 (12)

Source: FAO (2002b) for 1997–1999 data; FAO (1996) for all other data and 2010 projections.

salt, at a per-capita expense of a penny or two a week. Moreover, the costs of dealing with micronutrient deficiencies pale in comparison to the benefits to be enjoyed by healthier people. A far more ambitious undertaking is to eradicate food insecurity proper, which can be defined as the daily shortfall of 100–400 calories that some 800 million human beings experience chronically or frequently. Nearly all this group is counted among the 1.2 billion people categorized as extremely poor, in the sense that their daily earnings are 1 dollar or less.

Reported in Table 8.1 are broad trends and regional differences in the developing world, where all but 30–40 million of the food-insecure reside. The number of people who on average do not consume enough dietary energy for normal activity and good health has fallen steadily since the late 1960s, from more than 900 million throughout Africa, Asia, and Latin America around 1970 (column 1) to 777 million in the late 1990s (column 2). Relative to the total population of the developing world, the food-insecure portion has halved, from 35 percent of the total to 17 percent. This portion continues to decline, as does the entire number of human beings not getting enough calories.

By no means has progress toward greater food security been uniform. More than two in every five East and Southeast Asians consumed too few calories 35 years ago; nowhere was the incidence of undernourishment greater. Since then, the number of food-insecure people in the region has fallen by 282 million, largely because of great strides in China. Today, the incidence of undernourishment in East and Southeast Asia is one-in-ten. The current incidence is the same in Latin America and the Caribbean, where the size of the food-insecure cohort at the turn of the century was exactly what it had been three decades earlier. In the Middle East and North Africa, food security has improved. But as deterioration continues in some of the region's economies, the number of hungry people and the incidence of undernourishment are both rising.

The two parts of the globe of greatest concern are South Asia and Sub-Saharan Africa. After growing as rapidly as the rest of the population during the 1970s, the

Table 8.2 Daily per-capita calorie supplies during the 1990s

Region	1990–1992 (1)	1997–1999 (2)	Increase (%) (3)
East and Southeast Asia	2,647	2,899	9.4
South Asia	2,330	2,400	3.0
Latin America and the Caribbean	2,710	2,830	4.4
Middle East and North Africa	3,010	3,010	0.0
Sub-Saharan Africa	2,120	2,190	3.3
All developing countries	2,540	2,680	5.5

Source: FAO (2002b).

food-insecure cohort in India and neighboring countries has stabilized. Numbering 300 million, this cohort is larger than what one finds in any other part of the world. However, progress toward food security continues, in absolute and relative terms. In contrast, there is little positive news from south of the Sahara, where the number of food-insecure people nearly doubled during the last third of the twentieth century. It is generally expected that this number will continue going up about as fast as the region's population (Table 8.1, columns 4 and 5). Consequently, Sub-Saharan Africa will soon have not just the highest incidence of undernourishment, as it has had since the 1970s, but more food-insecure people than any other part of the world, including South Asia, as well.

Another perspective on food security is provided by Table 8.2. Since an adult male with a moderate physical work-load ought to consume about 2,800 calories per day, 2,200 calories per day are required by a woman engaged in moderate physical activity, and children (who are numerous in poor countries) need even fewer calories, it is clear that food supplies, including imports, are at least adequate in most developing regions. About 3,000 calories per day are available for every man, woman, and child in the Middle East and North Africa. During the 1990s, per-capita daily supplies went from 2,710 to 2,830 calories in Latin America and the Caribbean and from 2,647 to 2,899 in East and Southeast Asia.[1] Clearly, food availability comfortably exceeds nutritional requirements in each of these regions. Neither is there a shortfall in South Asia, where daily availability had risen to 2,400 calories per capita at the turn of the century.

A similar relative gain in per-capita supplies has occurred in Sub-Saharan Africa. However, much of this gain is accounted for by large relative increases in Angola, Mozambique, and Sierra Leone, where there has been a recovery from deep lows registered during civil wars. Food production per capita actually went down in most African nations during the 1990s. In a number of countries, local food availability is

---

[1]The 9.4 percent improvement in East and Southeast Asia largely reflects the increase of 12.2 percent that has happened in China. Obviously, progress of this magnitude, which far exceeds the 5.5 percent growth experienced in the developing world as a whole, explains much of the alleviation of global food insecurity during the late twentieth century.

less than the population's caloric requirements. Where this shortfall occurs, warding off hunger requires imports, some of which is likely to be food aid.

Just as food insecurity is more severe in some regions than in others, the incidence of undernourishment varies considerably within a specific population or community. As a rule, women are more vulnerable than men, even if females do most of the farming, as is often the case. They are especially at risk during pregnancy or while nursing their offspring, when they need to eat more to maintain energy levels and avoid anemia. Malnutrition is also severe among children. To a degree, it makes sense for an impoverished family to give smaller rations to its young, who need fewer calories than adults do. But between general poverty and inequitable allocation within households, countless African, Asian, and Latin American children are malnourished. According to Smith and Haddad (2000), 167 million youngsters under 5 years of age in developing countries – fully one-third of the age group – were undernourished in 1995. Even worse, there were 8.2 million more such children in 2000 than there had been 10 years earlier (Runge *et al.*, 2003, p. 19).

Something a little ironic about food insecurity is that it tends to be more acute precisely where crops and livestock are raised, rather than in teeming cities. As Barraclough (1991, p. 42) and many others report, hunger is severe among landless peasants, smallholders, and hired agricultural workers. Of course, the problem for the economically marginal in the countryside, like the problem for the smaller share of the urban population that is undernourished, is incomes that are so miserably low that a minimally adequate diet is unaffordable.

8.3 Achieving Food Security

Having defined key terms and identified the numbers of people affected, we now examine how hunger can be alleviated.

From time to time, food security has been characterized as something to which each and every human being is entitled. Such an entitlement is expressed, for example, in the International Declaration of Human Rights of 1948, as well as the International Covenant on Economic, Social, and Cultural Rights, which 145 governments agreed to in 1966. Adequate nourishment is also enshrined as a fundamental human right in the Rome Declaration on Food Security, which 182 countries endorsed during the World Food Summit of 1996. At the same meeting, world leaders pledged to reduce the number of undernourished people around the world to less than 400 million by 2015 – a goal that currently looks out of reach.

While broad declarations that food is a basic human right are meant well, one wonders what exactly is gained by statements of virtuous intent which, because of limited means or some other reason, no one is prepared to enforce. More telling is what the governments endorsing these statements actually accomplish, in terms of both food aid and alleviation of the poverty that makes adequate nourishment unaffordable.

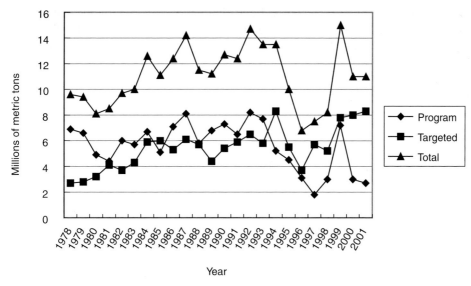

Figure 8.1 Program and targeted food aid, 1978–2000.
Source: World Food Program (2003).

Food Aid

When starving people flee to refugee camps, usually because of civil war or ethnic cleansing, food aid arrives as well. This humanitarian response is a point of pride for those making it possible. Americans, for instance, like to think of themselves as magnanimously caring for wretched cases on the far side of the world.

Food aid happens to be very small relative to the commodities that move commercially from country to country. For example, cereal donations, which comprise the larger part of the food given away at little or no cost to recipients, have averaged 12 million tons per annum since the middle 1980s (Figure 8.1) as annual grain exports have climbed above 230 million tons. Moreover, not all food aid is really humanitarian assistance. Non-targeted (or program) donations to the governments of poor countries are sold in local markets to provide funds (in the currency of the recipient country) for agricultural research and extension and other purposes. Much of this aid cannot be categorized as charity – in contrast to the majority of targeted donations, which are distributed to the hungry by non-governmental organizations (NGOs) through food-for-work, school-feeding, and direct distributions to the needy.

The case for humanitarian assistance is all but unassailable. In contrast, serious doubts have been raised about program food aid, which exceeded the targeted category up until the middle 1990s (Figure 8.1). To be specific, opponents have criticized non-targeted disbursements for being little more than an export subsidy, which distorts international trade and discourages production in recipient countries.

Program aid would not cause significant distortions if the commodities delivered to hungry people were purchased from local farmers. However, this usually does not happen. Instead, commodities are typically obtained in affluent settings where production exceeds consumption, such as North America and the European Union. Quite often, surpluses are a consequence of support prices and other policy-induced distortions in commodity markets.

If surplus grain from a rich donor nation is delivered to a poor country that imports cereals, as many recipients do, then a likely consequence is for imports to be displaced, thereby allowing recipients to save foreign exchange. Hoping to prevent commodity donations from crowding out trade, exporting nations, many of which are also donors, agree among themselves that food aid is not to be allowed if it causes imports to decline.[2] Enforcement of such an agreement means that program donations end up lowering prices in the recipient country and therefore displacing its domestic output, exactly as Schultz (1960) predicted in a classic critique of food aid. The overall effects of this intervention on economic well-being, then, are essentially the same as those observed if a country with a comparative advantage in grain subsidizes its cereal exports (Chapter 6). Producers' surplus (PS) grows and consumers' surplus (CS) diminishes in the donor country, with its incremental PS having a larger absolute value than its lost CS. Meanwhile, CS increases exceed PS losses in the recipient. From the gains in net economic value arising in both nations' commodity markets, one has to deduct reductions in well-being associated with increased taxation in the donor country, which is needed to underwrite program aid.

A comprehensive economic assessment of food aid must take into account other consequences. One of these is the risk a recipient nation runs of growing dependent on food donations, which vary according to production and market conditions in supplying countries. These changing conditions help to explain why international cereal donations fluctuate considerably, reaching 13 million tons in 1994–1995 then falling below 7 million in 1996–1997 and finally increasing to 15 million at the end of the century (Figure 8.1).

Of course, the benefits of development initiatives undertaken thanks to program assistance as well as food-for-work and school-feeding activities made possible by food aid also need to be considered. However, these benefits are sometimes exaggerated. Participants in food-for-work and school-feeding schemes are not always selected very carefully, which means that some of them would have worked, studied, and eaten even if no such schemes had been implemented. Another problem is that these activities as well as the development initiatives underwritten by program food aid are often not as cost effective as other sorts of foreign assistance. One reason for low cost effectiveness is corruption, which occurs for example when public officials pocket some of the proceeds of selling program donations in the domestic market. Another cost of corruption arises because administrative procedures must be put in place to limit peculation; these procedures can be cumbersome and expensive.

---

[2] Of course, compliance with a simple rule of non-declining imports could mask displacement of trade if aid caused deviation from an upward trend in a recipient's commercial purchases from other nations.

Since non-targeted food aid can distort international trade, discourage production in recipient countries, and foment corruption, its decline is not to be lamented. Conversely, recent increases in humanitarian assistance (Figure 8.1) are commendable.

Economic Growth and Lower Food Prices

Although it will continue to be offered, food aid cannot reach more than a few million people. If the vast majority of the 800 million human beings never benefiting from humanitarian assistance are to be adequately nourished, their earnings will have to rise, food will have to grow cheaper (not because of governmental meddling with market forces, but rather because of improvements in agricultural productivity), or a combination of the two.

Barring substantial redistribution of income, of the sort that is seldom attempted and is even less frequently successful, lifting food-insecure people out of extreme poverty requires economic growth. To examine the consequences of growth, Senauer and Sur (2001, cited in Runge et al., 2003, pp. 28–29, 209–213) have developed a model that combines statistical estimation of the linkage between per-capita earnings and daily calorie consumption with projections of GDP and population in various parts of the world. Also incorporated in the model are inter-regional differences in minimal energy requirements, which have to do with variations in body stature. So are increases that will happen over time in these requirements as demographic transition proceeds and, consequently, children become an ever smaller part of the population. A particular concern of the study is the portion of each region's population with earnings below the level at which, according to statistical analysis of the tie between living standards and calorie consumption, minimal energy requirements are barely met.[3]

The results obtained by Senauer and Sur (2001) are summarized in Table 8.3. If the GDP growth that the World Bank forecasts and the increases in population that the United Nations projects both materialize, the world's food-insecure population – defined as those with daily calorie consumption below the level required for light physical activity – will decline by nearly one-fourth by 2025 (column 1). There will be more progress in Asia, where GDP per capita is increasing rapidly, as well as Eastern Europe and the Former Soviet Union, where the food-insecure cohort is relatively small and not much economic expansion is needed for this cohort's nutritional status to reach minimal adequacy. In contrast, the number of food-insecure people in Latin America and the Caribbean will only go down by 8 percent and there will be no absolute change in the Middle East and North Africa, where growth in living standards is slated to be modest at best. Even worse, there

---

[3]These minimal requirements, which are consistent with nothing more than low levels of physical activity, currently range from 1,790 calories per day in South Asia, where short, thin people predominate, to 1,900 calories per day in Eastern Europe and the Former Soviet Union, where people tend to be taller and heavier. Thanks to aging of the population, minimal daily energy requirements will rise by the middle of this century, to 1,900 calories in Sub-Saharan Africa, 1,940 calories in the Middle East and North Africa as well as Asia, and 2,000 calories everywhere else (Senauer and Sur, 2001).

Table 8.3 Percentage changes between 2000 and 2025 in the number of people not consuming enough calories

Place	Current trends in GDP per capita (1)	Slower GDP growth (2)	Current trends in GDP per capita, but with pro-poor growth (3)	Current trends in GDP per capita, but with lower food prices (4)	Current trends in GDP per capita, but with pro-poor growth and lower food prices (5)
China	−71	−31	−80	−81	−88
Other East and Southeast Asia	−42	+16	−63	−63	−76
South Asia	−56	0	−71	−71	−81
Latin America and the Caribbean	−8	+52	−40	−40	−61
Middle East and North Africa	0	+72	−33	−36	−59
Eastern Europe and Former Soviet Union	−47	−27	−66	−66	−78
Sub-Saharan Africa	+52	+136	+22	+8	−26
Total	−23	+36	−43	−47	−65

Source: Senauer and Sur (2001).

will be 52 percent more food-insecure people south of the Sahara, where economic expansion is slow and human numbers are going up rapidly.

While a 23 percent decline in the number of hungry people between 2000 and 2025 is not particularly impressive, the consequences of slower growth – brought about because GDP expansion is only half of what the World Bank projects – would be appalling. Under this scenario, China and the Former Soviet Union and its erstwhile satellites would achieve modest progress toward food security, while there would be no change in the Indian Subcontinent. Everywhere else, the ranks of the food insecure would swell.

Economic development contributes substantially to food security if the rise in living standards is general, rather than being captured mainly by a small elite. While they do not examine broad-based development and its implications for hunger, Senauer and Sur (2001) do investigate the consequences of pro-poor growth, as they call it. In particular, they suppose that the share of each region's 2025 GDP going to the most destitute fifth of the population would increase by 50 percent, the share going to the next poorest quintile would rise by 25 percent, and, to accommodate this redistribution, the top quintile's GDP share would diminish. This scenario makes no allowance for the earning gains that would go to all, including the affluent, as the living standards of their poorer trading partners rose. Also unexamined are the GDP penalties that would result from taxation for the sake of redistribution. Regardless, the accelerated progress toward food security that Senauer and Sur (2001) project would be truly dramatic. Pro-poor economic expansion would cause the number of people in Asia and the former Soviet bloc who do not consume enough calories to fall by more than one-half. There would be declines of two-fifths in Latin America and the Caribbean, and one-third in the Middle East and North Africa. The cohort of food insecure Sub-Saharan Africans would go up between 2000 and 2025, but only by 22 percent instead of by one-half (Table 8.3, column 3).

Since households that are impoverished and food-insecure spend half or more of their meager earnings on edible goods, lower food prices, which result from productivity growth in agriculture and agribusiness, benefit them enormously. While Senauer and Sur (2001) refrain from forecasting prices, they investigate the consequences of 10 percent variation in per-capita calorie intake, as would result from changing prices. For example, the effects of cheaper food are almost exactly the same as those of the pro-poor scenario. An important difference is in Sub-Saharan Africa, where pro-poor growth alone would contain the increase in the food-insecure cohort to 22 percent while cheaper food alone would limit this increase to 8 percent (Table 8.3, column 4).

Needless to say, maximum progress against hunger would result from combining income growth for poor people and lower food prices. As reported in column 5 of Table 8.3, an integrated approach of this sort would cause most food insecurity to disappear in countries that have made a transition from communism as well as most of the developing world. There would even be progress south of the Sahara. Indeed, the only way to prevent more Africans from going hungry is by applying a full range of measures for improving their economic access to food.

8.4 The Food Security Synthesis and Economic Development

Focusing on the challenge of improving the economic access of impoverished people to available supplies of food, Tweeten (1999) has distilled a *food security synthesis*. Made up of a seven-step logical framework, this synthesis addresses fundamental causes of undernourishment as well as cures for the problem:

1. Transitory and chronic food insecurity is caused mainly by *poverty*. As already indicated, nearly all the underfed Africans, Asians, and Latin Americans identified in Table 8.1 have daily incomes of 1 dollar or less, and are therefore extremely poor. These people lack the buying power needed to overcome the frictions of time (e.g., unpredictable and unstable harvests from year to year) and space (e.g., local food shortages) that cause them to be food insecure.
2. Poverty is best alleviated through *economic development* that is broad based and sustainable. Altruism is commendable and plays a critical role in feeding members of a close-knit group. But in communities with little or nothing to share, there is little scope for altruism and the issue of redistribution is moot. Certainly, the linkage between economic expansion and progress toward greater food security is recognized by the FAO:

   "The need is for policy measures that address all aspects of food insecurity with a view to providing safety nets for the vulnerable and to creating the conditions that can lead to an eradication of endemic hunger. *This has to mean economic growth* [emphasis added] ... Improving the equitableness of the income distribution can only achieve so much (in countries with low and falling income), and, as seen time and again, will be strongly resisted by the potential losers. So growth is necessary, and against a background of economic growth, experience shows that it is easier, although never easy, to implement measures that increase equity, particularly if the growth is broadly based to include the agricultural sector (FAO, 1997, p. 3)."

3. The best way to pursue broad-based economic development is to follow the *standard model*. Described below, this model can be applied anywhere and provides a workable prescription for economic progress, thereby ensuring enough buying power for food self-reliance and food security.[4] The model is not prized for its ideology, but instead because it works. It is not one-size-fits-all, in the sense that each and every one of its elements must be implemented thoroughly. However, some of its key features are essential for a sound economy. These key features are shared by virtually every country that has experienced broad-based development.

---

[4]Food self-reliance emphasizes building up productivity in order to grow enough food domestically or to be able to purchase it abroad. It contrasts sharply with food self-sufficiency, which is a reckless policy if it compromises buying power so that a nation ends up not being able to afford imports if domestic production fails – as it is prone to do.

4. *Government or political failure* explains why some countries do not adopt enough components of the standard model to end poverty and food insecurity. Sometimes, economic illiteracy helps to explain why the standard model is not followed. However, political failure is far more often the main problem. Individuals and groups with power and authority frequently gain from policies that egregiously compromise the public interest. As these individuals and groups resist application of the standard model, direct costs (e.g., the inefficiencies created by state-owned enterprises (SOEs) and policy-induced distortions as well as the transaction costs associated with bureaucratic bloat) are created, as are indirect costs (relating to the effort expended to win favors from government). The magnitude of these costs as well as the sacrifices in national income that result from political failure are often obscured by political populism.
5. Political failure is inseparable from *institutional failure*. The ability of political leaders and bureaucrats to pursue inefficient policies, which harm overall well-being, has much to do with institutions of government that are weak, mismanaged, and corrupt – and consequently incapable of enforcing the rule of law and avoiding domination by special interests. It is no exaggeration to say that applying the standard model is inseparable from institutional reform.
6. Poorly structured and inadequate institutions often trace to *cultural factors*. Only a handful of economists have been willing to examine ties between culture and economic performance systematically (Landes, 2000; Sen, 2000). Nevertheless, the ties are of basic importance. For example, institutional failure is hard to avoid if few citizens are involved in political processes and if there is a general tolerance of government that is unrepresentative, corrupt, and incompetent. No less than anyone else, leaders are products of a nation's culture. If they view their position as a chance for personal aggrandizement, rather than an opportunity to serve the public, the rule of law is bound to be weak. Tribal animosities and other cultural characteristics militate against sociocultural change, of the sort required for economic progress.
7. The core challenge is *sociocultural change*. Achieving food security ultimately rests on the adoption of sociocultural norms that are conducive to economic progress. Two examples of such norms are the Protestant ethic (Weber, 1930; Tawney, 1966) and Confucist values (Johnston, 1966), both of which put a premium on honesty in interpersonal dealings, hard work, and enterprise. Likewise, much of Japan's economic success has been credited to "social and political attitudes that made economic development a priority bordering on an obsession" (Sowell, 1998, p. 345).

Sociocultural norms having an impact on development have been a recurring concern of economists, sociologists, and specialists in other disciplines. Safe to say, this subject will continue to command the attention of the finest minds in the social sciences for a long time to come.

The Standard Model

As observed at the end of the preceding chapter, economic progress requires that human, natural, and other resources be complemented by a culture of trust that is sustained by institutions that enforce the rule of law. Once the cultural and institutional underpinnings of broad-based development are satisfied, a set of economically sound prescriptions for public policy tends to be followed. To describe this set of prescriptions, Williamson (1990) coined the phrase, "Washington Consensus." As the same individual subsequently admitted, this phrase, which has come to be widely used, has the unfortunate connotation of a program imposed by the International Monetary Fund (IMF) and World Bank, both of which are headquartered in the US capital (Williamson, 2000). The standard model, which is a term Tweeten (1999) proposes for a broader set of prerequisites for economic progress, does not share this disadvantage.

Numerous studies cited by Tweeten (1999) provide statistical evidence of the standard model's success. Anecdotal evidence is also compelling. For example, the East Asian tigers, Chile, and Ireland have all enjoyed impressive economic gains by adopting critical elements of the standard model, which is not the "end of economics" and will have to be revised as new experiences are evaluated. However, it promises the overall economic well-being essential to end hunger in any poor country willing to embrace it.

An under-girding principle of the standard model is that things that the market does best should be left to markets – in particular, the competitive allocation of most goods and services – and the government should focus on doing what the marketplace neglects or is incapable of performing. One fundamental task for the state is to provide public goods (Chapter 3), which the market economy does not supply if left to its own devices. Something else the marketplace cannot do for itself is to establish the institutional framework without which the market economy simply does not exist. If government performs the latter function and supplies public goods, then competitive markets can be counted on to determine what, when, where, and how to produce the vast majority of goods and services. Not merely unnecessary, governmental manipulation of prices or output levels in markets is usually undesirable (Tweeten, 1999).

A critical part of the standard model for economic development that is broadly based and environmentally sustainable is sound public administration, which reinforces the institutional framework for efficient markets:

- *Security, stability, and order*: Courts must interpret laws and administer justice evenhandedly. Under the rule of law, contracts can be made and acted on, which allows commercial exchange and investment to take place.
- *Property rights*: Ownership that is exclusive as well as tradable is another prerequisite for the market economy. Private investment, for example, is all but impossible if property rights are weak, in part because the collateral needed for loans does not exist in a legal sense. Also, ownership rights encourage investment because the value of improvements accrues to the owner (or his or her heirs) when

a property is sold. Property rights need to be enforced consistently by courts, of course. Furthermore, a system of delineating and registering these rights is required.
- *Competition*: The full benefits of market exchange are realized when there are many buyers and sellers. At the very least, government must avoid protecting monopolies. More than that, competition ought to be promoted, through the application of antitrust laws. Where a natural monopoly exists, because the presence of two or more firms would cause economies of scale to be lost, regulation is appropriate. Otherwise, free trade is often the best way to prevent domestic firms from exercising monopoly power.
- *Honesty and competence in government*: Although some economists used to be complacent about peculation – regarding the impacts of bribery as little different from the consequences of taxation, for example – the toll on economic progress associated with corruption in the public sector is now widely appreciated (Sachs, 1997). To contain corruption, transparency and accountability in government dealings are required. So are merit hiring, competitive pay, and civil-service protection, which are also needed for a competent and politically neutral bureaucratic workforce. A free press, which is a core element of democracy, also serves as a brake on governmental malfeasance.

Along with providing a market economy's institutional underpinnings, the state needs to avoid macroeconomic policies that distort incentives and hinder investment:

- *Fiscal responsibility*: Chronic deficits, which arise because government is unwilling to live within its means, can be sustained for a while by borrowing – rising public indebtedness, in other words. But a point is often reached at which the state prints money to settle its bills, which ignites inflation. Inflation erodes the value of private savings, thereby reducing investment. To be sure, borrowing that permits governmental spending in excess of tax revenues might be justified if this is required for investments in public goods. A case can also be made for counter-cyclical fiscal policy – that is, deficits during recessions and surpluses at other times. But chronic deficits, which are the hallmark of fiscal irresponsibility, are by all means to be avoided.
- *Monetary restraint*: Sound fiscal policy makes the task of sound monetary policy, which creates price stability, much easier. The complete absence of inflation is not really ideal, since deflation (i.e., a prolonged decline in the general price level) is almost always accompanied by a slowdown in economic activity.[5] Instead, modest inflation (i.e., inflation of no more than 3 percent) has merit. This is normally accomplished by increasing the money supply a little more rapidly than growth in GDP, with appropriate adjustments for foreign exchange

---

[5] As prices fall, consumers postpone purchases whenever they can, in the hope of getting a bargain in the future. Also, business profits are squeezed, insofar as labor contracts keep wages from falling along with prices. Firms are likely to respond to this by laying off workers. Slumping sales and unemployment occurred on a large scale during the Great Depression, when deflation persisted for a number of years.

and direct investment flows. Monetary authorities who are largely independent of immediate political concerns are most likely to follow this guideline, although their ability to do so is compromised if government is fiscally irresponsible.
- *Appropriate taxation*: One of the greatest challenges facing governments in the developing world is the collection of taxes required even for minimal public services. The economic disincentives of taxation are reduced if marginal rates are low and the tax base encompasses a large part of the economy. But it is also true that progressive taxation minimizes the burden on the poor. Successful governments tax "bads" (e.g., tobacco, alcohol, and emissions) instead of "goods," like investment and exports. Also, user fees charged for electricity, irrigation water, and other things provided by the public sector cover the costs of these goods and services. Taxes on sales, value added, and property distort the economy less than do taxes on exports and corporate profits.

Just as sound macroeconomic policy creates the price stability that causes markets to flourish, a liberal trade policy promotes efficiency by encouraging everyone to specialize according to comparative advantage:

- *Openness to trade and investment*: Although some countries have succeeded at protecting infant industries from foreign competition until comparative advantage was achieved, this strategy has serious pitfalls. Rather than taking advantage of barriers to imports to bring costs down below international prices, firms often grow indolently dependent on protection, and therefore lobby for its continuation. Any hint that protection will turn out to be permanent instead of temporary is apt to lock a country into a path of inefficient industrialization and arrested growth.
- *A realistic foreign exchange rate*: Efficient transmission of international prices to domestic markets hinges on a proper rate of exchange between the national currency and the currencies of trading partners. As indicated in Chapter 6, currency over-valuation, which discourages exports and encourages imports, occurs frequently in the developing world. To deal with this problem, Fischer (2001) recommends two solutions. One is floating (or flexible) exchange rates, with currency values determined in the marketplace. The other, which particularly suits a small country that trades mainly with a larger nation, is adoption by the former of the latter's currency. Illustrative of the latter approach is dollarization, which Panama underwent in 1904 and which Ecuador and El Salvador have done more recently.

Rather than trying to develop by manipulating exchange rates and other market forces, developing countries need to exploit the opportunities proliferating in a globalizing world by raising productivity. Granted that productivity gains in the private sector are impressive, the government's role of providing public goods is also important:

- *Infrastructure*: Commercial activity is facilitated by all-weather roads, communication networks, and so forth. Likewise, seaports, airports, and the

like are needed for international trade. While private firms can administer many such facilities efficiently, much of this infrastructure is best classified as public goods, which means that governmental financing is unavoidable.
- *Agricultural research and development*: As noted in Chapter 3, investment in the scientific and technological base for crop and livestock production yields high returns. However, this investment is deficient in the developing world. Whereas spending on agricultural research and development amounts to 2–3 percent of the value of agricultural output in affluent nations, the ratio is closer to 1/2 percent in Africa (Pinstrup-Andersen, 2002). If this imbalance is not redressed, African farmers, who comprise the majority of the labor force and produce half of GDP in many countries south of the Sahara, will not benefit from genetic engineering and other biotechnologies, which show promise of raising yields and reducing pressure on threatened ecosystems.
- *Education*: Universal access to schools is essential for development. The penalty is sizable for denying access to women or ethnic minorities, for example. Since the social pay-off of elementary education is especially high, universal primary schooling, financed by government, is a priority for food security and development.
- *Public sanitation and health*: Along with education, public investment in human capital involves attention to water and wastes. Otherwise, parasites and bacteria impair food utilization and sap vitality, even interfering with the physical and mental development of the very young. Human capital is also built up through networks of health clinics, at which immunization, vitamin supplements, knowledge about the prevention of Human Immunodeficiency Virus/Acquired Immunodeficiency Syndrome (HIV/AIDS) and other diseases, family planning services, and guidance for pre- and postnatal care are provided.
- *Environmental quality*: Yet another public good is the quality of air, water, and other natural resources. As emphasized in Chapter 5, conservation depends on public policy. For one thing, correcting market failure requires intervention by the state. Also, research and development allows for production of agricultural commodities and other goods and services to go up without damaging the environment. For example, achieving higher yields on land well suited to agriculture reduces the incentives to cultivate fragile lands.
- *Food and income safety nets*: In just about any part of the world, emergency food aid is made available in the wake of environmental disasters, civil disturbances, and other such events. Also, many governments offer food and other sustenance to those constituents who are either unable to provide for themselves or who lack support from family and other private sources. Economic analysis is needed to determine how changing the height and breadth of the social safety net is likely to affect labor effort and risk-taking, which are needed of course to create the largesse distributed to the less fortunate. That said, it must be recognized that the safety net's specific dimensions are never exclusively an economic concern. Among other things, these dimensions reflect a sense of human community, as this is felt at the local, national, and global levels.

The standard model and its various elements do not fit easily into any single ideological category. It certainly is not state centered, at least as a Marxist would define the term. If public administration is generally sound and if appropriate policies are in place to deal with problems like environmental market failure, competitive markets can and ought to be treated as if they are on autopilot. But neither is the standard model libertarian, in the sense that government is expected to wither away as all its functions are taken over by private entrepreneurs who respond to market demands. To the contrary, the state is uniquely suited to administering the marketplace and supplying public goods.

While the preceding list of prescriptions define proper roles for government as well as markets, it does not really amount to a comprehensive and precise blueprint for the sort of economic progress needed for everyone to be food secure. Experience shows that the exact balance struck between the state and the marketplace in one country can differ from the balance chosen somewhere else, even though development is happening in both places. Also, converting the list of prescriptions into a blueprint would require that questions of sequencing be addressed. For example, Tweeten and McClelland (1997, pp. 1–31) recommend beginning with the reform of policies that hinder economic performance and create food insecurity while investing in elementary education and roads and other basic infrastructure. They also point out that supporting the improvement of agricultural technology is very important. As Watkins and von Braun (2003) observe, three out of every four of the world's 1.2 billion extremely poor people live in rural areas. Also, 90 percent of the food consumed in developing nations is produced domestically. But in spite of these realities, not to mention the high returns to public investments in agricultural research and development (Chapter 3), the neglect of agriculture, which Schultz (1964) criticized four decades ago, continues.

Sequencing is a major concern for development agencies. In its structural adjustment initiatives, the IMF offers loans to countries in financial distress in exchange for the trimming of fiscal deficits, adoption of realistic exchange rates, and related changes in public policy. Likewise, the World Bank, which specializes in the financing of development projects with a longer-term pay-off, sometimes attempts to tie its lending to policy reform. But for various reasons, this sort of "conditionality" has proven difficult to implement (Easterly, 2001, pp. 115–120).

When spent in the right ways, foreign assistance can help a nation break out of the "poverty trap" that exists when incomes are too low to pay for investments in infrastructure, technology transfer, and education, which are needed to raise living standards. For example, a country where a large segment of the population engages in subsistence farming, and consequently is impoverished and food insecure, can benefit enormously from agricultural research and development financed by an organization like the World Bank. However, just as there are beneficial and important uses of foreign aid, there are many inappropriate ways that this assistance has been used in the past, including the underwriting of subsidies and other policies for distorting markets. This is one reason why the contributions that foreign aid has made to development often have been disappointing (Easterly, 2001, pp. 35–44).

Although foreign assistance has its place, it will not be the primary driver of economic progress in Africa, Asia, and Latin America because the amounts available are modest. Currently, just over $50 billion are provided by affluent countries and multilateral institutions, such as the World Bank. This sum is dwarfed by international capital flows (Hausler, 2002). Adoption of standard model policies would speed economic growth in no small part by attracting more foreign direct investment (FDI) to poor countries from rich nations. In theory, the relative abundance of labor and natural resources and relative scarcity of capital in developing countries should create high returns to investment and massive capital inflows. But aside from China and a few other exceptions, poor countries in recent years have pursued policies that keep returns low and consequently repel FDI. Reforming these policies would stimulate the growth needed to alleviate poverty, hunger, and environmental degradation.

It is important to keep in mind that no poverty trap applies to many elements of the standard model for broad-based economic progress and food security. Different from many of the investments that foreign aid can make possible, trade liberalization and other reforms in economic policy do not require large expenditures by government. Initiatives like strengthening the judiciary and related agencies, so that enforcement of contracts and property rights becomes more reliable, are not free. However, the expenses are modest in absolute terms and, of even greater importance, exceeded by the benefits of strengthening the marketplace's institutional framework. Thus, there is no financial barrier to adoption of a large part of the standard model.

Finally, the most effective contribution that wealthy countries can make to economic progress and food security in Africa, Asia, and Latin America has nothing to do with foreign aid. As emphasized in Chapter 6, trade restrictions in the former countries impinge significantly on exports from less fortunate parts of the world. Removing these restrictions would create a substantial impetus for adoption of the standard model and, hence, improvements in living standards around the globe.

8.5 The Standard Model, Communitarian Values, and Economic Equity

In light of the great debate over the relative merits of capitalism, socialism, and various intermediate alternatives that dominated intellectual life since before the Second World War, the coalescence that happened in the late twentieth century about the meaning of sound economic policy as well as the prosperity it creates is remarkable. Call this coalescence the Washington Consensus, the standard model, or anything else, the range of opinions about how to pursue development is much narrower than it used to be (Yergin and Stanislaw, 1998). Reflecting this coalescence is the accord that now exists concerning the problem of food insecurity. In particular, there is wide agreement that the policy reforms and other initiatives required for economic progress are also needed to alleviate hunger and vice versa. Regardless of which broad aim is chosen, the same standard model applies (Tweeten, 1999).

As the latter steps of Tweeten's (1999) food security synthesis emphasize, cultural norms ultimately determine whether or not the standard model is applied. Among these norms are enterprise, thrift, hard work, and honesty in personal dealings (Weber, 1930; Johnston, 1966; Tawney, 1966). However, there are other cultural norms, including what one might call communitarian values – a sense of community, in other words. These values, which influence our thinking about food security, determine how the standard model is adopted. At an extreme, communitarian impulses can even block the model's adoption entirely.

By no means is a sense of community economically irrational, either for individuals or for entire groups and populations. In many parts of the world, the natural environment is harsh and unpredictable and, consequently, human survival has been tenuous. In such settings, sharing is a well-entrenched tradition. If one family's harvest fails this year, it can count on getting something to eat from neighbors or distant relatives, who have enough to spare. This option, which has obvious value, is kept alive by sharing what one has when other folks who are down on their luck come calling. Barring a catastrophe, which leaves everyone hungry at the same time, individuals survive from one year to the next and the community remains intact.

Regardless of environmental and other conditions that created values like sharing in the first place, communitarian impulses can have a down side in terms of economic growth – especially when these impulses apply to financial resources, not just food. For example, someone with savings – thanks to his or her thriftiness, which is clearly a cultural norm well suited to capitalist expansion – might be viewed mainly as a source of largesse by traditionally minded members of his or her family and community. The thrifty person then faces a choice. To continue amassing funds, he or she must strain, perhaps sever, his or her closest human ties. The alternative is to give up on capital accumulation. Inasmuch as many people choose the latter alternative, investment is hampered and, as a result, living standards stagnate.

A sense of community is alive and well even in affluent places with a long history of capitalism. It is even reasonable to suppose that prosperity, which capitalist development has made possible, makes the application of communitarian values easier, although acting on communitarian impulses often has a negative impact on economic growth. Sweden is a good example. Like many of its neighbors, Scandinavia's leading nation has a generous and comprehensive social safety net, which helps to explain its low Gini coefficient. The decline in Swedish GDP per capita during the 1990s (Chapter 7) suggests that striving for minimal economic inequality might reduce GDP. However, high average incomes in the country also indicate that the trade-off between growth and equity can be eased by taxing consumption so as to finance human capital formation, which raises living standards and reduces inequality simultaneously. Insofar as possible, it is also important to avoid high marginal tax rates, which penalize the firms and individuals that constitute the main engine of economic expansion.

While potential trade-offs between growth and equity must be kept in mind, it is also true that the distribution of productive assets can be so skewed that economic expansion is depressed (Deininger and Squire, 1997). In particular, redistribution of farm real estate sometimes "becomes the only option for improving rural livelihoods rapidly and substantially," even though governments usually are reluctant to undertake land redistribution because it is socially divisive and often provokes violence (Barraclough, 1991, p. 130). A better way to reduce asset inequality is to build up other sorts of productive wealth for the poor, especially their human capital. This approach makes sense in Sub-Saharan Africa, which is impoverished and agriculturally dependent. Even in that part of the world, human capital comprises three-fifths of all assets, while agricultural land and other environmental wealth make up just 10–20 percent (Dixon and Hamilton, 1996).

As indicated in the preceding section, human capital formation and other productivity-enhancing investments are key components of the standard model for economic progress and food security, not just a good way to reduce asset inequality. So is appropriate taxation, which is needed to pay for these investments. However, increased taxation, especially when it enables government to deviate from the standard model, diminishes economic expansion.[6]

Communitarian values and desires to reduce inequality are strong, so advocates of the standard model are likely to end up frustrated if they simply ignore these values and desires. Runge *et al.* (2003, p. 6) are not off the mark when they characterize the alleviation of poverty and hunger as a public good, in the sense that failure to do so constitutes an indictment of the prevailing economic order in the eyes of many and therefore dims the prospects for policies that facilitate economic expansion. The challenge, then, is one of balance, with communitarian and egalitarian impulses being harnessed to maintain safety nets that are both adequate and do not impinge on broad-based development, which as stressed in this chapter is essential to raise the living standards of the food insecure.

Study Questions

1. The FAO's definition of food security emphasizes economic access to an adequate diet. Explain this emphasis, taking into account other dimensions of human nourishment.
2. In what parts of the developing world has food security improved and in what regions is food insecurity still a severe problem?
3. Distinguish between program and targeted food aid and assess the likely impacts of each on imports and domestic production in recipient countries.

---

[6]For example, Tanzi and Schuknecht (1995) have concluded that economies with the slowest growth in public spending from the 1960s to 1990s were more efficient and innovative and experienced faster expansion in employment.

4. What combination of economic expansion and lower food prices would be needed to lower the incidence of food insecurity appreciably in Sub-Saharan Africa?
5. Is there any substantial conflict between trying to achieve broad-based economic development and pursuing wider food security?
6. Compare and contrast the elements of the Washington Consensus and the elements of the standard model.
7. What are the potential inconsistencies between the standard model and communitarian values?

9

A Synopsis of Regional Trends in the Global Food Economy

Up to this point, the book's content has been thematic, relating to the overall demand and supply of food, agriculture's environmental impacts, the sector's role in globalization and economic development, and food security. Examination of these trends puts differences among various parts of the world in sharp relief. Malnourishment, for example, has been reduced substantially in East and Southeast Asia – defined here not to include Japan, where practically everyone is well fed. In contrast, the incidence of food insecurity is very high in South Asia and Sub-Saharan Africa. Also, forests are being lost at a faster pace in the latter region than anywhere else, whether rich, poor, or in between.

From this chapter on, the book is organized geographically. The study of inter-regional differences and similarities enhances one's understanding of the challenges facing the food economy, such as conserving the natural resources on which agriculture depends while consistently providing adequate nourishment for everyone – not just the four out of every five human beings who currently are avoiding extreme poverty. In this chapter, broad regional trends and characteristics are compared and contrasted. The next six are about the world's affluent nations, Asia, Latin America and the Caribbean, the Middle East and North Africa, Eastern Europe and the Former Soviet Union, and Sub-Saharan Africa, respectively. Examined in each of these chapters are national trends in living standards and the incidence of poverty, population, and human fertility, as well as agriculture and environmental resources.

9.1 Economic Growth and Income Distribution

Clearly apparent in Table 9.1 is the economic expansion that various parts of the world experienced during the last third of the twentieth century. Not limited by any means to affluent settings, GDP growth (column 1) has had much to do with increased specialization and trade, as reflected in the table by export trends (column 6). Furthermore, economic expansion has been driven by the buildup of productive assets, which include fixed capital (column 5). At the same time, capital formation has been made possible by a rise in living standards.

Table 9.1 Long-term trends in economic growth for major regions of the world 1965–2003 and current levels of income per capita and income inequality

Region	Total GDP growth (average annual percent) 1965–1999 (1)	Total population growth (average annual percent) 1965–1999 (2)	Per capita GDP growth (average annual percent) 1965–1999 (3)	Per capita GDP growth (average annual percent) 2000–2003 (4)
1. High-Income Nations	3.2	0.8	2.4	1.5
2. East and Southeast Asia	7.4	1.8	5.6	5.6
3. South Asia	4.7	2.2	2.4	3.5
4. Latin America and Caribbean	3.5	2.1	1.4	−0.2
5. Middle East and North Africa	3.0	2.8	0.1	2.0
6. Eastern Europe and Former USSR	0.7	0.8	0.0	5.0
7. Sub-Saharan Africa	2.6	2.7	−0.2	1.2

Region	Gross fixed capital formation (average annual percent) 1965–1999 (5)	Exports of goods and services (average annual percent) 1965–1999 (6)	Income per capita (PPP $) 2001 (7)	Income inequality (Gini coefficient) 1990s (8)
1. High-Income Nations	–	5.5	26,650	–
2. East and Southeast Asia	9.7	10.1	3,790	38
3. South Asia	5.3	7.2	2,570	34
4. Latin America and Caribbean	1.9	6.0	6,900	50
5. Middle East and North Africa	–	–	5,430	36
6. Eastern Europe and Former USSR	–	–	6,320	33
7. Sub-Saharan Africa	0.1	2.4	1,750	47

Source: Maddison (2001), pp. 183–185; World Bank (2001b), pp. 26, 72; World Bank (2003a), p. 16; World Bank (2005).

Regional Trends in GDP Per Capita

Nowhere have investment, increases in trade, and growth of GDP coincided more than in East and Southeast Asia. In terms of increased exports, formation of fixed capital, and economic expansion, the region's performance has been unmatched. Also, economic expansion has outstripped increases in human numbers (Table 9.1, column 2) in China (which is so large that its national trends heavily influence trends for the entire region), the four tigers (Hong Kong, Singapore, South Korea, and Taiwan), and other countries. Accordingly, annual increases in GDP per capita in East and Southeast Asia averaged 5.6 percent from 1965 through 1999 (column 3), which caused living standards to double every 12 years. In spite of the region's economic miracle, as it is often described, growth has flagged at times, such as during the financial crisis of 1997. However, the upward trajectory continues, as indicated by average annual growth in GDP per capita of 8.0 percent from 2000 through 2003 (column 4).

Productive capacity has been accumulating and exports have been increasing in India and neighboring countries as well. With economic expansion accelerating in South Asia after 1980, living standards have improved. In Latin America and the Caribbean, average yearly increases between 1965 and 1999 in fixed capital (1.9 percent) and exports (6 percent) compared poorly with Asian trends. Furthermore, more emphasis has been placed on enhancing consumption, at the expense of savings and capital accumulation. As a result, economic growth in the region has been modest compared to what has been achieved in Asia, where trade and investment have grown quickly.

In three parts of the world, GDP per capita has stagnated since the middle 1960s. Data on export growth and fixed capital formation are unavailable for the Middle East and North Africa as a whole, although one supposes that neither has shown much improvement. Another problem in this region has been chronic currency overvaluation (Chapter 6), brought about because oil exports have provided the means to keep exchange rates stable in spite of domestic inflation. Often called "Dutch Disease" (Perkins et al., 2001, pp. 643–651), this condition has throttled development outside the petroleum sector, which employs a small workforce and purchases little from the local economy[1]. Yet another brake on economic progress relates to military expenditures – expenditures that as a portion of GDP are three to four times the levels observed in other parts of the world. Typically rationalized by the threat from Israel, purchases of planes, tanks, and other hardware derive mainly from the desire of undemocratic regimes to cling to power and obviously detract from the investment required for GDP growth, which has been about the same as demographic expansion in the Middle East and North Africa since the middle 1960s.

In Eastern Europe and the Former Soviet Union, living standards were stagnant during the last third of the twentieth century, even though demographic trends in

---

[1] The first economists to study Dutch Disease and its impacts were from the Netherlands, which experienced a stronger currency and various economic distortions during the 1970s due to sharply higher prices for the country's natural gas exports.

the region differed little from those of richer nations. Communism, which the USSR adopted before the First World War ended and then imposed on captive lands behind the Iron Curtain after the defeat of Nazi Germany, clearly failed as a form of economic organization (Aslund, 2002, pp. 20–69). Compounding the problem in Russia and elsewhere has been mismanagement of the transition after communism's collapse in the early 1990s (Chapter 14). But since the transition is mostly complete, GDP per capita is rising again.

In Sub-Saharan Africa, rapid population growth has complicated the task of raising the standard of living, although it is also true that fixed capital formation and export growth have been sluggish.

Income Distribution Differences and Economic Convergence

Just as average incomes (Table 9.1, column 7) and trends in GDP per capita vary among different parts of the world, income inequality, as indicated in column 8 by Gini coefficients (Chapter 7), is not the same everywhere. At one end of the Kuznets Curve (Chapter 7) are wealthy countries with modest inequality. At the other end are poor regions, such as South Asia, where Gini coefficients tend to be low. The regional coefficient for the Middle East and North Africa is not very high either, although this figure does not reflect pronounced inequality in Saddam-era Iraq, Saudi Arabia, and other oil-exporting nations for which no measures of income distribution are available.

The highest regional Gini coefficient is for Latin America and the Caribbean. In part, this reflects the fact that, with a few exceptions, GDP per capita is higher in that region than in Africa and Asia; thus, many nations in the Western Hemisphere are near the peak of the Kuznets Curve, at which greater inequality is to be expected. However, the concentration of income in relatively few hands has been a notable feature of the region's economy for centuries.

A clear outlier is Eastern Europe and the Former Soviet Union. Inequality in Russia, where the post-communist transition often has been troubled, is now at Latin American levels (Chapter 14); nevertheless, the region's Gini coefficient is very low. Sub-Saharan Africa is another outlier, though in a less positive sense. With GDP per capita as low as it is for the region, one would expect to observe a Gini coefficient under 40. In fact, the worst of all combinations occurs south of the Sahara, with total earnings being very modest and distributed in a highly skewed fashion. Indeed, measures of economic inequality are about the same as in Latin America and the Caribbean.

On numerous occasions, economists have tested a hypothesis, called convergence, about trends in GDP per capita. The hypothesis states that living standards in various countries come together – not because of a deterioration of average earnings in better off places, but rather as a consequence of rising incomes in poor settings (Perkins *et al.*, 2001, pp. 64–67). Although statistical analysis suggests that convergence is only "conditional," in large part because of sluggish growth in places like Sub-Saharan Africa, global inequality has actually declined since the middle 1960s. This is mainly explained by sizable increases in GDP per capita in East and Southeast Asia as well as South Asia, two regions that are about as populous as the rest of the world combined and that were largely destitute a couple generations ago. Since regional Gini coefficients have not been pushed above 40 as economic expansion has taken place in Asia, economic convergence of a sort has happened (Bourguignon and Morrisson, 2002).

9.2 Population Dynamics

When obliged by the facts to admit that GDP per capita – or at least GDP – has gone up, critics of the economic perspective on human progress often fall back on the argument that positive trends in conventional economic indicators do not really reflect an improvement in the well-being of people. But when they make this contention, the critics are not on firm ground. As pointed out by Johnson (2000), many non-economic indicators of well-being, such as average food consumption and life expectancy, have been improving markedly in poor countries, quite often in the absence of comparable growth in incomes. Since non-economic indicators are of fundamental importance, it is in a sense comforting that the convergence hypothesis applies more to these variables than to GDP per capita – not the reverse.

Increased Human Longevity

Convergence of a very encouraging sort is happening for life spans in most of the world. Thanks mainly to diminished mortality among infants and small children (Chapter 2), life expectancy at birth went up by 4 years in East and Southeast Asia, 5 years in Latin America and the Caribbean, and fully 9 years in South Asia and the Middle East and North Africa (where, as already indicated, there has been no appreciable change in GDP per capita) during the last two decades of the twentieth century. Each of these increases equaled or exceeded the gains in affluent nations. In light of what life expectancies at birth were in 1980, relative progress has been greater in the four developing regions than in wealthier parts of the world (Table 9.2, columns 1 and 2).

Although newborns in Asia, Latin America and the Caribbean, and the Middle East and North Africa can now expect to live 63–70 years, scope for improvement remains in these places. In India and surrounding nations, for example, more than half the children are malnourished (Table 9.2, column 3). Also, living conditions are crowded and impoverished, as indicated by the high incidence of tuberculosis (column 4). As deficiencies of this sort are remedied, more infants and small children will survive and proceed to live longer than their ancestors.

There are exceptions to the trend toward longer life spans. In Eastern Europe and the Former Soviet Union, life expectancy at birth was a little less than 70 years in 1980 and remained at about that level at the turn of the century. Within the region, improvements have occurred in Eastern Europe. At the same time, there

Table 9.2 Regional indicators of mortality, health, and poverty

Region	Life expectancy at birth (years)		Prevalence of childhood malnutrition (percentage of population under 5 years)* 1992–1998	Incidence of tuberculosis (per 100,000) 1997
	1980	1999		
	(1)	(2)	(3)	(4)
1. High-Income Nations	74	78	–	–
2. East and Southeast Asia	65	69	22	150
3. South Asia	54	63	51	193
4. Latin America and Caribbean	65	70	8	81
5. Middle East and North Africa	59	68	15	67
6. Eastern Europe and Former USSR	68	69	8	75
7. Sub-Saharan Africa	48	47	33	267

Region	Prevalence of HIV (percentage of adults) 1999	Poverty levels (percentage of population living on $1 per day or less)	
		1987	1998
	(5)	(6)	(7)
1. High-Income Nations	–	–	–
2. East and Southeast Asia	0.22	26.6	15.3
3. South Asia	0.56	44.9	40.0
4. Latin America and Caribbean	0.58	15.3	15.6
5. Middle East and North Africa	0.03	4.3	1.9
6. Eastern Europe and Former USSR	0.18	0.2	5.1
7. Sub-Saharan Africa	8.38	46.6	46.3

Note: *The figures in this column account for wastage (weight for age) as well as stunting (height for age).
Source: World Bank (2001b), pp. 112, 116; World Bank (2001c), pp. 23, 277.

has been a steep decline, especially for men, in Russia and the Ukraine, due to the lack of meaningful employment during the difficult transition from communism, widespread alcoholism, and other factors (Powell, 2002).

While increasing death rates for some demographic cohorts have prevented life spans from going up in the Former Soviet Union, much more severe threats to life have actually caused people to die younger in Sub-Saharan Africa. As emphasized in Chapter 15, the toll taken by malaria and a host of other tropical diseases is enormous in the region. The incidence of tuberculosis is very high as well. In addition, one in every three African children is malnourished.

Most diseases carry off the young, the old, and the infirm – groups that are also more apt than adolescents and young and middle-aged adults to be poorly fed. Since the late 1970s, however, a different sort of mortal threat has emerged south of the Sahara, especially in East and Southern Africa. Spread mainly through sexual intercourse, HIV/AIDS mainly afflicts teenagers and adults under the age of 50 years or so. The devastation resulting from the wide spread of this disease is hard to exaggerate. Economies shrink and, as soldiers are infected, military power is diminished. Even the passing of skills and knowledge from one generation to the next is interrupted, as children are orphaned and as their teachers die. Needless to say, health services, which are barely able to deal with diseases that have long plagued the region, are overwhelmed, even when cheap generic drugs that control the symptoms of AIDS are available.

Mainly because the incidence of HIV/AIDS is elevated (Table 9.2, column 5), life expectancy at birth has gone down, not up, in Sub-Saharan Africa since 1980. As reported in Chapter 15, some countries have managed to bring down infection rates. One of these is Uganda, where a combination of abstinence education (especially for pre-teen and adolescent girls) in Catholic and other religious schools and the government's vigorous promotion of condom use is proving successful (WHO, n.d.). But elsewhere, there has been less progress. For Botswana, life expectancy at birth would have been 74 years in 2010 if the disease had not appeared, but now may fall as low as 27 years by the same date because the rate of infection has risen very high; the forecasts for Zimbabwe are similar. During the first decade of the twenty-first century, life expectancy at birth might decline by 10 percent for the region as a whole (Lamptey et al., 2002).

Reduced Human Fertility

As emphasized in Chapter 2, diminishing mortality sets in motion a demographic transition. To begin with, growth in human numbers accelerates, since the decline in mortality is not matched immediately by a fall in the number of births. But sooner or later, people recognize that the threat of mortal disease has receded – for infants and small children, among others – and respond by having fewer children. The resulting decline in the total fertility rate (TFR), which can be regarded as synonymous with family size, eventually causes the population to stop increasing.

Although the reduced incidence of mortality triggers the demographic transition, which is not entirely over until distribution of the population among age cohorts has stabilized (Lee, 2003), lower death rates are not the only reason for diminished fertility. The alleviation of poverty is also a causal factor. Changes in the incidence of extreme poverty (Table 9.2, columns 6 and 7) are generally consistent with trends in GDP per capita. The percentage of people in East and Southeast Asia living on $1 per day or less fell by nearly one-half between 1987 and 1998; there was a decline in South Asia as well. In contrast, this percentage did not vary in either Sub-Saharan Africa or Latin America and the Caribbean. The Middle East and North Africa, where the incidence was already below 5 percent in the late 1980s, managed to reduce extreme poverty during the 1990s, even though average living standards were not going up very rapidly. In contrast, the rising incidence of extreme poverty in Eastern Europe and the Former Soviet Union is proof that the post-communist transition has been far from perfect in that region.

As living standards rise and people emerge from poverty, more resources are devoted to the formation of human capital. Especially when females benefit from this investment, TFRs move toward or even below the replacement level, of 2.1 births per woman. As reported in Table 9.3, literacy, which is a fundamental indicator of human capital, increased during the last decade of the twentieth century throughout the developing world – even south of the Sahara. However, the numbers reported in the table's first four columns reflect pronounced inequalities between males and females. Latin America and the Caribbean are exceptional in this regard, as are Eastern Europe and the Former Soviet Union. Also, the gap between female and male illiteracy in East and Southeast Asia is falling, although it remains sizable. Elsewhere, the gap is huge. Illiteracy rates for women are nearly double the rates for men in the Middle East and North Africa, Sub-Saharan Africa, and South Asia. In all three of these regions female rates are close to or over 50 percent. This not only hinders the economic empowerment of women, as is needed to reduce human reproduction, but holds back expansion of the economy as a whole, which suffers from being denied the full talents and capabilities of half the adult population.

The future state of human capital in different parts of the world is indicated by data reported in columns 5–10 of Table 9.3. Outside of Sub-Saharan Africa, gross enrollment rates, which equal the number of students attending primary, secondary, or tertiary (i.e., university level) school divided by the relevant age group,[2] have risen – dramatically so in some places. It is becoming normal for children to attend secondary school, and even to earn a high school diploma, in Latin America and the Caribbean as well as the Middle East and North Africa, which are both highly urbanized. The same holds for East and Southeast Asia, even though much of that

---

[2] Since students above the normal age range can, for instance, attend primary school, gross enrollment rates can exceed 100 percent. An alternative measure is the net enrollment rate, which equals the share of the relevant age cohort enrolled in school and which therefore is always less than 100 percent. While gross enrollment rates have been estimated for practically every nation, net rates are unavailable for a number of countries.

Table 9.3 Regional indicators of human capital

Region	Males		Females	
	1990	1999	1990	1999
	(1)	(2)	(3)	(4)
(A) Adult illiteracy rates (percentage aged 15 years and over)				
1. High-Income Nations	–	–	–	–
2. East and Southeast Asia	13	8	29	22
3. South Asia	41	34	66	58
4. Latin America and Caribbean	14	11	17	13
5. Middle East and North Africa	33	25	59	47
6. Eastern Europe and Former USSR	2	2	6	5
7. Sub-Saharan Africa	40	31	60	47

Region	Primary		Secondary		Tertiary	
	1980	1997	1980	1997	1980	1997
	(5)	(6)	(7)	(8)	(9)	(10)
(B) Gross enrollment ratios (percentage of relative age groups)						
1. High-Income Nations	102	103	87	106	36	62
2. East and Southeast Asia	111	119	44	69	4	8
3. South Asia	77	100	27	49	5	7
4. Latin America and Caribbean	105	113	42	60	14	17
5. Middle East and North Africa	87	95	42	64	11	16
6. Eastern Europe and Former USSR	99	100	86	–	31	32
7. Sub-Saharan Africa	81	78	15	27	1	–

Source: World Bank (2001b), pp. 88, 96.

region is not very prosperous and a large portion of the population is rural. South Asia, which in spite of recent growth still has very low living standards, has made impressive strides, with gross enrollment at the primary and secondary levels reaching 100 percent and 49 percent, respectively, in 1997. In contrast, there appear to be widening inequities south of the Sahara. Secondary enrollment rates, almost entirely in urban areas, increased between 1980 and 1997. Meanwhile, primary rates actually fell, due mainly to educational deterioration in the countryside. This is one of several facets of the socioeconomic marginalization that rural Africa is suffering.

Although economic expansion is difficult in the face of persisting and high illiteracy, human capital formation does not guarantee economic progress (Easterly, 2001, pp. 71–84). In Eastern Europe and the Former Soviet Union, educational spending was considerable, including at the tertiary level, under the communists. However, this investment did not result in the region's living standards converging toward western levels, which means that the economic returns to education were negligible under communism. The absence of market incentives kept GDP per capita stagnant, even for an adult population with high school diplomas and university degrees.

If higher female literacy truly reflects greater economic empowerment for women, the chances are very good that human fertility will fall. Already low in 1980, TFRs in wealthy nations show few signs of increasing (Table 9.4, columns 1 and 2). Since the average Chinese woman delivers fewer than two children, the TFR for East and Southeast Asia stands at the replacement level of 2.1 births per woman – a level that is being approached quickly in Latin America and the Caribbean. In spite of low living standards, human fertility in India and the surrounding region plummeted during the 1980s and 1990s. The same thing happened in the Middle East and North Africa, even though the socioeconomic status of women leaves much to be desired in many parts of the region. A decline has even occurred in Sub-Saharan Africa, although the regional TFR still exceeds five births per woman.

In places like Sub-Saharan Africa, elevated human fertility, in general, coincides with a high fertility rate for adolescents (Table 9.4, column 3), in particular. As a rule, this reflects a low average age of marriage and family formation. The situation is considerably different in East and Southeast Asia, where women tend to marry and start having babies at an older age; as a result, the region's adolescent fertility rate is about equal to the rate for wealthy countries. Of course, women who desire small families because they are urbanized, have put poverty behind them, can take advantage of post-primary education, and regard infant mortality as a remote risk act on that desire by using contraceptives. In contrast, a large portion of the women who live south of the Sahara are poor and uneducated and many of them reside in rural areas, where a large family can be advantageous. Under these circumstances, contraceptive prevalence is low (column 4).

Natural Increase

As pointed out in Chapter 2, human numbers do not stop growing once mortal disease has been contained and women are having just enough children to replace themselves and their male partners. Instead, population growth continues because of demographic momentum. However, the plummeting of human fertility that has taken place in so many parts of the world since 1980 has led to a sharp deceleration of natural increase, which is the difference between the crude birth rate (CBR) and the crude death rate (CDR) and which comprises all but a tiny fraction of population growth in most regions (Table 9.4, columns 5–8).

In two sets of countries, the difference between birth and death rates is negligible. One of these is the group of high-income nations, where TFRs have been below the replacement level for more than a generation. With CBRs barely above CDRs, immigration from poorer settings accounts for roughly half of all population growth. Without this immigration, which mainly adds to the working-age cohort, the age dependency ratio (Table 9.4, columns 9 and 10), which equals the number of people who are either below or above normal working age divided by the remainder

Table 9.4 Regional population trends, 1980s and 1990s

Region	TFR (births per woman)		Adolescent fertility rate (births per 1,000 women aged 15–19 years) 1999	Contraceptive prevalence rate (percent of women aged 15–49 years) 1990–1999
	1980	1999		
	(1)	(2)	(3)	(4)
1. High-Income Nations	1.8	1.7	25	75
2. East and Southeast Asia	3.0	2.1	27	57
3. South Asia	5.3	3.4	110	49
4. Latin America and Caribbean	4.1	2.6	73	59
5. Middle East and North Africa	6.1	3.5	52	52
6. Eastern Europe and Former USSR	2.5	1.6	39	64
7. Sub-Saharan Africa	6.6	5.3	130	21

Region	Population growth (average annual percent)		CDR (per 1,000) 1999	CBR (per 1,000) 1999	Age-dependency ratio (dependents/working-age population)	
	1980–1990	1990–1999			1980	1999
	(5)	(6)	(7)	(8)	(9)	(10)
1. High-Income Nations	0.6	0.6	9	12	0.5	0.5
2. East and Southeast Asia	1.6	1.3	7	18	0.7	0.5
3. South Asia	2.2	1.9	9	27	0.8	0.7
4. Latin America and Caribbean	2.0	1.7	7	23	0.8	0.6
5. Middle East and North Africa	3.1	2.2	7	26	0.9	0.7
6. Eastern Europe and Former USSR	0.9	0.2	11	12	0.6	0.5
7. Sub-Saharan Africa	2.9	2.6	16	40	0.9	0.9

Source: World Bank (2001c), pp. 108, 179; World Bank (2001b), p. 46.

of the population, would be even higher than it now is. The other part of the world where natural increase is coming to a halt is Eastern Europe and the Former Soviet Union, in which human fertility has dropped far below the replacement level and life spans are shorter for some parts of the population.

In most of the developing world, population growth, though still greater than 1.0 percent per annum, is clearly diminishing. Although the TFR for East and Southeast Asia reached the replacement level at the turn of the twenty-first century, the region continues to experience increasing human numbers mainly because of demographic momentum. During the next few decades, the population of China and nearby countries will age, which will result in a lower CBR and a slight increase in the CDR. The age dependency ratio of East and Southeast Asia already looks like that of an affluent place, although children are a slightly larger segment of the dependent group. Similarly, stabilization of human numbers is not too far off for Latin America and the Caribbean. In South Asia and the Middle East and North Africa, family size fell dramatically during the late twentieth century. But since TFRs remain above three births per woman, these two regions still have several years to go before population growth will be driven exclusively by demographic momentum.

As in so many other demographic and economic categories, Sub-Saharan Africa stands apart. Since declines in human fertility have been modest and since the toll taken by HIV/AIDS and other diseases has been so staggering, natural increase has only eased off slightly since 1980 and annual population growth is still well above 2 percent. Also, the age-dependency ratio, which is the highest in the world, indicates that demographic momentum will be substantial for many years to come – for the simple reason that children, not the elderly, comprise the vast majority of the dependent population. HIV/AIDS might cut back on demographic expansion. However, it is more likely that natural increase will continue for many more decades.

9.3 Agriculture's Response to Demand Growth

As observed in Chapter 3, growth in the demand for food almost always used to lead to a geographic expansion of farming. But during the last century or two, land has become a progressively less important factor of agricultural production. In recent decades, increases in demand, which are clearly influenced these days by income trends although population continues to be the main driving variable, have caused farmers to increase their use of fertilizer, irrigation, machinery, and other non-land inputs.

Changes in the Mix of Inputs

In any particular setting, changes in the input mix undertaken to raise farm output depend on the relative scarcities of different factors of production (Hayami and Ruttan, 1970). Where labor is much scarcer than land, mechanization (i.e., increased use of tractors and other equipment, along with the fuel to drive this machinery) occurs, thereby increasing what each agricultural worker produces. Elsewhere, land is scarcer than labor, which encourages irrigation as well as increased use of fertilizer, improved seeds, and other purchased inputs. Yields (i.e., output per hectare) go up as a result.

These two paths to increased production can be discerned from regional patterns of factor use. Since rural population density is low (Table 9.5, column 1), mechanization, which is measured by the number of tractors per square kilometer of arable land (columns 2 and 3), is far advanced in high-income nations as well as Eastern Europe and the Former Soviet Union. Although still high in the latter region, tractor use has fallen as communist-era subsidies have been cut and increased exposure to market forces has obliged farmers to use inputs more efficiently. Outside of the Caribbean and Central America, rural population densities are not very high south of the US border, which has led to intensive use of machinery in many countries. One consequence of mechanization in Latin America has been rapid deforestation (column 4), because an individual with a tractor and other equipment can clear land faster than a farmer using a machete and fire.

Elsewhere in the developing world, rural population densities are higher, which implies that land is relatively scarce and incentives are strong to irrigate, fertilize, and use high-yielding varieties. More than one-third of all the arable land in Asia is irrigated (Table 9.5, column 5). Also, fertilizer use has increased in China, India, and neighboring countries (columns 6 and 7). Indeed, with fertilizer applications having eased off in affluent nations and fallen dramatically in Eastern Europe and the Former Soviet Union, East Asia now has the highest application rate in the world. Between widespread irrigation and intensive use of chemical inputs, the region's average cereal yield is high, exceeded only by that of high-income countries (column 8). Except for a few places, virtually all land that is well suited to crop production is being farmed and, with the intensification of agriculture that is happening in the region, deforestation is abating throughout Asia.

Input use in the Middle East and North Africa, where no important opportunities exist for agricultural extensification, is a little anomalous. Rural population density there is nearly identical to South Asia's and not much lower than that of East and Southeast Asia, which suggests that yield-enhancing technology should be adopted. But while the irrigated portion of arable land in the region is high, fertilizer application rates are not. Furthermore, there are more tractors per square kilometer of farmland in the Middle East than in any other group of countries, save Eastern Europe and the Former Soviet Union as well as affluent nations. All of this is probably explained by extraordinarily high population densities and correspondingly low mechanization in some places (e.g., the Nile River Valley) and highly mechanized (and subsidized) farming in oil exporters, like Saudi Arabia.

In agriculture, as in economic performance and the demographic transition, Sub-Saharan Africa lags the rest of the developing world. Although its rural areas are not as densely populated as the Asian and Middle Eastern countryside, machinery use, which was limited in the early 1980s, is on the decline. Irrigation development remains at an incipient stage, and fertilizer applications are below

Table 9.5 Regional trends in the use of land and other agricultural inputs

Region	Rural population density (people per sq. km of arable land) 1998	Agricultural machinery (tractors per sq. km of arable land)		Change in forest area (annual percent) 1990–2000
		1979–1981	1996–1998	
	(1)	(2)	(3)	(4)
1. High-Income Nations	175	3.9	4.3	0.1
2. East and Southeast Asia	691	0.6	0.7	−0.2
3. South Asia	537	0.3	0.9	−0.1
4. Latin America and Caribbean	252	1.0	1.2	−0.5
5. Middle East and North Africa	534	0.6	1.2	0.1
6. Eastern Europe and Former USSR	125	2.2	1.7	0.1
7. Sub-Saharan Africa	369	0.2	0.2	−0.8

Region	Irrigated percentage of cropland 1996–1998	Fertilizer consumption (kg per ha of arable land)		Cereal yield (kg per ha) 1998–2000
		1979–1981	1996–1998	
	(5)	(6)	(7)	(8)
1. High-Income Nations	11.2	131	126	4,002
2. East and Southeast Asia	37.1	116	233	2,982
3. South Asia	40.8	36	98	2,274
4. Latin America and Caribbean	13.7	59	81	2,488
5. Middle East and North Africa	36.2	42	70	1,411
6. Eastern Europe and Former USSR	10.4	145	34	2,407
7. Sub-Saharan Africa	4.2	16	14	1,120

Source: World Bank (2001b), pp. 128, 136, 140, 132.

levels required to maintain soil fertility. As emphasized in Chapter 5, impoverished farmers south of the Sahara subsist by mining nutrients in one setting, clearing bush and forests by hand somewhere else, and then repeating the cycle. Rapid deforestation in the region is symptomatic of this way of life.

Production Trends

As indicated at the beginning of Chapter 3, temperate zones are very good settings to produce corn, wheat, and other grains, which directly or indirectly comprise much of the human diet. In the world's high-income nations, which mainly have temperate environments, mechanization (to economize on the employment of scarce labor), fertilization, and (where appropriate) irrigation have been drawn on to raise yields to levels unmatched in any other part of the world. In places like Australia and North America, growth in human numbers (Table 9.6, column 1) has been outstripped by increases in food supplies (column 2), generally, and livestock output (column 3), specifically.

Eastern Europe and parts of the Former Soviet Union also enjoy the agricultural advantages of a temperate environment. However, these advantages were negated by the imposition of collective farming – an imposition from which the region is still recovering. Production per hectare is little more than half the yields achieved in wealthy nations. While there is considerable potential for agricultural recovery in places like the Ukraine, the urgency of this task is reduced by the absence of demographic expansion.

Having benefited substantially from the Green Revolution, Asia is another setting where food supplies have increased faster than the population, with the greatest difference between the two trends in China and the surrounding region. In addition, the agricultural sector has responded to the growing demand for livestock products,

Table 9.6 Regional trends in agricultural output during 1990s

Region	Population growth (%) (1)	Food production growth (%) (2)	Livestock production growth (%) (3)
1. High-Income Nations	6	13	10
2. East and Southeast Asia	14	56	98
3. South Asia	21	26	36
4. Latin America and Caribbean	19	31	32
5. Middle East and North Africa	25	34	37
6. Eastern Europe and Former USSR	2	–	–
7. Sub-Saharan Africa	30	25	14

Source: World Bank (2001c), p. 279; World Bank (2001b), p. 36.

resulting from growth in GDP per capita. Aside from a few impoverished and densely populated places in the Caribbean Basin, hunger in the Western Hemisphere has little to do with agricultural limitations. The resources available for crop production are considerable, and not fully exploited, and intensification has occurred. Consequently, yields are not much below those of East and Southeast Asia and increases in food and livestock output comfortably exceed population growth. Trends are similarly encouraging in the Middle East and North Africa, which is another region where the Green Revolution has had a positive impact.

The situation in the world's second-largest continent is alarming. Human numbers continue to shoot up south of the Sahara. There is very little irrigation and fertilization, so yields are abysmal, and farmers are encroaching rapidly on forests and other habitats. Sub-Saharan Africa is the only part of the world where the population is racing ahead of food supplies. Furthermore, GDP per capita has been stagnating or declining in all but a few places. Hence, incentives to increase livestock production have been weak.

9.4 Summary

Although the food economies of various parts of the world are linked, primarily through international trade, demand and supply trends in each region are distinctive in at least some respects. In places like the United States and Western Europe, where living standards are high and natural increase is winding down, overeating is a more serious problem than hunger. Agriculture, which employs a small share of the labor force because of mechanization and attractive employment options off the farm, is highly productive in many settings, thanks to temperate conditions and the use of improved seeds and other purchased inputs. Affluent countries are the source of most grain exports and all but a tiny share of food aid.

The developing world's stellar performer has been East and Southeast Asia. Rates of savings and investment are high in the area, and an export-oriented approach to industrialization and development has been pursued (Chapter 7). Very low in the middle 1960s, regional GDP per capita has risen at a fast clip since then, which has had a substantial impact on the consumption of food, generally, and of livestock products, specifically. Human capital formation has accelerated and increases in population have decelerated. Growth of non-agricultural sectors has driven up the opportunity costs of labor, capital, and other factors of production used on farms. Agriculture has responded by intensifying, although it is becoming internationally uncompetitive in some of the region's wealthier countries.

Although South Asia is much poorer than East and Southeast Asia, trends in the former region are encouraging. Thanks mainly to a marked decline in TFRs, human numbers are not going up nearly as quickly as they were a generation ago. Nowadays, regional growth in GDP per capita is exceeded only by that of East and Southeast Asia. In addition, agricultural intensification has occurred, during the Green Revolution and since then, thereby alleviating food insecurity.

In many ways, the economic performance of Latin America and the Caribbean has been disappointing. When compared to what other parts of the developing world possess, the region's endowment of natural resources is abundant. Illiteracy is low and population growth is not particularly high. Moreover, living standards south of the US border in the middle of the twentieth century were the envy of Africa and Asia and are still higher today. More than being a simple case of economic convergence, the failure of Latin America and Caribbean to match the growth in GDP per capita that Asia has experienced is largely a consequence of profligate monetary and fiscal policies, protectionism, low savings rates, and pronounced inequality. However, favorable environmental conditions have allowed agricultural output to grow faster than food demand in much of the Western Hemisphere.

In the Middle East and North Africa as well as Eastern Europe and the Former Soviet Union, living standards have not improved for a generation or more. In the former region, rapid population growth has put a brake on increases in GDP per capita. But in both places, economic expansion has been held back by sizable military expenditures and the distortion of incentives by autocratic states reluctant (to say the least) to give market forces free rein. Something positive to say about the two regions is that extreme poverty has been largely avoided and, in Eastern Europe, income inequality has been and remains modest because education is widely available. Also, the former USSR and its satellites have considerable potential for agricultural recovery.

As emphasized throughout this chapter, the trends are much less encouraging in Sub-Saharan Africa, where living standards have slipped below those of South Asia. Population growth is unparalleled, as is deforestation. In contrast, agricultural development is stymied in many parts of the continent and it is difficult to envision an end to widespread food insecurity, poverty, and illness. On top of everything else, HIV/AIDS is devastating the region.

Poverty and food insecurity, which are inextricably linked, often coincide with environmental degradation, insofar as people who are poor and malnourished are often concentrated where land and other resources are susceptible to erosion and other forms of degradation. As reported in columns 1–3 of Table 9.7, small portions of the populations of wealthy nations, Eastern Europe and the Former Soviet Union, and even Latin America and the Caribbean live in fragile environments. The shares for East, Southeast, and South Asia are much higher. Higher still are the shares for the Middle East and North Africa (38 percent) and Sub-Saharan Africa (39 percent). Especially in the latter region, development is a considerable challenge, one that involves improving the livelihoods of rural people while simultaneously safeguarding the natural resources they must now exploit for survival's sake.

Exacerbating the challenge in Africa have been war and corruption as well as crime on a scale that can cause civil society to collapse. The region suffers from a long history of acute ethnic divisions, a colonial legacy of borders that neglected these divisions and of poor preparation for nationhood, as well as the emergence of repressive regimes after independence. As a result, state legitimacy does not really exist in many places. The rate at which lives have been lost to armed conflict in

Table 9.7 Human occupation of fragile environments and history of civil conflict, by region

Region	Population in 2000 (millions) (1)	Population on fragile lands by region	
		Number (millions) (2)	Share of total (%) (3)
(A) Regional distribution of people living on fragile land			
1. High-Income Nations	850.4	94	11.1
2. East and Southeast Asia	1,856.5	469	25.3
3. South Asia	1,354.5	330	24.4
4. Latin America and Caribbean	515.3	68	13.1
5. Middle East and North Africa	293.0	110	37.6
6. Eastern Europe and Former USSR	474.7	58	12.1
7. Sub-Saharan Africa	658.4	258	39.3
Other	27.3	2	6.9
Total	6,030.1	1,389	24.7
Total less high-income nations	5,179.7	1,295	26.9

Region	Total civil conflict deaths per year 1944–1996 (4)	Total homicides per year (5)
(B) Civil conflict and reported homicides (per 100,000 people)		
1. High-Income Nations	10.3	2.3
2. East and Southeast Asia	10.1	0.6
3. South Asia	7.4	1.9
4. Latin America and Caribbean	3.2	4.5
5. Middle East and North Africa	9.4	1.2
6. Eastern Europe and Former USSR	3.4	1.8
7. Sub-Saharan Africa	30.3	7.4

Source: World Bank (2003), pp. 61, 155.

Sub-Saharan Africa from 1944 through 1996 has been triple the rate for high-income nations and East and Southeast Asia (Table 9.7, column 4) – even though these were the main settings for the Second World War and the Korean and Vietnam Wars. Likewise, Africa's homicide rate (column 5) is more than 50 percent above Latin America's, which in turn is double the rate for rich countries. Homicide is often a more severe problem in cities and towns. In contrast, most of Africa's wars – the decades long insurrections in Angola, Mozambique, and Sudan, for example, as well as the ancient vendettas plaguing Burundi and Rwanda – have been fought primarily in the countryside. The horrific consequences for rural people and the food economy are obvious.

Study Questions

1. Are trends in GDP per capita in different parts of the world consistent with the hypothesis of economic convergence?
2. Is there any evidence that life expectancies at birth in various parts of the world are converging?
3. Relate regional differences in female illiteracy to regional variation in human fertility?
4. Explain regional differences in mechanization and intensification in terms of varying factor endowments?

10

Affluent Nations

The classic novel, *Anna Karenina*, begins with the observation that happy families are all alike, but every distressed family is unhappy in its own way (Tolstoy, 1992, p. 1). The same can be said of entire countries. Nations in which large segments of the population are below, at, or barely above the international poverty line (i.e., per-capita income of $2 per day) are heterogeneous. While most are tropical, a few are in the temperate latitudes. Culture and historical experience vary widely. Of particular concern to this book, living standards and food security are improving in some parts of the developing world while other parts remain mired in poverty, with much of the population frequently or chronically hungry. In contrast, nations where most people are comfortably middle-class closely resemble one another. Without exception, most goods and services are allocated in markets, political democracy is well established, and few people go hungry. Aside from a handful of special cases – commercial city states with highly educated populations, such as Hong Kong and Singapore, and tiny places with enormous deposits of petroleum or natural gas, such as Brunei, Kuwait, and Qatar – all developed countries are located far from the equator.[1] Nearly all are western, the most important exception being Japan. But although its heritage is manifestly non-European, the cultural and institutional antecedents for commercial exchange and political democracy are no less entrenched there than in the West.[2]

As reported in the first part of this chapter, GDP per capita has not stopped rising in affluent parts of the world. However, family size has plummeted, thereby causing human numbers to stabilize, or even shrink. To say the least, food shortages are a remote concern, either because domestic output exceeds domestic consumption or because food imports are easily paid for with non-agricultural exports. Instead of

---

[1] As indicated in Chapter 3, temperate zones have various advantages over tropical settings, including the diminished incidence of communicable diseases and a winter for cutting into populations of agricultural pests. Admonishing development specialists not to forget that climate and other geographic factors are important, Landes (1998, pp. 17–22) points out the myriad environmental advantages enjoyed by Western Europe, where temperate isotherms extend farther from the equator than in any other part of the world.

[2] The same observations apply to Hong Kong, Singapore, South Korea, and Taiwan, which are discussed in the next chapter.

worrying about undernourishment, most people have good reason to fret about eating too much. Furthermore, government policies create incentives for excessive farm production, which weighs heavily on the world's economy.

10.1 Standards of Living

Living standards in places like North America and Western Europe do not merely exceed those of the developing world. Rather, differences of at least an order of magnitude are the rule. Moreover, aside from Asia, which in recent decades has enjoyed the greatest relative improvement in real GDP per capita, the record of advanced, capitalist nations – average increases of 2.4 percent per annum from 1965 through 1999 – is superior to those of other regions. Living standards in Eastern Europe and the Former Soviet Union at the turn of the twenty-first century were no better than what these had been 35 years earlier. The Middle East and North Africa stagnated as well. Even worse, average income deteriorated south of the Sahara during the last third of the twentieth century. The modest improvement experienced by Latin America and the Caribbean fell short of what OECD members achieved (Table 9.1).

As indicated in Table 10.1, annual increases in real GDP per capita generally have been between 1.0 and 2.5 percent in the developed world. Three exceptions during the 1980s were Japan, the United Kingdom, and New Zealand, with growth rates of 3.4, 3.0, and 0.7 percent, respectively (column 1). In Britain, Prime Minister Margaret Thatcher pressed to end policies, including inflationary monetary management, that were discouraging private enterprise. Gross investment (column 4) accelerated as a result, to the benefit of living standards. In contrast, reform did not take place in New Zealand until the middle of the decade, as indicated by the country's having the highest rate of inflation (column 3) in the OECD. Since the risks and uncertainties created by inflation discourage capital formation and since export growth (column 2) was disappointing, economic expansion was anemic.

During the 1980s, Japan seemed to be yielding new lessons about growth and development. Its industrial policy, with the Ministry of International Trade and Investment (MITI) guiding financial resources to firms and sub-sectors supposedly poised for take-off, was extolled as a model for the United States (Thurow, 1992, p. 37), where emphasis was placed on bringing down tax rates and easing the regulatory burden on business during Ronald Reagan's Presidency. However, the Japanese economic paradigm subsequently lost its luster. After soaring during the 1980s, prices for real estate and other assets collapsed during the 1990s. This crippled a number of large banks, which were heavily burdened with bad debts in part because industrial policy had tempted politicians to meddle with banking decisions (Neely, 1999). Due to Japan's financial crisis, gross investment was disappointing during the last decade of the twentieth century (column 8), even though there was no inflation (column 7) and even though Japanese households consistently have saved large shares of total earnings. In the absence of capital formation, growth in

Table 10.1 Patterns of economic growth for selected OECD countries, 1980–2003

	A. Average annual rate of growth 1980–1990 (%)				B. Average annual rate of growth 1990–1999 (%)					C. Average annual rate of growth 2000–2003 (%)	
Country (ranked by column 1)	GDP per capita (1)	Exports (2)	Inflation (3)	Gross investment (4)	Country (ranked by column 5)	GDP per capita (5)	Exports (6)	Inflation (7)	Gross investment (8)	Country (ranked by column 9)	GDP per capita (9)
Japan	3.4	4.5	1.7	5.7	Australia	2.6	7.9	1.6	6.1	New Zealand	2.5
United Kingdom	3.0	3.9	5.7	6.4	Denmark	2.4	3.8	1.7	4.8	United Kingdom	2.3
Spain	2.6	5.7	9.3	5.7	United States	2.4	9.3	1.8	7.0	Spain	2.3
Denmark	2.3	4.3	5.6	3.7	The Netherlands	2.1	4.8	2.1	1.5	Australia	2.0
Italy	2.3	4.1	10.0	2.0	Spain	2.0	10.9	4.0	−0.5	Canada	2.0
Canada	2.1	6.3	4.5	4.9	United Kingdom	1.9	6.0	2.9	1.8	Sweden	2.0
Germany	2.1	–	–	2.4	New Zealand	1.7	5.4	1.5	8.1	France	1.5
United States	2.1	4.7	4.2	4.4	Canada	1.2	8.8	1.3	2.6	USA	1.3
Sweden	2.0	4.3	7.4	4.2	France	1.2	4.9	1.5	−3.2	Italy	1.3
Australia	1.9	6.9	7.3	3.0	Germany	1.1	4.1	2.0	0.5	Denmark	1.0
France	1.8	3.7	6.0	2.6	Japan	1.1	5.1	0.1	1.1	Germany	1.0
The Netherlands	1.7	4.5	1.6	2.3	Sweden	1.1	8.3	2.1	−2.2	The Netherlands	0.5
New Zealand	0.7	4.0	10.8	4.4	Italy	1.0	7.2	4.1	−1.0	Japan	−0.3

Source: World Bank (2000/2001), pp. 278–279 and 294–295; World Bank (2005).

GDP per capita has been sluggish since 1990 (columns 5 and 9), in spite of export performance (column 6) within the normal range for OECD members.

Along with Japan, a few countries have moved up or down noticeably in growth-rate rankings shown in Table 10.1. Australia, for example, has made a positive change, thanks to sharply lower inflation and accelerated investment. Meanwhile, economic expansion has flagged in some of the leading economies in continental Europe, including France, Germany, and Italy. Spain's experience has been different. The country joined the European Common Market – the precursor to the European Union (EU) – in the early 1980s. Since then, its GDP per capita has risen faster than average incomes in other EU members. This squares well with the qualified, or conditional, view of economic convergence. To be specific, among a group of countries sharing the same characteristics – in this case, membership in the same common market – living standards in the poorer countries rise toward those of wealthier members of the group (Chapter 9).

By OECD standards, the United States has enjoyed robust growth in GDP per capita since the early 1980s, when it suffered its sharpest economic downturn since the Great Depression. With inflation contained, gross investment increased at a good clip during the 1980s and accelerated during the 1990s, when there was a boom in high technology. The country's low savings rate has been compensated for by highly efficient capital markets, which reliably direct funds to their highest and best use, as well as sizable inflows of money from Japan and other countries where economic expansion was slow as the twentieth century drew to a close. Funds are attracted to the United States in part by the rule of law, which protects all investors – foreign and domestic alike. Although GDP per capita has not improved as quickly since 2000, the world's leading economy enjoys faster relative growth than many other developed nations to this day.

There is also more economic inequality in the United States than in other OECD members, its Gini coefficient being 40 or so (World Bank, 2004). In part, this traces to failing schools in inner cities and other impoverished settings and gaps in the social safety net. Economic inequality also reflects a heterogeneous population, growing out of the nation's openness to immigrants – many of whom lack English fluency and other skills required for remunerative employment. It must also be remembered that any Gini coefficient is a static measure – specifically, a snapshot of inequality at one particular date. Thus, it does not necessarily reveal much about social mobility – which is normally measured in terms of changes in annual earnings during a single person's lifetime or income changes within a family from one generation to the next. Social mobility is constrained in places like Japan, which has a Gini coefficient of 25, because children are sorted at a young age between those who will eventually get into university and people whose education will end in technical school. Education in the United States is much more flexible, even allowing for adults whose performance in secondary school was mediocre to earn university degrees. This obviously enhances social mobility.

In any event, inequality in the United States that is high by OECD standards still compares favorably with the yawning gaps between the rich and the poor in Latin

America and the Caribbean as well as Sub-Saharan Africa, not to mention Russia and a number of other countries outside the developed world.

10.2 Population Dynamics

Living standards that are the envy of the rest of the world are matched in the OECD by long life spans and the rarity of infant deaths. More than 99 percent of all the babies born in affluent nations live beyond their first birthday, and life expectancies at birth are around 75 years for men and 80 years or more for women (World Bank, 2004). Instead of succumbing to infectious diseases, as do many Africans, Asians, and Latin Americans, most people in places like the United States die of cardiovascular disease and cancer. While much of this mortality is an inevitable consequence of old age, some of it is linked to bad habits – smoking, of course, as well as overeating.

Due to the progress made against mortal illness, the demographic transition in countries belonging to the OECD is very far advanced. By 1980, human fertility (Table 10.2, column 1) had already fallen below the replacement level throughout the developed world, Spain and a few other nations being minor exceptions. Since then, total fertility rates (TFRs) have risen a little in the United States (where an influx of immigrants inclined to have larger families has had an impact), Denmark, and the Netherlands. But in all other countries listed in Table 10.2, family size in 2002 (column 2) was the same as or lower than it had been two decades earlier. In Spain, for example, the TFR plummeted after 1980, reaching 1.3 births per woman at the turn of the century.

Human fertility that is low, declining, or both is to be expected where infant mortality is limited, living standards are very high (column 3) and a clear majority of the population is urban (column 4). More than anything else, small family size is an outcome of economic opportunities for women that are generally comparable to those enjoyed by men. As reported in columns 5 and 6, the female portion of the OECD workforce is rising and currently approaches 50 percent in a number of countries. In addition to driving total and per-capita GDP to levels that could not be achieved if women were denied equal economic status, female empowerment raises the opportunity cost of bearing and raising children. Recognizing these costs, the vast majority of women of childbearing age use contraceptives, regardless of what some religious leaders have to say on the subject. Indeed, contraceptive use is so prevalent that documentation is not readily available for much of the OECD.

Due to demographic momentum, crude birth rates (CBRs) still slightly exceed crude death rates (CDRs) (Table 10.3, columns 3 and 2, respectively) in most nations where GDP per capita is over $20,000. Australia, New Zealand, and the United States have the highest rates of natural increase (column 4) in the OECD. In contrast, a generation or more of TFRs consistently below the replacement level have been enough to cause birth rates to slip below death rates in Germany and Italy. Other countries are about to join this group. One of these is Spain, where (as in Italy) most

Table 10.2 Human fertility and its determinants in selected OECD countries, 1980 and 2002

Country (ranked by column 3)	Total fertility rate		Income per capita, 2002 PPP ($)	Urban % of total population 2002	Female % of labor force	
	1980 (1)	2002 (2)	(3)	(4)	1980 (5)	2002 (6)
United States	1.8	2.1	35,750	78	41.0	46.2
Denmark	1.5	1.7	30,940	85	44.0	46.5
Canada	1.7	1.5	29,480	79	39.5	46.0
The Netherlands	1.6	1.7	29,100	90	31.5	40.9
Australia	1.9	1.8	28,260	91	36.8	44.0
Germany	1.4	1.4	27,100	88	40.1	42.4
Japan	1.8	1.3	26,940	79	37.9	41.7
France	1.9	1.9	26,920	76	40.1	45.3
Italy	1.6	1.3	26,430	67	32.9	38.7
United Kingdom	1.9	1.7	26,150	90	38.9	44.3
Sweden	1.7	1.6	26,050	83	43.8	48.1
New Zealand	2.0	1.9	21,740	86	34.3	45.2
Spain	2.2	1.3	21,460	78	28.3	37.5

Source: World Bank (2004).

Table 10.3 Natural increase in selected OECD countries, 2002

Country (ranked by column 1)	Income per capita, 2002 PPP ($)	CDR, 2002 (per 1,000 population)	CBR, 2002 (per 1,000 population)	Rate of natural increase 2002 (%)	Average annual population growth 1980–2002 (%)
	(1)	(2)	(3)	(4)	(5)
United States	35,750	9	14	0.5	1.1
Denmark	30,940	11	12	0.1	0.2
Canada	29,480	7	11	0.4	1.1
The Netherlands	29,100	9	12	0.3	0.6
Australia	28,260	7	13	0.6	1.3
Germany	27,100	10	9	−0.2	0.2
Japan	26,940	8	9	0.1	0.4
France	26,920	10	13	0.3	0.5
Italy	26,430	11	9	−0.2	0.1
United Kingdom	26,150	10	11	0.0	0.2
Sweden	26,050	11	11	0.0	0.3
New Zealand	21,740	7	14	0.7	1.0
Spain	21,460	9	10	0.1	0.4

Sources: World Bank (2002a), pp. 48–50; World Bank (2004).

people are identified as Roman Catholics. Denmark, Sweden, and the United Kingdom are also on the verge of demographic contraction. So is Japan.

With human numbers stabilizing in the developed world, its share of the global population is falling. Equal to 15 percent in 2000, this share is projected to reach 11 percent of the global total (10 billion or so) midway through the twenty-first century (UNPD, 2003). Furthermore, as natural increase converges toward (or perhaps below) zero, demographic transition in wealthy nations will have to do mainly with shifts in the age structure of the population, with the elderly making up an ever-larger share of the total. To be specific, low TFRs combined with longer life spans are expected to raise the age-dependency ratio from 0.5 in 2000 to 0.7 in 2050 (UNPD, 2003).

The demographic flux unleashed if human fertility is low for a generation or more leaves no socioeconomic reality unaffected. More is spent on health care, obviously. If retired people use pension income and accumulated savings to buy imported goods, the trade balance deteriorates. Also, foreign workers might have to be recruited, so that pensions for an aging population are adequately funded from tax revenues. Immigration is readily accomplished in North America and Australia, which have long traditions of assimilating newcomers from various parts of the world. The same cannot be said, though, of many parts of Europe. There, the arrival of foreigners, which inevitably alters the national identity, sometimes creates social tensions. To this day, there are relatively few immigrants in Japan, where national identity and ethnicity are tightly intertwined.

10.3 The Food Economy

As with every other sector, the food economy is profoundly altered by the achievement of high living standards and the maturing of the population. For one thing, demand for environmental quality is positively tied to average incomes. Accordingly, the discharge of agricultural wastes into rivers, lakes, and streams is circumscribed. Similarly, natural landscapes that provide scenic views and other amenities are less likely to be converted to cropland and pasture. For that matter, "multifunctionality" – by which is meant that rural holdings are not simply a source of commodities, but of environmental and lifestyle values as well – becomes a part of the farm policy lexicon.

No less than land and other natural resources used by agriculture, farm labor is affected by affluence and changes in the age structure of the population. As highlighted in Chapter 3, the opportunity cost of agricultural employment has been driven up in places like the United States because of high earnings in the rest of the economy. Responding to these earnings, many farmers choose not to work solely at crop and livestock production; part-time farming has become so common in the OECD that non-agricultural wages and salaries comprise the majority of earnings for a large number of farm households. Lucrative opportunities outside of agriculture also strengthen incentives for mechanization, which in turn diminishes the demand for farm operators and workers. As producers react to these incentives, the average size of agricultural holdings increases, which drives up the capital requirements of farming – in particular, the investment in real estate and other assets needed to produce commodities at a low cost while simultaneously supporting an operator and his or her family.[3] Also, since every commercial farmer now feeds scores of other people, agriculture now employs a small portion of the total labor force. In addition, by reducing the hard, physical labor that used to characterize farming, mechanization and related technological change increases the involvement of women in commodity production. Mirroring economy-wide trends, the female share of the farm workforce has risen in recent decades.

In addition to being driven by supply-side factors, the transformation of OECD food economies is a consequence of shifting patterns of food consumption. As observed in Chapter 2, income growth has a positive impact on sales of animal products and fruits and vegetables. Demand also increases for ready-to-eat items. Along with depending on the opportunity cost of time, which clearly is very high in an affluent setting, consumption of processed and prepared products requires the use of microwaves and other appliances, which poor households in the developing world cannot afford but that just about every OECD household possesses. Another feature of consumption in a rich country relates to food safety. Where most people are poor, safety concerns are by no means absent, although these tend to be muted. In contrast, middle-class consumers have little patience for products that do not meet exacting standards of quality. At an extreme, some of them adhere strictly to the so-called precautionary (or "better-safe-than-sorry") principle, as indicated by their categorical rejection of genetically modified organisms (GMOs).[4]

Aside from avoiding GMOs, some consumers try to guarantee food safety by eating organic products. However, as pointed out by Rutgers University food scientist Joseph D. Rosen, who organized a symposium on food safety at the August 2004 meeting of the American Chemical Society, there is no scientific evidence that organic food is any healthier than conventional fare (Truth about Trade and Technology, 2004). A safeguard used by many more consumers is to purchase name brands offered by companies that go to great lengths to maintain the reputations of their products. Expenditures are required on these companies' part to maintain quality as edible commodities are processed and move through marketing channels. In light of these costs, brand-name products usually fetch higher prices than substitute goods.

---

[3] The substantial capital requirements of a farm of economic size has obliged individual operators to take advantage of leasing and other alternatives to conventional ownership.

[4] An extreme example of the precautionary principle in action is the pressure that European non-governmental organizations (NGOs) applied to Zambia in 2002 to reject a shipment of genetically modified corn, which the United States had donated to help alleviate the hunger that three million Zambians were suffering at the time. For the NGOs, the value of improved nourishment that food aid would have created in the recipient country was outweighed by the risk that some of the donated grain would be planted and that some of the resulting harvest would find its way to Europe. Reluctant to defy anti-GMO activists, who might have lobbied to block Zambian commodity exports at a later date, the African nation acceded to their demands and the donation was not accepted (Carroll, 2002).

As the demands of affluent and maturing customers for processed and prepared foods of invariably high quality are catered to, agribusinesses that specialize in this sort of service become more important. In the United States, for example, farmers comprise one-tenth of the food economy's workforce. Another 10 percent supplies production agriculture with fertilizer, seeds, and pesticides, machinery, and other inputs. Fully 80 percent is employed beyond the farm gate, storing, processing, transporting, and retailing food. Some populists regard agribusinesses' dominance of the food economy as exploitative. However, claims that firms that process and market commodities incur unnecessary costs and absorb food receipts that rightfully belong to growers are mostly unfounded. As Persaud and Tweeten (2002) contend, agribusinesses are a vital and productive part of the food economy in developed countries. The share of consumer spending on edible goods that goes to these enterprises reflects the value added that they create, in terms of supplying food in the form, time, and location that consumers desire. As stressed in Chapter 3, agribusiness development reflects specialization and trade, which benefit everyone – farmers and consuming households alike. It is no accident that limited development of this sort coincides in much of the African, Asian, and Latin American countryside with agricultural self-sufficiency, poverty, and hunger.

Agricultural Subsidies and Protectionism

For good reason, Gardner (2002, pp. 271–277) emphasizes the linkage between elevated farm incomes, on the one hand, and elevated compensation outside of agriculture, on the other. As already mentioned, this linkage, which applies throughout the OECD, explains part-time farming, mechanization, and other supply-side adaptations. There is another consequence of high earnings outside of agriculture, however, which is that governments often prop up farmers' earnings.

Since moving commodities from one country to another is not particularly expensive, farmers' earnings cannot be shored up without agricultural protectionism. No prior training in economics is needed to appreciate that domestic commodity price supports (Chapter 4) that exceed international market levels would unleash a torrent of imports if there were no tariffs or quantitative restrictions (Chapter 6). It is hardly surprising, then, that agriculture's producer subsidy equivalent (PSE), which measures the portion of total farm receipts attributable to all forms of governmental support (including price premiums associated with trade restrictions and governmental regulation of commodity markets as well as various kinds of payments), tends to be high where the nominal protection coefficient (NPC), which is found by dividing domestic prices by international prices, is well above unity. By the same token, a low PSE often goes hand in hand with negligible differences between domestic and international prices.

A few OECD members meddle little with farm earnings and agricultural trade. One of these exceptions is Australia, which has a strong comparative advantage in the production of various products and which chooses not to undermine its advocacy of free trade in the WTO by distorting its own food economy. The country's PSE fell from 9 percent in the late 1980s to 5 percent at the turn of the century, as state support for farmers went from $1.3 billion to a little under $1.0 billion per annum. Only equal to 1.05 between 1986 and 1988, the NPC is now slightly over 1.00. Likewise, New Zealand, where farmers benefit as the global food economy moves toward freer trade, has a PSE of just 1 percent and an NPC barely above unity (Table 10.4).

Canada and the United States also have a comparative advantage in agriculture, although each of these countries distorts prices more and offers a higher level of support to farmers. Since the late 1980s, the former nation has made greater progress than its southern neighbor toward a lower PSE and NPC. From 1999 through 2001, Canadian prices were 13 percent above world levels and government support, which declined during the 1990s, amounted to 18 percent of farm receipts. Meanwhile, governmental support has increased in the United States, where price distortions averaged 16 percent and the PSE was 23 percent at the turn of the century (Table 10.4).

Elsewhere in the OECD, there is more policy-induced distortion of the food economy. In the EU, state support as a share of total farm receipts has gone down since the late 1980s, as have differences between domestic and international prices. However, the PSE was still 36 percent and the NPC remained at 1.38 from 1999 through 2001. Even more severe are wealthy European nations outside the EU that, in the absence of protection, would import nearly all their food. One such nation is Norway, where two-thirds of farm receipts come from the state (which takes in billions of dollars annually thanks to offshore oil production) and domestic food prices are 150 percent above world levels. Another is Switzerland, with a PSE of 70 percent and NPC of 2.76. Both these countries are in the same position as Japan, which is an efficient producer of various non-farm goods and services but has no meaningful comparative advantage in agriculture. Rice producers and other growers in East Asia's wealthiest economy stay in business only because three-fifths of their receipts come in one way or another from the state and because farm commodity prices in domestic markets are 2½ times world prices.

To put PSEs and NPCs in Western Europe and Japan into perspective, consider the situation in Poland, which joined the EU in 2004, and Turkey, which aspires to do so. During the 1990s, as Poland was harmonizing its agricultural policies with those of Germany and other nations to the west, its PSE rose from 4 to 12 percent and the country's NPC increased a little, to 1.14. Governmental support of farm earnings and the gap between domestic and international prices have gone up in Turkey as well.[5] These distortions remain modest by EU standards, which means that Polish and Turkish farmers are not as sheltered from international market forces. But if recent trends toward higher PSEs and NPCs continue, taxpayers and consumers in the two countries will be footing a higher bill.

---

[5] Note that Turkey's PSE was nearly equal to that of the United States from 1999 to 2001: 21 versus 23 percent. But because payments to farmers in the latter country are more decoupled from production, price distortion, as measured by the NPC, is lower. This illustrates the fact that PSEs and NPCs are not perfectly correlated.

Table 10.4 State support of farm incomes and agricultural protectionism

Country or region	Units	1986–1988	1999–2001
Australia			
Total producer support	$US million	1,285	947
Producer subsidy equivalent	Percent	9	5
Nominal protection coefficient	Ratio	1.05	1.01
New Zealand			
Total producer support	$US million	476	67
Producer subsidy equivalent	Percent	11	1
Nominal protection coefficient	Ratio	1.02	1.01
Canada			
Total producer support	$US million	5,667	3,930
Producer subsidy equivalent	Percent	34	18
Nominal protection coefficient	Ratio	1.40	1.13
United States			
Total producer support	$US million	41,839	51,256
Producer subsidy equivalent	Percent	25	23
Nominal protection coefficient	Ratio	1.19	1.16
European Union			
Total producer support	$US million	93,719	99,363
Producer subsidy equivalent	Percent	42	36
Nominal protection coefficient	Ratio	1.87	1.38
Norway			
Total producer support	$US million	2,628	2,274
Producer subsidy equivalent	Percent	66	66
Nominal protection coefficient	Ratio	3.38	2.50
Switzerland			
Total producer support	$US million	5,063	4,480
Producer subsidy equivalent	Percent	73	70
Nominal protection coefficient	Ratio	3.85	1.25
Japan			
Total producer support	$US million	49,498	51,980
Producer subsidy equivalent	Percent	62	60
Nominal protection coefficient	Ratio	2.51	2.42
Poland			
Total producer support	$US million	528*	1,676
Producer subsidy equivalent	Percent	4*	12
Nominal protection coefficient	Ratio	1.00*	1.14
Turkey			
Total producer support	$US million	2,779	6,522
Producer subsidy equivalent	Percent	14	21
Nominal protection coefficient	Ratio	1.15	1.25
OECD			
Total producer support	$US million	238,936	248,302
Producer subsidy equivalent	Percent	38	33
Nominal protection coefficient	Ratio	1.58	1.35

Note: *Data for 1991–1993.
Source: OECD (2002).

As reported in Chapter 6, liberalizing the global food economy would create substantial benefits – especially for OECD consumers, who would pay lower prices for commodities purchased from more efficient producers in other parts of the world, and OECD taxpayers, who as indicated at the bottom of Table 10.4 provide a quarter trillion dollars annually in farm subsidies. Producers in less fortunate parts of the world would also benefit, as international prices rose and as they gained access to European, Japanese, and North American markets. As also noted in this book's chapter on globalization and agriculture, progress in the WTO and its predecessor, the GATT, toward freer trade in farm commodities often has been disappointing. Indeed, more is sometimes accomplished through regional trade agreements. For example, the EU would have been unable to afford previous levels of farm subsidies and agricultural protectionism once it expanded to include Poland and other Eastern European nations with large farming sectors. Anticipation of this problem might help to explain recent drops in the PSE and NPC in the EU (Table 10.4). Similarly, declining protectionism in Canada and the United States might be driven in part by expectations of a Free Trade Area of the Americas (FTAA). These initiatives toward freer trade at the regional level might lead to further progress at the WTO, which would be best for the world as a whole.

Production Technology and Trends in Output

The generous subsidies received by farmers in affluent nations have not shielded them from forces of change. During the twentieth century, OECD agriculture became industrialized: technology driven, capital intensive, and with standardized production procedures. In recent years, it also has acquired post-industrial features as individual farmers have complemented production operations with service activities. On many farms, marketing, utilization of information systems, and asset management in the home office creates at least as much profit as labor performed in the field, shop, or barn.

In addition, agriculture has been affected by environmental concerns. Instructive in this regard are patterns of fertilizer use in Western Europe. Among the EU members listed in Table 10.5, application rates (columns 2 and 3) went up during the 1980s and 1990s only in Spain, which had a low rate by regional standards 25 years ago, and the United Kingdom, where the increase was small. Elsewhere in the region, there was an appreciable decline – significantly, a decline that went along with much higher yields. Environmental regulation, aimed at protecting water quality, had an effect in some places. But elsewhere, a falling NPC has been the main driving force. Demanded by taxpayers and consumers, diminished incentives for excessive farm output have had an environmental pay-off as well.

Application rates have fallen in Japan, where farmers started using more fertilizer for the sake of higher yields during the nineteenth century (Chapter 3), and have remained stable in the United States. In contrast, there have been sizable increases in Australia and Canada, which had the lowest application rates in the OECD from 1979 to 1981. The fastest growth has occurred in New Zealand, which

Table 10.5 Agricultural inputs in selected OECD countries, 1979–2001

Country	Rural population density (people per sq. km) 2001 (1)	Fertilizer use (kg per ha of arable land)		Irrigated land (% of cropland)		Agricultural machinery (tractors per sq. km of arable land)	
		1979–1981 (2)	1999–2001 (3)	1979–1981 (4)	1999–2001 (5)	1979–1981 (6)	1999–2001 (7)
United States	37	109	110	10.8	12.6	2.5	2.6
Denmark	35	245	152	14.5	19.5	7.1	3.7
Canada	14	42	55	1.3	1.6	1.4	1.6
The Netherlands	184	862	476	58.5	59.9	22.4	16.4
Australia	3	27	48	3.5	4.7	0.8	0.6
Germany	86	425	237	3.7	4.0	13.4	8.7
Japan	603	413	316	56.0	54.7	27.2	46.0
France	78	326	237	7.2	13.4	8.4	6.9
Italy	232	229	208	19.3	24.2	11.2	19.7
United Kingdom	109	319	325	2.0	1.8	7.4	8.6
Sweden	55	165	106	2.4	4.2	6.2	6.2
New Zealand	36	188	532	5.2	8.6	3.5	5.0
Spain	69	101	167	14.8	20.1	3.4	6.7

Sources: World Bank (2004).

as already indicated has a comparative advantage in agriculture and neither protects nor subsidizes its farmers.

While agriculture is capital intensive throughout the OECD, the specific form of farm-related investment differs from place to place. Dry conditions in many parts of Italy and Spain explain irrigated percentages of cropland (Table 10.5, columns 4 and 5) that are above OECD norms – though well below corresponding percentages in dry parts of the developing world. The Netherlands is not by any means an arid nation. However, three-fifths of its arable land is irrigated because on-farm water management, undertaken mostly for proper drainage, is intense. Japan's long history of agricultural intensification explains why more than half the country's farmland is irrigated.

Although agriculture is highly mechanized throughout the affluent world, the ratio of tractors to agricultural land (Table 10.5, columns 6 and 7) varies considerably, mainly because of different natural resource endowments. At one extreme, land suited to agriculture is very scarce in Japan and the Netherlands. Consequently, average farm size is modest and tractors and other machinery are correspondingly of small scale, though numerous. At the other extreme are Australia, Canada, and the United States, where land is abundant and agricultural equipment tends to be large-scale and tractor-to-land ratios are lower.

Starting in the late 1800s, use of machinery, improved seeds, fertilizer, pesticides, and other inputs aside from labor and land increased substantially in Europe, Japan, and North America. This technological shift caused yields to multiply. Since agricultural land use remained stable, production growth resulted entirely from this intensification (Chapter 3). In spite of stable or diminished fertilization in many countries, per-hectare output continued to go up during the last two decades of the twentieth century (Table 10.6, columns 1 and 2). Evidently, new ways were still being found to produce more output from available resources.

Of course, yields vary, depending on resource endowments, and it is a mistake to suppose that a nation can compete internationally only if its agriculture is very intensive. Consider Australia and Canada, both of which have a comparative advantage in cereal production. In each country, average per-hectare output that is low by OECD standards is more than offset because land is not very scarce. Topography strongly favors mechanization and economies of size are considerable. In contrast, Japan is not internationally competitive because it is difficult in spite of high yields to cover the opportunity costs of land, which is extremely scarce, and other farm inputs.

Japan's lack of comparative advantage explains why food production fell there during the 1980s and 1990s (Table 10.6, columns 3 and 4). A decline also occurred in Sweden, where farmers find competing with foreigners difficult. For similar reasons, there was no appreciable change in Britain and Italy. For several other countries listed in Table 10.6, however, the index of food production has increased. As was true earlier in the twentieth century, productivity-driven growth in food supplies has outstripped increases in demand. The latter increases have been modest both because human numbers are not going up very rapidly and because the

Table 10.6 Agricultural and food output in selected OECD countries, 1979–2002

Country (ranked by income per capita, 2002 PPP $)	Cereal yield (kg per ha)		Food production index (1989–1991 = 100)		Agricultural productivity (value added per worker – 1995 $)	
	1979–1981	2000–2002	1979–1981	2000–2002	1979–1981	1999–2001
	(1)	(2)	(3)	(4)	(5)	(6)
United States	4,151	5,770	94.5	122.5	20,672	52,307
Denmark	4,040	6,025	83.3	106.0	19,350	60,054
Canada	2,173	2,544	79.7	123.5	16,002	43,235
The Netherlands	5,696	7,341	86.5	98.4	24,360	58,915
Australia	1,321	1,740	91.3	138.8	20,872	36,151
Germany	4,166	6,587	91.4	97.1	9,119	32,843
Japan	5,252	6,150	94.1	91.6	17,378	32,005
France	4,700	7,149	93.6	104.3	19,318	58,177
Italy	3,548	4,940	101.4	102.3	11,090	26,717
United Kingdom	4,792	6,862	92.2	92.4	20,326	33,634
Sweden	3,595	4,681	100.6	96.0	20,865	39,422
New Zealand	4,089	6,286	90.7	135.2	16,637	28,389
Spain	1,986	3,200	81.9	120.1	7,556	21,955

Source: World Bank (2004).

income elasticity of demand for food is generally low where GDP per capita is elevated (Chapter 2).

Finally, productivity growth in agriculture has driven up value added per worker in the sector (Table 10.6, columns 5 and 6). Since farming must compete for labor with other sectors, GDP per capita, which is a good indicator of the opportunity cost of labor dedicated to crop and livestock production, is positively related to value added per agricultural worker. Note, for example, the relatively low numbers in columns 5 and 6 for New Zealand and Spain, where living standards are lower than in other countries listed in the table, and the relatively high numbers for the United States, Denmark, and the Netherlands.

10.4 Dietary Change and Consumption Trends

Transformed in various ways because of elevated living standards, technological improvement, and capital intensity, OECD agriculture also has been greatly influenced by the large appetite that most affluent people have for livestock products. This impact arises because multiple plant-derived calories must be fed to cattle, chickens, and other domesticated animals for each calorie of meat, eggs, etc. eaten by people. The former inputs include grain, soybean products, and other feed components raised by farmers or derived from crops. Forage, produced on rangeland and pasture, is another source of plant-derived calories for livestock, especially cattle and sheep. But regardless of how domesticated animals are fed, use of land and other resources is affected if demand for livestock products is elevated, as it is in the world's wealthy countries.

Rask (1991) has developed a methodology that takes dietary composition and its resource impacts into account in the analysis of food consumption and production. Prices of different elements of the human diet are not used because of the distortions caused by protectionism and other policies. Instead, everything that people consume is expressed in cereal equivalents.[6] This methodology, though it involves a large number of calculations, is conceptually straightforward. However, it has a pair of limitations. One is the implicit assumption that all plant-derived calories eaten by livestock are equally scarce, even though the forage that cattle and other domesticated animals graze does not have the same value as corn and other grain provided at feedlots. The other limitation is that the plant-derived calories required to produce one calorie's worth of livestock products are assumed to be invariable. Since forage production and the feeding efficiency of livestock both improve over time, Rask's (1991) methodology, which is used here and in the regionally focused chapters that follow, tends to overstate agricultural use of natural resources.

Using FAO data on food consumption in three time periods – 1979–1981, 1989–1991, and 1997–1999 – and converting all food and beverages into annual cereal equivalents, Rask has documented variation over time in OECD diets (Table 10.7). The United States is unusual in that intake of plant as well as animal products (columns 1 and 2) is trending upward, albeit slightly. In other countries where per-capita consumption (column 3) has gone up, most of the change has related to animal products. Illustrative in this regard is Denmark, where the average person was eating about the same amount of plant products in the late 1990s as he or she did two decades earlier. Meanwhile, per-capita consumption of animal products rose from 1.40 to 1.59 cereal-equivalent tons per annum. Similar changes happened in the Netherlands, Italy, and Spain. Relative increases in the consumption of eggs, meat, and milk were even more pronounced in Japan, although this consumption continues to make up a smaller portion of the human diet than in other countries with comparable living standards.

---

[6] Cereal equivalents for 1 cal of human consumption of selected foods (Rask, 1991) follow.

Cereals (excluding beer)	1.0000	Bovine meat	19.8000
Starchy roots	0.2648	Mutton and goat meat	19.0000
Pulses	1.0998	Pig meat	8.5000
Oil crops	0.9070	Poultry meat	4.7000
Vegetable oils	2.7479	Other meats	12.0000
Fruits (excluding wine)	0.1475	Animal fats	3.1000
Vegetables	0.0766	Milk (excluding butter)	1.2000
Treenuts	0.7955	Butter and ghee	21.0000
Sugar and sweeteners	1.1161	Eggs	3.8000
Alcoholic beverages	0.2450		

Changes in per-capita intake of animal products are likewise the larger part of consumption trends where these trends are declining. For example, reductions in the meat, eggs, milk, etc. that Canadians eat have exceeded the small increases in their consumption of plant products. The same is true of France, Germany, Sweden, and the United Kingdom, as well as Australia and New Zealand. The latter two nations are highly efficient producers of grain and livestock, which keeps prices low and encourages consumption.

In most of the OECD, food self-sufficiency – defined simply in Table 10.7 (column 4) as domestic production (expressed in cereal equivalents) divided by consumption – has been achieved, is increasing, or both. Japan is an exception, as are Italy and Sweden. None of these nations has any reason to worry about food supplies since, as emphasized in this chapter, commodity imports are easily paid for by selling non-agricultural products to foreigners. Indeed, a more serious imbalance in much of the OECD is a self-sufficiency ratio that is too high. An excessive ratio may be above or below unity. Regardless, the gains from efficient specialization and trade are being foregone wherever domestic farmers are producing too much, invariably because of governmental meddling with market forces.

The Obesity Problem

Domestic farm output is not the only thing that is too high in many rich countries. Aside from Japan, where per-capita consumption of plant and animal products is just 1.10 cereal-equivalent tons per annum (or 3.01 kilograms per day), daily food intake is between 4.25 and 6.36 cereal-equivalent kilograms in the other OECD members listed in Table 10.7. Needless to say, consumption levels in this range are well above what the average African, Asian, or Latin American eats on a typical day. Moreover, these levels are excessive relative to nutritional requirements, which causes a large part of the population to be overweight.

Since excessive weight is a worldwide problem, not just an issue in rich countries, a full discussion of causes, impacts, and solutions is deferred until the last chapter of this book, which addresses the future of the food economy and in which the scope of discussion is global rather than regional. Only noted here are the problem's dimensions in the OECD.

One perspective on the obesity epidemic, as some prefer to call it, is to count the number of people whose body mass index (BMI) – equal to a person's weight in kilograms divided by the square of his or her height in meters – exceeds 30.[7] The portion of Americans over the age of 18 years categorized as obese according to this criterion has gone up in recent decades, from 13 percent in the early 1960s to 14 percent in the late 1970s, to 22 percent in the early 1990s, and 30 percent at the turn of the twenty-first century. Even more worrying is the rising incidence of children

---

[7] To calculate the BMI using English units of measurement, weight in pounds is multiplied by 704.5 and then divided by the square of height in inches. The ideal BMI for adults of both sexes is 20–22. Anyone with a BMI over 25 is categorized as overweight and, as already noted, the cutoff for obesity is 30.

Table 10.7 Food consumption and production trends in selected OECD countries, 1979–1999

Country	Years	Plant product consumption per capita (in cereal-equivalent tons) (1)	Animal product consumption per capita (in cereal-equivalent tons) (2)	Total food consumption per capita (in cereal-equivalent tons) (3)	Total food self-sufficiency* (4)
United States	1979–1981	0.26	1.84	2.11	0.99
	1989–1991	0.29	1.81	2.10	0.99
	1997–1999	0.31	1.84	2.15	1.02
Denmark	1979–1981	0.23	1.40	1.62	2.39
	1989–1991	0.23	1.63	1.86	2.34
	1997–1999	0.25	1.59	1.84	2.36
Canada	1979–1981	0.22	1.71	1.93	1.19
	1989–1991	0.24	1.60	1.84	1.07
	1997–1999	0.26	1.53	1.79	1.15
The Netherlands	1979–1981	0.24	1.45	1.69	1.57
	1989–1991	0.26	1.46	1.72	1.78
	1997–1999	0.24	1.58	1.82	1.92
Australia	1979–1981	0.23	2.35	2.58	1.74
	1989–1991	0.24	2.24	2.49	1.66
	1997–1999	0.25	1.98	2.23	2.07
Germany	1979–1981	0.26	1.67	1.92	0.97
	1989–1991	0.26	1.65	1.91	1.01
	1997–1999	0.27	1.45	1.72	0.98
Japan	1979–1981	0.26	0.68	0.93	0.73
	1989–1991	0.26	0.82	1.08	0.67
	1997–1999	0.26	0.85	1.10	0.56

(*Continued*)

Table 10.7 (Continued)

Country	Years	Plant product consumption per capita (in cereal-equivalent tons) (1)	Animal product consumption per capita (in cereal-equivalent tons) (2)	Total food consumption per capita (in cereal-equivalent tons) (3)	Total food self-sufficiency* (4)
France	1979–1981	0.24	2.05	2.29	1.05
	1989–1991	0.25	2.02	2.27	1.03
	1997–1999	0.25	1.92	2.17	1.08
Italy	1979–1981	0.32	1.38	1.70	0.78
	1989–1991	0.31	1.54	1.85	0.76
	1997–1999	0.31	1.54	1.85	0.78
United Kingdom	1979–1981	0.22	1.48	1.71	0.76
	1989–1991	0.25	1.38	1.62	0.81
	1997–1999	0.26	1.29	1.55	0.81
Sweden	1979–1981	0.22	1.51	1.73	0.99
	1989–1991	0.22	1.42	1.65	0.99
	1997–1999	0.24	1.46	1.70	0.94
New Zealand	1979–1981	0.21	2.57	2.78	3.34
	1989–1991	0.23	2.21	2.44	3.44
	1997–1999	0.24	2.07	2.32	3.60
Spain	1979–1981	0.27	1.12	1.38	0.96
	1989–1991	0.28	1.35	1.63	0.95
	1997–1999	0.28	1.53	1.81	1.00

Notes: *Total food self-sufficiency equals total domestic food production divided by total domestic food consumption.
Source: Derived from FAOSTAT Agriculture Data, commodity balances and food supply data from 1979 to 1999 (www.fao.org).

with an elevated BMI, who tend to struggle with their weight their entire lives. Whereas 4 percent of the US cohort between the ages of 2 and 17 years were obese in the middle and late 1960s, 14 percent were so categorized in 1999 and 2000 (Rashad and Grossman, 2004). In contrast, the number of Americans who do not eat enough is much smaller; according to a survey carried out in December 2002 by the US Department of Agriculture (USDA), only 3 percent of US households chronically consume too little food (Nord et al., 2003).

Another perspective on weight problems is provided by data on mortality and health-care costs. According to the US Centers for Disease Control and Prevention (CDC), as many as 400,000 Americans died in 2000 because of poor diets and lack of exercise; this was only 35,000 less than the number of their countrymen whose lives were claimed the same year by tobacco-related illness (Song, 2004). At present, annual health-care expenditures traced to smoking still exceed those associated with obesity. However, the latter will soon move past the former. The reversal will occur in part because the incidence of obesity is rising as the portion of the adult population that smokes is falling. Overeating will also become the most costly US health problem because people whose excess weight makes them ill tend to cling to life longer than those who die from cancer, emphysema, and other diseases related to tobacco use.

In light of the mounting human and economic toll of overeating, proposals for state action to deal with the problem are being debated actively. Improved food labeling, to help consumers who need to watch their weight arouses little controversy, as does education about the health consequences of weighing too much. Sweden has banned food advertising during television programs aimed at youngsters; nevertheless, the incidence of childhood obesity is not appreciably different from the incidence in places similar to Sweden where such advertising is not restricted (Duncan, 2003). No country has attempted to deal with the market failure that arises because the medical expenses resulting from obesity are neglected entirely by farmers, agribusinesses, restaurants, and other food providers and not fully internalized entirely by heavy people, themselves. In the United States, for example, a sizable portion of these expenses are paid by the government's Medicare and Medicaid programs. An obvious corrective intervention by the state would be similar to a tax on pollution (Chapter 5). A charge on edible products laden with fat and sugar, for example, would counteract the long-term trend toward ever-cheaper food that has contributed substantially to weight gain in the United States (Lakdawalla and Philipson, 2002; Cutler et al., 2003).

Safe to say, the issue is far from settled. Indeed, it is not farfetched to envision a time when overeating might command more attention of policy-makers than food insecurity – not just in the OECD, but around the world.

10.5 Summary

With due allowance for differences in geography and (especially in Japan's case) cultural antecedents, a number of generalizations apply to the world's rich nations.

In each of them, standards of living are historically unprecedented. Moreover, additional improvement is all but guaranteed since trade continues to expand and savings are routinely invested in human, material, and technological capital. Natural increase is becoming a thing of the past. Throughout the OECD, there is little immediate prospect that human fertility will rise above the replacement level, mainly because not much prevents women from taking advantage of the economic opportunities available in an affluent society. Since families with one or two children are the rule, demographic contraction and aging of the population can be forestalled only by accelerated migration from poorer settings, which Japan and some parts of Europe are reluctant to accept.

No less than any other sector, the food economy is greatly affected by generalized affluence and the rising predominance of the middle aged and elderly. High earnings outside of agriculture create overpowering incentives for mechanization, thereby causing farm employment to decline and the average size of holdings to increase. The gradual demise of traditional family farms is indicated not just by reductions in the agricultural workforce, but by its maturation as well. As of 2002, the average age of US farm operators was 55 years and on the rise. Since an individual operator must harness assets worth millions of dollars to produce commodities like corn and soybeans efficiently, there will be fewer commercial farms in the future and many of these will owe their existence to generous parents who have passed the family homestead to a son or daughter, to the proprietor's engagement of capital markets (even including the use of equity financing), or to both (Tweeten and Hopkins, 2003). Since TFRs have gone down in rural areas no less than in cities, much of the workforce aside from farm operators themselves will comprise younger immigrants from poorer nations.

Throughout the OECD, farming has been protected from international competition and subsidized. As reported in this chapter, PSEs and NPCs have not budged in Japan and a couple of rich European nations outside the EU. However, protectionism has all but disappeared in Australia and New Zealand and seems to be following a downward trend in the EU and North America. Further modification of government policies that distort the food economy is likely, both because freer trade serves broad, national interests and because of mounting budgetary pressures. The impulse for reform also has to do with environmentalism. Where the population is stable (or declining) and growing older, demand for clean waterways is bound to outweigh the desire to stimulate commodity output with price supports, import quotas, and the like. When the choice between the two is stark, the former demand ought to trump the latter desire – even when large producers, who capture most of the gains from subsidies and protection, lobby hard for the largesse coming their way.

Due to policy-induced distortions, exports have been excessive from some parts of the developed world, most notably Western Europe, and imports have been choked off in other places, such as Japan. To be sure, reform will correct inefficiencies. Nevertheless, reform may not greatly alter patterns of agricultural trade between the OECD and the rest of the world. Between demographic stabilization and low-income elasticities for food, demand for imported commodities is unlikely to grow appreciably in wealthy nations. Meanwhile, the continuation of technological progress in agriculture, including in biotechnology, will drive up supply. A significant portion of the additional production is bound to be exported, especially from those OECD members with a comparative advantage in agriculture.

Thanks to specialization and trade among farmers and agribusinesses, which is facilitated by the infrastructure, scientific advances, and other public goods provided by government, the food economy supplies abundant output to consumers at a low cost. With food insecurity all but eradicated in the OECD, other issues command more attention. One of these is fine-tuning of the safety net, so that work incentives are maintained as people are fed who otherwise would be unable to afford an adequate diet. Another issue is environmental protection, which has considerable appeal for people who are secure about the basics of life. By the late 1900s, the primary concern related to the food economy had become excessive consumption. No consensus has been reached yet about dealing with obesity, although this is certainly a less painful problem than hunger.

Study Questions

1. How do per-capita GDP trends and income distribution in the United States compare with income trends and distribution in other OECD nations?
2. What are the chances that population contraction will be avoided in Western Europe and Japan? In the United States?
3. Why is agricultural protectionism declining in the EU?
4. Explain the linkage between agricultural mechanization and per-capita GDP in the OECD?
5. What is the general trend in the consumption of livestock products in affluent nations?
6. What are the trends in childhood obesity in the United States?

11

Asia

Defined in this chapter and throughout this book not to include Japan, which is affluent, or any part of the Middle East or the Former Soviet Union, Asia has been the setting for intensive farming and great civilizations for millennia. It is also home to half the human population. The world's two largest nations are Asian: China, with nearly 1.3 billion inhabitants, and India, where more than 1 billion people live. So are the fourth, sixth, and seventh most populous countries: Indonesia (209 million), Pakistan (141 million), and Bangladesh (133 million), respectively (World Bank, 2003b, pp. 14–15).

Domination by outside imperial powers having come to an end by the middle of the twentieth century, two of the largest conflicts between communists and their opponents were fought in Korea and Vietnam in the decades immediately following the Second World War. During this same period, average incomes were low and poverty widespread throughout Asia, and the urgency of development was universally recognized. However, more than a few observers were pessimistic. Paul Ehrlich, for one, held out no hope for the region, even predicting that India would soon succumb to a famine of historically unprecedented proportions (Ehrlich, 1968, pp. 38–41).

As dire warnings of this sort were being issued, a Green Revolution was under way in India, China, and neighboring lands. Thanks to the introduction of high-yielding varieties of rice and other crops, food production has more than kept pace with consumption since the 1960s. With the specter of hunger receding in many parts of the region, accelerated development has occurred. As a result, the global center of economic gravity, which moved across the Atlantic from Europe to the North America during the twentieth century, shows signs of another shift, this time across the Pacific.

11.1 Trends in GDP Per Capita

Since the late 1970s, Asia has set a standard for economic growth matched by no other region (Chapter 9). Increases in GDP per capita have been particularly rapid in China and South Korea (Table 11.1). Also, yearly growth during the last two decades of the twentieth century averaged 4 percent in Indonesia, Singapore, and

Table 11.1 Patterns of economic growth in Asia, 1980–2003

	(A) 1980–1990 (%)				(B) 1990–1999 (%)					(C) 2000–2003 (%)			
Country (ranked by column 1)	GDP per capita	Exports	Inflation	Gross investment	Country (ranked by column 5)	GDP per capita	Exports	Inflation	Gross investment	Country (ranked by column 9)	GDP per capita	Economic freedom index* 2002	Political freedom status* 2001–2002
	(1)	(2)	(3)	(4)		(5)	(6)	(7)	(8)		(9)	(10)	(11)
1. China	8.6	19.3	5.9	11.0	1. China	9.6	13.0	8.2	12.8	1. China	2.5	3.55	NF
2. South Korea	8.2	12.0	6.1	11.9	2. Vietnam	6.3	27.7	16.8	25.5	2. Vietnam	2.3	3.85	NF
3. Thailand	5.9	14.1	3.9	9.4	3. Singapore	6.1	–	1.6	8.5	3. South Korea	2.3	2.50	F
4. Hong Kong	5.7	14.4	7.7	4.0	4. Myanmar	5.1	7.5	25.9	14.7	4. Thailand	2.0	2.40	F
5. Singapore	5.0	–	1.9	3.7	5. South Korea	4.7	15.6	5.8	1.6	5. Cambodia	2.0	2.60	NF
6. Indonesia	4.3	2.9	8.5	7.0	6. India	4.3	11.3	8.6	7.4	6. India	2.0	3.55	F
7. India	3.7	5.9	8.0	6.5	7. Sri Lanka	4.1	8.4	9.7	6.2	7. Hong Kong	1.5	1.35	–
8. Pakistan	3.6	8.4	6.7	5.9	8. Malaysia	3.8	11.0	5.0	6.2	8. Bangladesh	1.3	3.70	PF
9. Malaysia	2.5	10.9	1.7	2.6	9. Laos	3.8	–	22.9	–	9. Laos	1.3	4.55	NF
10. Sri Lanka	2.6	4.9	11.0	0.6	10. Thailand	3.5	9.4	4.6	–2.9	10. Indonesia	1.0	3.35	PF
11. Vietnam	2.5	–	210.8	–	11. Bangladesh	3.2	13.2	3.9	7.0	11. Sri Lanka	1.0	2.80	PF
12. Nepal	2.0	3.9	11.1	1.8	12. Indonesia	3.0	9.2	14.4	5.1	12. Malaysia	0.5	3.10	PF
13. Bangladesh	1.9	7.7	9.5	1.4	13. Nepal	2.4	14.3	8.6	5.7	13. Mongolia	–0.3	2.90	F
14. Mongolia	2.5	–	–1.6	1.7	14. Cambodia	2.0	–	28.7	–	14. Nepal		3.40	PF
15. Myanmar	–1.2	1.9	12.2	–	15. Hong Kong	1.8	8.4	5.2	6.3	15. Pakistan		3.30	NF
16. Philippines	–1.6	3.5	14.9	–2.1	16. Pakistan	1.5	2.7	10.7	2.1	16. Singapore		1.55	PF
					17. Mongolia	0.9	9.6	8.4	4.1				
					18. Philippines	–1.2	–	66.5	–				

Note: *The economic freedom index ranges from 1.00 (completely free markets) to 5.00 (completely repressed markets). See Heritage Foundation (2002). Political freedom status is F (free), PF (partly free), and NF (not free). See Freedom House (2002).
Source: World Bank (2000/2001), pp. 278–279, 294–295; World Bank (2005).

Thailand. Since the World Bank ceased reporting data for Taiwan after the Republic of China (with its capital in Taipei) was obliged in 1972 to surrender its UN seat to the People's Republic of China (with its capital in Beijing), the country is not included in Table 11.1. Be that as it may, the Taiwanese economy has been one of Asia's best performers, with GDP per capita rising by 5–6 percent per annum during the 1980s and 1990s (Council for Economic Planning and Development, 1999, p. 1).

In light of recessionary conditions throughout Latin America and the Caribbean, the Middle East, and Africa during the 1980s, the improvement in living standards that occurred the same decade in most of Asia (Table 11.1, column 1) was impressive. The reasons for this success are indicated in columns 2–4. By and large, exports were rising at a fast clip. Rates of capital formation were also high – in large part because inflation was kept in check. Exports and investment drove GDP per capita higher in the 1990s as well (columns 5–8). Likewise, improvements have been sustained since the turn of the twenty-first century (column 9).

In countries belonging to the OECD, elevated living standards generally go hand in hand with economic and political freedom. The same linkage holds in Asia, in the sense that the region's freest economies (i.e., the four East Asian tigers: Hong Kong, Singapore, South Korea, and Taiwan) are all affluent and, as has been demonstrated in South Korea and Taiwan, prosperous people press hard for democratic freedoms (columns 10 and 11). But in India, democracy is well established in spite of poverty. Furthermore, GDP per capita has shot up rapidly in China and Vietnam even though governmental restraints on economic freedom are serious in each country. If these restraints are not removed, the sort of growth enjoyed recently could well prove difficult to sustain.

As can be seen in Table 11.1, increases in GDP per capita have varied over time, with some nations falling back relative to their neighbors between the 1980s and 1990s and others improving. The uncertainty associated with the 1997 transfer from British to Chinese rule caused growth to decelerate in Hong Kong. The decline in regional rankings for South Korea, Thailand, and Indonesia during the 1990s was largely the result of a major financial crisis that erupted in 1997, when foreign investors' loss of confidence in the financial markets of these and other countries led to widespread capital flight and severe currency devaluations (Perkins *et al.*, 2001, pp. 556–564). After the 1980s, economic expansion in Pakistan was held back by political turmoil. Among countries that have come up in the rankings is Vietnam, which has moved away from central planning and has invested heavily in human capital and physical infrastructure. On the export side, it has moved past Colombia to become the world's second largest supplier of coffee, with foreign sales sizable enough to depress that commodity's price in international markets.

For many years, three economies have been among Asia's poorest performers. Mongolia and Nepal are geographically isolated, which obviously hinders trade. Remoteness is not a problem for the Philippines, which once was better off than most of the region. Instead, that country has suffered long from poor governance and corruption, which underlie stagnation in much of Latin America, the Middle East, and Sub-Saharan Africa. In contrast, the Chinese economy has expanded consistently and rapidly, which is a major reason why GDP per capita has been growing faster in Asia than in other parts of the world. A closer look at China is obviously warranted, as is examination of the other Asian colossus, India, where growth in GDP per capita picked up as the twentieth century drew to a close.

China

In a transformation triggered by agricultural reform, the world's largest country emerged from economic dormancy during the 1980s. Introduced by Premier Deng Xaioping in 1978, the Household Responsibility System (HRS) allowed peasants to acquire long-term leases on land and, with time, to decide what to produce and where to sell their crops and livestock. Preceded by 18 years of collective agriculture, this liberalization caused farm output to shoot up, more than meeting national food requirements (Naughton, 1995, pp. 138–142). Within a few years, market reforms were spreading to other parts of the economy, with rapid industrial expansion in enterprise zones along the coast. Besides textiles and food products, in which an emerging manufacturing economy usually has a comparative advantage, electronic goods and machinery of various sorts were (and are still) produced.

China's economic accomplishments are put into perspective by the dismal record during the late twentieth century of Russia and other remnants of the old USSR. One disadvantage of the Former Soviet Union has been the entrenched legacy of socialist agriculture. Farming was collectivized, with atrocious loss of life, soon after Stalin seized power in the late 1920s (Chapter 7). Sixty years later, in the early 1990s, farmers consequently lacked first-hand knowledge of free markets, not to mention traditional farming practices, when the USSR disintegrated. As emphasized in Chapter 14, agriculture was still a bleak feature of the economic landscape in the Former Soviet Union at the turn of the twenty-first century. In contrast, Chinese peasants endured collective agriculture for a little less than two decades, which was not enough to eradicate their familiarity with markets and traditional practices. Having been coerced into joining village-level communes around 1960, peasants quickly exploited the opportunities created by the HRS. Another clear advantage of China not shared with the Former Soviet Union is the large and affluent population of ethnic Chinese living in Taiwan, Hong Kong, Singapore, and other parts of Southeast Asia that is willing to invest in the ancestral homeland. In no small measure, the tens of billions of dollars made available by these people have driven industrial development, export growth, technological improvement, and solidification of the private sector in China (Ahmad, 2004).

Remarkable though the country's recent progress has been, enormous challenges remain. Rapid industrial expansion along the coast has created glaring regional disparities. Wage and salary discrepancies are also increasing, with private entrepreneurs, managers, and the employees of foreign businesses being paid much more than others. In addition, there is tremendous inefficiency among state-owned enterprises (SOEs), which dominate the steel industry, vehicle production, ship-building, mining, energy, and transport. The mounting losses of state-owned firms are being covered by loans from government banks; this has, in turn, made the Chinese banking sector insolvent, with non-performing loans amounting to about half of all debt held by the sector (Kynge, 2002; Righter, 2002). The weakness of financial institutions bodes ill for future capital formation. So does the lack of a legal system adapted to a market economy (Ahmad, 2004). To date, the small- and medium-sized firms that have accounted for much of China's economic expansion have not been severely hampered by deficiencies in the rule of law. But unless the enforcement of contracts and property rights improves, investment is unlikely to continue increasing at current rates.

Doubts have been expressed about the pace of economic expansion in China. As a rule, rising unemployment (which the country is suffering) and mounting SOE losses are signs of recession, not of an economy in which average income is going up each year by 7, 8, or 9 percent. Moreover, provincial authorities admit that past estimates of local economic activity cannot be reproduced, which suggests that national figures that are based on those estimates might well have been exaggerated. Kurlantzick (2002) contends that the true annual rate of increase in GDP per capita is 4–6 percent, not 7–9 percent.

Even a growth rate in the lower range is well above rates sustained since the late 1970s in all but a handful of countries. Also, there is little doubt that poverty has been reduced significantly, even as income inequality has increased along regional and occupational lines. SOE losses have to be contained and reform of the financial sector is essential. Corruption and other problems of governance must be faced. In light of these challenges, a debate has arisen at the highest levels of government concerning the proper direction of economic reform (Kynge, 2002). It is not unreasonable to expect the outcome of this debate to be positive, in terms of continued economic expansion.

India

The other continent-sized nation in Asia, India has accomplished a rare feat in the developing world by maintaining a functional democracy since achieving independence in 1947. For most of this period, socialist policies that the Labor Party championed in the old colonial power after the Second World War were embraced in the former crown jewel of the British Empire. Protectionist barriers and an intrusive regulatory regime were established and banks and major industries were nationalized. Not surprisingly, GDP grew slowly up to 1980 (Ahuluwalia, 2002).

Even though performance of the national economy as a whole was middling during the 1950s, 1960s, and 1970s, India was able to achieve a breakthrough in agricultural productivity. It was the first developing country to experience the Green Revolution, with impressive increases in rice and wheat yields that drove food costs down for landless and urban populations (Perkins *et al.*, 2001, pp. 596–604). Agricultural development contributed directly to declines in the incidence of extreme poverty from 1970 onwards, especially in rural areas (Datt and Ravallion, 2002).

During the 1980s, gradual liberalization of India's closed, regulated economy was occurring, which led to the growth performance documented in the left-hand panel of Table 11.1. Reform accelerated during the last decade of the twentieth century. For example, average import duties declined from 72 percent in 1991 and 1992 to 25 percent in 2001, when all remaining quantitative restrictions on trade were removed (Ahuluwalia, 2002). In addition, some SOEs were privatized and licensing requirements, which formerly stifled the private sector, were lifted (Long, 2004). Consequently, the annual rate of increase in GDP per capita rose above 4 percent.

Liberalization is still a work-in-progress in India. Still among the highest in the developing world, the country's tariffs are nearly double the levels in China and Southeast Asia. Also, privatization has been slow; SOEs still account for 35 percent of industrial value added. In addition, poor infrastructure and modest incentives for foreign investment limit exports and growth (Ahuluwalia, 2002). A more encouraging trend has been the increase of relatively well-paid jobs in information technology and remote business services outsourced from wealthy places like the United States. Here, a comparative advantage has been created because tens of thousands of young Indians have the education and excellent English that these jobs require (Long, 2004).

While India has demonstrated that a democratic and pluralist nation can open its economy to world trade, with positive effects on living standards and the incidence of poverty, continued growth depends on not abandoning the path of economic reform on which the country clearly has embarked. Likewise, the progress that liberalization has brought to India (and other countries) must not be jeopardized by protectionism in other parts of the world, as advocated by economic populists in prosperous nations.[1]

11.2 Population Dynamics

Before 1980, a large segment of the Asian population was destitute. But thanks to economic progress in China, India, and a number of other countries, hundreds of millions were subsequently lifted out of poverty. As pointed out in Chapter 9, global inequality declined during the last two decades of the twentieth century mainly because of the region's economic expansion (Bourguignon and Morrisson, 2002). This expansion also has facilitated the rapid adoption of health and sanitation technology, which along with improved nutrition has diminished the threat posed by mortal disease.

---

[1] Demagogues and critics in affluent places have vastly exaggerated the threat posed by outsourcing and disregarded its benefits for the world as a whole, including in countries belonging to the OECD (Drezner, 2004).

Reduced Human Fertility

The demographic change set in motion as life expectancy at birth goes up is far enough along in many Asian nations that human fertility is either approaching or has fallen to or below the replacement level of 2.1 births per woman (Table 11.2, columns 1 and 2). As in other affluent settings, total fertility rates (TFRs) are minimal in Hong Kong, Singapore, and South Korea. If current rates are sustained in China, where parents with more than one child have faced severe penalties, and Thailand, where no such penalties were ever applied, demographic stabilization will be achieved within a few decades.

As in other parts of the world, average income (Table 11.2, column 3) obviously has an impact on human fertility. The number of births per woman has fallen below 2.5 in nearly every country with a GDP per capita above $3,000 (column 3), with Malaysia and the Philippines being modest exceptions. There is good reason to expect that, as economies continue to expand, fertility will fall into this range in poorer countries, which actually experienced the greatest TFR declines between 1980 and 2002. In India, where average income is still under the $3,000 threshold, the number of births per woman went from 5.0 to 3.1 during these 22 years. Even though GDP per capita has not risen above $2,000 in Vietnam, the country's TFR fell by more than one-half in the same period and is currently below the replacement level.

Economic expansion tends to go hand in hand with urbanization (Table 11.2, column 4) and the latter change correlates with diminished human fertility. These linkages certainly hold for Asia. Something interesting about the region, though, is that living standards have risen and TFRs have fallen dramatically in at least a few countries where most of the population is still rural. Sri Lanka, Thailand, and Vietnam are cases in point. So are China and India. A lesson here is that, where development has benefited the countryside and not just cities, rural families have responded by having fewer children, just as urban households have done.

Another factor influencing human fertility, of course, is female economic empowerment, as reflected by literacy and remunerative employment. Due to a history of severe sex discrimination, male illiteracy is appreciably lower than the rate for women throughout Asia, the Philippines being the sole exception (Table 11.2, columns 5 and 6). The gap between the sexes is pronounced and total illiteracy is especially high in South Asia (aside from Sri Lanka) and Cambodia and Laos, both of which are poverty-stricken. In contrast, female illiteracy is not particularly high in countries where GDP per capita has risen above the $3,000 threshold, which means that the vast majority of women are sufficiently qualified to join the labor force. The female portion of the labor force has been rising, especially in nations that are urbanized and relatively prosperous (columns 7 and 8). This has driven up the opportunity cost of raising children, thereby causing more women to do what is required to limit fertility. Their resulting choices are reflected by rates of contraceptive prevalence (column 9) in places like Thailand, not just China, that are comparable to what one observes in rich nations.

Table 11.2 Social indicators for Asia in 2002

Country (ranked by income per capita 2002)	Total fertility rate		Income per capita 2002 ($PPP)	Urban % of total population in 2002	Adult illiteracy rate 2002 (% people 15 and over)		Females % of labor force		Contraceptive prevalence rate 1990–2002 (% women 15–49)
	1980	2002			Males	Females	1980	2002	
	(1)	(2)	(3)	(4)	(5)	(6)	(7)	(8)	(9)
Hong Kong	2.0	1.0	27,490	100	–	–	34.3	37.2	–
Singapore	1.7	1.4	24,910	100	3	11	34.6	39.2	–
South Korea	2.6	1.5	16,960	83	–	–	38.7	41.8	–
Malaysia	4.2	2.8	8,500	59	8	15	33.7	38.3	–
Thailand	3.5	1.8	6,320	20	5	9	47.4	46.2	72
China	2.5	1.9	4,520	38	5	13	43.2	45.2	83
Philippines	4.8	3.2	4,450	60	7	7	35.0	38.0	47
Sri Lanka	3.5	2.1	3,460	23	5	10	26.9	36.9	–
Indonesia	4.3	2.3	3,070	43	8	17	35.2	41.2	57
India	5.0	2.9	2,650	28	–	–	33.7	32.5	52
Vietnam	5.0	1.9	2,000	25	6	13	48.1	48.7	79
Cambodia	5.7	3.8	1,970	18	19	41	55.4	51.5	24
Pakistan	7.0	4.5	1,960	34	47	71	22.7	29.5	28
Bangladesh	6.1	3.0	1,770	26	50	69	42.3	42.5	54
Laos	6.7	4.8	1,660	20	23	45	–	–	25
Nepal	6.1	4.2	1,370	13	38	74	38.8	40.5	39
Myanmar	4.9	2.8	N/A	29	11	19	43.7	43.4	–

Source: World Bank (2004).

Natural Increase

Now that TFRs have reached or are approaching the replacement level in so much of Asia, natural increase in the region is being driven largely by demographic momentum. Crude death rates (CDRs) will probably go down in a handful of the poorest countries – Cambodia, Laos, Myanmar, and Nepal – in the next few decades. But elsewhere, the death rate is at or close to its nadir (Table 11.3, column 2) and will therefore remain stable or even creep up as the population ages.

By regional standards, natural increase (Table 11.3, column 4) is elevated in Malaysia and the Philippines. As already observed, these are the only two nations where GDP per capita has risen above $3,000 and yet human fertility remains noticeably above 2.1 births per woman. In all other countries that have broken through this income threshold, low TFRs have caused the gap between crude birth rates (CBRs) and CDRs to narrow. As a result, annual rates of natural increase are now under 1.5 percent – well below average yearly population growth since 1980 (column 5). Rates in this range are observed even in some poorer Asian nations, such as Myanmar and Vietnam.

Table 11.3 Population dynamics in Asia for 2002

Country (ranked by column 1)	Income per capita 2002 ($PPP) (1)	Crude death rate 2002 (per 1000 population) (2)	Crude birth rate 2002 (per 1000 population) (3)	Rate of natural increase per 1,000 population (%) (4)	Average annual population growth 1980–2000 (%) (5)
Hong Kong	27,490	5	7	0.2	1.5
Singapore	24,910	5	11	0.6	2.5
South Korea	16,960	7	12	0.5	1.1
Malaysia	8,500	5	22	1.7	2.6
Thailand	6,320	8	15	0.7	1.3
China	4,520	8	15	0.7	1.3
Philippines	4,450	6	26	2.0	2.3
Sri Lanka	3,460	6	18	1.2	1.4
Indonesia	3,070	7	20	1.3	1.7
India	2,650	9	24	1.5	2.0
Vietnam	2,000	6	19	1.3	1.9
Cambodia	1,970	12	27	1.5	2.8
Pakistan	1,960	3	26	2.3	2.6
Bangladesh	1,770	8	28	2.0	2.1
Mongolia	1,710	6	23	1.7	1.8
Laos	1,660	12	36	2.4	2.5
Nepal	1,370	10	32	2.2	2.3
Myanmar	N/A	12	23	1.1	1.7

Source: World Bank (2002a), pp. 48–50; World Bank (2004).

With natural increase decelerating dramatically in Asia, the region's population will undoubtedly peak during the next few decades. This is the main reason to anticipate that the human population as a whole will stabilize this century (Chapter 2). However, the ending of population growth, as will soon occur in countries like China and Thailand, does not mean that demographic issues will disappear. To the contrary, these countries must soon face the same problem that currently confronts Japan and other wealthy nations – paying pensions and otherwise tending to an aged cohort that makes up a sizable and growing part of the total population. Needless to say, caring for the elderly will not be easy in places where there is little prospect of living standards reaching Japanese levels (Eberstadt, 2000). For example, limiting couples to one child makes little sense in China now that the ratio of retirees to workers is rising.

11.3 Agricultural Development

Since agriculture and civilization have such a long history in China, India, Java (Indonesia's most populous island), and other Asian settings, population density is elevated in many parts of the region. True, there are Caribbean Islands as well as East African highlands where the number of rural inhabitants per square kilometer of farmland is very high. Nevertheless, overall population density is much lower in Latin America and the Caribbean and Sub-Saharan Africa than in Asia. The same holds for the Middle East and North Africa, even though Egypt's Nile River Valley is one of the world's most crowded agricultural landscapes.

Opportunities to increase crop and livestock production by increasing the geographic domain of agriculture are very limited in Asia. Even where such opportunities exist, on some Indonesian islands for example, there are adverse environmental impacts of agricultural extensification (Chapter 5). Between these impacts and the direct expenses farmers must incur to clear away trees and other vegetation, no major increase in agricultural land use is anticipated in the region (Chapter 3).

Intensified Production

High population densities in Asia have induced the kind of agricultural intensification characteristic of land-scarce settings described in Chapter 9. As indicated in Table 9.5, fertilizer application rates in the region exceed those of other parts of the developing world, with the average rate in China and neighboring countries more than 10 times that for Sub-Saharan Africa for a whole. Another fundamental aspect of intensification in Asia is irrigation. As reported in the same table, the irrigated portion of arable land in the region even exceeds the irrigated portion in the Middle East and North Africa, where water is a limiting factor of crop production for obvious climatic reasons.

Rural wages being low in China, India, and many other countries, incentives for agricultural mechanization are not overwhelming. Despite an increase during the

1980s and 1990s, the number of tractors per square kilometer of arable land remains below levels in Latin America and the Caribbean – where machinery use facilitates agricultural extensification – and the Middle East and North Africa (Table 9.5). As indicated in the last chapter, this indicator of mechanization is much higher in wealthy nations, where rural labor is scarce relative to other factors of production. In Asia, where land rather than labor is the limiting factor, intensification is not driven primarily by increased use of tractors and other equipment, which has little effect on production per hectare, but instead by fertilization, irrigation, and the planting of crop varieties that respond to increased applications of nutrients and water. As stressed in Chapter 3, the latter path to intensification creates yield growth.

In a region as large and diverse as Asia, patterns of agricultural input use are bound to vary. These patterns are reported in Table 11.4, in which countries are ranked from richest to poorest – with Hong Kong and Singapore being omitted because these two small former colonies of Great Britain lack enough land for farming to be important. Average living standards and the rate of fertilizer application were positively correlated around 1980 (column 2). While still apparent, the correlation has since become less pronounced (column 3). In nations like the Philippines, the application rate around 1980 was modest by regional standards but subsequently went up dramatically. Also, there are poor countries, such as Vietnam, where high population densities in rural areas (column 1) have a lot to do with intensive nutrient applications. Mongolia is like the Former Soviet Union in that applications have declined since the fall of communism. Finally, Thai farmers could easily afford to fertilize more, although the high natural fertility of the land many of them cultivate allows them not to do so.

Columns 4 and 5 of Table 11.4 underscore the importance of irrigation in Asia. Malaysia, where precipitation is abundant, has a negligible network. So do Mongolia, where water scarcity hinders irrigation development, and Cambodia, which lacks the financial resources needed for a large-scale build-up of canals and other infrastructure. In addition, the irrigated portion of arable land in China declined in the late twentieth century, in part because of agricultural extensification in rain-fed areas but also because environmental limits on irrigated agriculture were being reached. Elsewhere in the region, an expansion has occurred. The irrigated portion of arable land has gone up dramatically in Bangladesh. The same change has happened in Vietnam, in part because war-related damage to infrastructure has been repaired. Expansion even has occurred in Pakistan, where nearly three out of every four arable hectares were irrigated around 1980.

A modest region-wide trend toward mechanization is indicated by the number of tractors per square kilometer of farmland (Table 11.4, columns 6 and 7). One exception to this trend is Mongolia, where capital-intensive agriculture of the kind favored in the old USSR had failed. Another is Sri Lanka, where mechanization had surpassed levels observed anywhere else in the region around 1980 and subsequently declined. One more exception is China, where there were fewer tractors per square kilometer around 2000 than there had been from 1979 to 1981. Evidently, the HRS has discouraged inefficient use of farm equipment, which was characteristic

Table 11.4 Agricultural inputs in Asia, 1979–2001

Country (ranked by income per capita 2002)	Rural population density (people per km² arable land) 2001	Fertilizer use (kg per ha)		Irrigated land (% of cropland)		Agricultural machinery (tractors per km² of arable land)	
		1979–1981	1999–2001	1979–1981	1999–2001	1979–1981	1999–2001
	(1)	(2)	(3)	(4)	(5)	(6)	(7)
South Korea	491	392	454	59.6	60.4	0.1	11.1
Malaysia	554	427	670	6.7	4.8	0.8	2.4
Thailand	326	18	112	16.4	27.1	0.1	1.5
Philippines	564	64	134	12.8	14.6	0.2	0.2
China	561	149	256	45.1	36.3	0.8	0.7
Sri Lanka	1,607	180	277	28.3	33.6	1.4	0.9
Indonesia	591	65	124	16.2	14.4	0.0	0.4
India	460	35	107	22.8	32.2	0.2	0.9
Vietnam	923	30	341	25.6	37.6	0.4	2.5
Pakistan	438	53	136	72.7	81.6	0.5	1.5
Mongolia	87	8	3	6.7	4.8	0.8	0.4
Bangladesh	1,228	46	166	17.1	49.6	0.1	0.1
Laos	495	4	11	13.3	18.2	0.1	0.1
Cambodia	274	5	0	5.8	7.1	0.1	0.1
Nepal	668	10	26	22.5	36.2	0.1	0.2

Source: World Bank (2004).

of the socialistic communes established around 1960. Currently, there are more tractors per square kilometer of arable land in India.

Mechanization has proceeded the furthest in relatively wealthy nations. In South Korea (and Taiwan), a small portion of the workforce is employed in agriculture and the opportunity cost of farm labor reflects the high wages that can be earned in the industrial and service sectors. As in Japan and other countries belonging to the OECD, tractors (generally of small size) and other equipment are substituted for labor wherever possible. Similarly, agriculture is mechanized in Malaysia, which is more prosperous than its neighbors. Also, better living standards have led to a sharp increase in the ratio of tractors to agricultural land in Thailand. In most other nations, however, equipment use, though it has increased, remains limited, even by the standards of the developing world as a whole.

Intensification, Extensification, and Output Growth

With fertilizer being applied at higher rates and a considerable investment being made in irrigation infrastructure, agriculture has intensified significantly in the world's most populous continent, both in China and other parts of East Asia and in India and its South Asian neighbors. For all major crops other than fruit and vegetables, annual increases in planted area between 1970 and 2000 (Table 11.5A and B, column 2) were considerably less than annual yield growth (column 3).

Table 11.5 Trends in crop area, yield, and output in Asia, 1970–2000

	2000 Production (million tons) (1)	Average Annual Change, 1970–2000 (%)		
		Area (2)	Yield (3)	Production (4)
East Asia				
(A) Crops				
Rice	344	0.4	1.8	2.2
Vegetables	313	4.4	1.5	6.0
Roots and tubers	239	0.1	1.2	1.3
Maize	127	1.0	2.7	3.8
Fruits	105	4.7	1.2	5.9
Wheat	100	0.1	4.0	4.1
Oil crops	41	2.2	3.7	5.8
Other cereals	15	−3.5	1.4	−2.2
(B) Animal products				
Total meat	74	–	–	6.9
Total eggs	26	–	–	7.7
Total milk	16	–	–	6.1
Total wool	0.3	–	–	2.8
South Asia				
(A) Crops				
Rice	184	0.5	2.0	2.5
Wheat	98	1.4	2.8	4.3
Vegetables	71	1.7	1.2	3.0
Fruits	40	3.0	1.2	4.3
Pulses	15	0.3	0.2	0.5
Maize	14	0.4	1.0	1.6
Millet	10	−1.7	0.7	−1.0
Oil crops	10	1.3	1.4	2.6
Sorghum	10	−1.6	0.7	0.5
(B) Animal products				
Total milk	105	–	–	4.2
Total meat	8	–	–	3.2
Total eggs	2	–	–	6.2
Total wool	1	–	–	0.9

Source: Dixon and Gulliver with Gibbon (2001), pp. 182, 183, 228, 230.

Obviously, intensification was the primary driver of output growth, which except for minor grains was rapid during the same three decades (column 4).

As fertilization and irrigation have increased, production per hectare has improved. Of all the nations listed in Table 11.6, Malaysia was the only one with little growth in cereal yields (columns 1 and 2) during the 1980s and 1990s. This exception arose because the country's agricultural sector lacks comparative advantage in rice

Table 11.6 Agricultural and food output in Asia, 1979–2002

Country (ranked by income per capita 2002)	Cereal yield (kg per ha)		Arable and permanent cropland (1,000 ha)		Food production index (1989–1991 = 100)		Agricultural productivity (value added per worker $)	
	1979–1981	2000–2002	1980	1999	1979–1981	2000–2002	1979–1981	2000–2002
	(1)	(2)	(3)	(4)	(5)	(6)	(7)	(8)
South Korea	4,986	6,118	2,196	1,899	77.5	132.3	3,765	14,251
Malaysia	2,828	3,132	4,310	7,605	55.6	142.1	3,939	6,912
Thailand	1,911	2,654	17,970	18,000	79.7	123.5	616	863
Philippines	1,611	2,692	9,920	10,050	86.1	137.1	1,381	1,458
China	3,027	4,845	99,200	135,361	60.8	185.9	161	338
Sri Lanka	2,462	3,520	2,147	1,900	98.1	117.2	642	725
Indonesia	2,837	4,141	19,500	30,987	63.1	123.6	604	748
India	1,324	2,390	169,130	169,700	68.2	131.8	269	401
Vietnam	2,049	4,375	6,055	7,350	62.5	171.4	–	256
Pakistan	1,608	2,266	20,320	21,880	66.3	152.7	416	716
Mongolia	573	751	1,182	1,322	88.1	91.9	994	1,444
Bangladesh	1,938	3,312	9,145	8,440	79.3	138.3	232	318
Laos	1,402	3,140	880	955	70.3	186.4	–	621
Cambodia	1,006	1,978	3,046	3,807	48.9	152.0	–	422
Nepal	1,615	2,178	2,330	2,968	65.4	135.8	156	203
Myanmar	2,521	3,453	10,023	9,961	88.2	176.5	–	–

Source: World Bank (2004), FAO Production Yearbook (2000), pp. 3–13; FAO Production Yearbook (1982), pp. 45–56.

and other grains and instead produces tree crops like palm oil and rubber efficiently. Measures to raise cereal yields have not been very rewarding and, as in South America, the capital intensity of agriculture has facilitated agriculture's geographic expansion. As indicated in columns 3 and 4, Malaysia has experienced more extensification than any other Asian nation.[2]

Besides Malaysia, with its three-quarters increase in farmed area, just a few countries in the region have experienced agricultural extensification of any consequence since 1980. The two most important are China and Indonesia, with growth of 36 and 59 percent, respectively. Agricultural land use also increased by 25 percent in Cambodia and 27 percent in Nepal. In Indonesia, extensification exceeded growth in per-hectare output of cereals. In Nepal, the two were comparable. In contrast, yield increases have contributed significantly more to increased grain production in China and Cambodia. The latter two countries are much closer to the norm in the rest of Asia, which is that increases in food supplies result almost entirely from higher yields.

Thanks mainly to yield-driven growth, the food production index (Table 11.6, columns 5 and 6) more than doubled during the 1980s and 1990s in Cambodia, China, Laos, Malaysia, Pakistan, and Vietnam. In countries such as these, output growth significantly exceeded demographic expansion, which means of course that per-capita food supplies have increased. Another aspect of the development of Asian agriculture is that value added per worker in the sector has gone up in recent decades (Table 11.6, columns 7 and 8). This basic indicator of productivity remains low where population density is high and GDP per capita is modest, in spite of large relative increases registered by China, India, and other nations in recent decades. In contrast, value added per agricultural worker has risen appreciably in Malaysia, as it has done in other places where agriculture is capital-intensive. In South Korea, where as already observed well-paid jobs in non-agricultural sectors have driven up the opportunity cost of farm labor, this productivity measure has reached a very high level. In addition, value added per agricultural worker is inflated in that country for exactly the same reason that it is in nearby Japan. To be specific, an obvious lack of comparative advantage for South Korean and Japanese agriculture is masked by agricultural protectionism, which keeps domestic commodity prices well above international values.

11.4 Dietary Change, Consumption Trends, and Food Security

As demographic expansion decelerates, food demand trends in Asia are being driven less by simple increases in the number of mouths to be fed and more by

---

[2] Whenever rapid extensification coincides with stagnating yields, one might suppose that the former has something to do with the latter. In particular, the spread of farming to less fertile areas is bound to diminish production per hectare, all else remaining the same. But, to repeat, trends in cereal yields are an inaccurate indicator of overall trends in agricultural yields in Malaysia since that country lacks comparative advantage in grain production.

other forces. Among the latter are changes in the age composition of the population. For example, children, who need to eat less than young- and middle-aged adults, are becoming less numerous. At the same time, there are more elderly people, whose nutritional requirements are relatively modest. Something else that has a substantial effect on food demand in Asia is income growth. As pointed out in Chapter 2, income elasticities of demand for livestock products are, by and large, greater than income elasticities for other foods. Thus, improved living standards cause the former to become a more important part of the human diet.

Changes in the composition of Asian diets have been identified using FAO data on food consumption and Rask's (1991) methodology for converting these data into cereal equivalents, as explained in Chapter 10. As indicated in Table 11.7, consumption of plant-derived products (column 1) did not change dramatically during the 1980s and 1990s. Noticeable increases occurred in China, Indonesia, India, Myanmar, and Nepal. Elsewhere, however, there were small increases, no change at all, or (in Hong Kong, Sri Lanka, and South Korea) modest declines. Growth in per-capita consumption, then, was almost entirely the result of the rising importance of livestock products (column 2). In terms of cereal equivalents, goods like meat and eggs account for large shares of total food consumption in Hong Kong, Malaysia, and South Korea. In all but a few nations where GDP per capita is under $3,000, most calories consumed by humans continue to be plant-derived, even though relative increases in livestock consumption during the last two decades of the twentieth century exceeded relative growth in plant-derived calories. Only in Bangladesh, which has the lowest per-capita consumption of food in the region, have the livestock and plant-derived portions of the human diet remained steady.

Particularly where people were eating more livestock products in the late 1990s than they did 20 years earlier, food self-sufficiency – defined simply in Table 11.7 (column 4) as domestic production (expressed in cereal equivalents) divided by domestic consumption (column 3) – has diminished. In South Korea, the ratio of production to consumption slid from 0.79 in 1979–1981 to 0.73 in 1997–1999. During the same period, there were declines in Malaysia (1.17–0.98), Thailand (1.33–1.23), and the Philippines (0.81–0.68). The ratio went up markedly in Cambodia and Vietnam, which were still recovering from war-related devastation in the late 1970s. Pakistan is a special case in that per-capita livestock consumption is significantly above levels in nearby countries with similar living standards and it has become a little more (not less) food self-sufficient even though a lot more animal products are being eaten. Another special case is Sri Lanka, which has experienced much less dietary change although its ratio of production to consumption has deteriorated significantly. Elsewhere, the ratio changed little during the 1980s and 1990s.

Where accelerated GDP growth is happening, especially in non-agricultural sectors, a self-sufficiency ratio below 1.00 is not a great concern. There is no need to worry about food supplies in Hong Kong and South Korea, for example, because these countries can easily afford the imports needed to close gaps between national production and domestic consumption. By the same token, feed grains can be

Table 11.7 Food consumption indicators in Asia, 1979–1999

Country (ranked by income per capita 1999)	Years	Plant product consumption per capita (in cereal-equivalent tons) (1)	Animal product consumption per capita (in cereal-equivalent tons) (2)	Total food consumption per capita (in cereal-equivalent tons) (3)	Total food self-sufficiency* (4)
Hong Kong	1979–1981	0.26	1.14	1.40	0.48
	1989–1991	0.27	1.33	1.60	0.44
	1997–1999	0.24	1.55	1.79	0.25
South Korea	1979–1981	0.32	0.28	0.60	0.79
	1989–1991	0.31	0.48	0.80	0.69
	1997–1999	0.31	0.74	1.04	0.73
Malaysia	1979–1981	0.27	0.45	0.72	1.17
	1989–1991	0.26	0.60	0.86	1.35
	1997–1999	0.27	0.74	1.01	0.98
Thailand	1979–1981	0.24	0.33	0.57	1.33
	1989–1991	0.22	0.37	0.59	1.28
	1997–1999	0.25	0.42	0.66	1.23
Philippines	1979–1981	0.23	0.33	0.56	0.81
	1989–1991	0.24	0.36	0.60	0.74
	1997–1999	0.23	0.42	0.65	0.68
China	1979–1981	0.25	0.16	0.41	0.97
	1989–1991	0.27	0.32	0.59	0.98
	1997–1999	0.29	0.61	0.91	0.97
Sri Lanka	1979–1981	0.26	0.15	0.41	0.79
	1989–1991	0.24	0.15	0.39	0.70
	1997–1999	0.25	0.18	0.44	0.65

Country	Period			
Indonesia	1979–1981	0.25	0.12	0.37	0.95
	1989–1991	0.29	0.16	0.46	1.02
	1997–1999	0.32	0.19	0.51	0.97
India	1979–1981	0.22	0.15	0.38	0.97
	1989–1991	0.26	0.19	0.45	1.00
	1997–1999	0.26	0.22	0.48	1.01
Vietnam	1979–1981	0.23	0.16	0.39	0.93
	1989–1991	0.24	0.23	0.46	1.04
	1997–1999	0.26	0.30	0.56	1.07
Pakistan	1979–1981	0.22	0.35	0.57	0.70
	1989–1991	0.23	0.44	0.67	0.73
	1997–1999	0.24	0.51	0.74	0.76
Bangladesh	1979–1981	0.22	0.10	0.32	0.93
	1989–1991	0.23	0.10	0.33	0.91
	1997–1999	0.24	0.11	0.35	0.89
Laos	1979–1981	0.23	0.15	0.38	0.97
	1989–1991	0.23	0.18	0.41	0.99
	1997–1999	0.23	0.25	0.48	1.02
Cambodia	1979–1981	0.19	0.09	0.28	0.82
	1989–1991	0.20	0.20	0.40	0.97
	1997–1999	0.21	0.22	0.43	0.96
Nepal	1979–1981	0.20	0.26	0.46	1.00
	1989–1991	0.27	0.29	0.55	0.99
	1997–1999	0.25	0.29	0.53	0.99
Myanmar	1979–1981	0.26	0.17	0.43	1.08
	1989–1991	0.29	0.16	0.46	0.96
	1997–1999	0.31	0.18	0.49	1.00

Note: *Total food self-sufficiency equals total domestic food production divided by total domestic food consumption.
Source: Derived from FAOSTAT Agriculture Data, commodity balances and food supply data from 1979 to 1999 (www.fao.org).

purchased from other nations so that the growing demand for animal products can be satisfied. The situation is different, though, in poorer countries. If agricultural imports are not affordable and food aid is not available in sufficient quantities, then food insecurity is a concern if agricultural production consistently lags behind food consumption. Pakistan, for one nation, is illustrative in this regard.

In most countries where food self-sufficiency went up or down a little or did not change at all, per-capita livestock consumption has not grown very much, remains modest, or both. However, the situation is different in China, where the self-sufficiency ratio stood at 0.97 as the HRS was being implemented and remained at exactly the same level two decades later, on the eve of the twenty-first century. No one argues that failure to achieve a higher degree of self-sufficiency heralds impending famine in Asia's largest nation. However, Brown (1994) has contended that rapid economic growth in China could cause the country's agricultural imports to balloon, thereby driving up global commodity prices; the end result, he adds, would be aggravated food insecurity in poorer parts of the world, most notably Sub-Saharan Africa.[3] Other analysts discount this possibility. In a detailed statistical analysis of food demand trends in China, Huang *et al.* (2000) find no reason to expect a major increase in imports of feed grains and other commodities, since consumption growth is not expected to exceed increases in production by a wide margin. Thus, the price run-up that Brown (1994) foresees is not likely.

Finally, we note that frequently used measures of food security generally correlate in Asia with improved diets. These measures are provided in Table 11.8, in which countries are ranked according to food consumption per capita (column 1). In places with the highest average consumption, infant mortality (column 2), childhood malnutrition (column 3), and the incidence of poverty (column 4) are all negligible – even where the Gini index of income concentration (column 5) is elevated, as it is in Hong Kong. As always, there are some interesting exceptions to the general tendency. One is Sri Lanka, which is justly famous for its public health and family planning services – as reflected in infant mortality and poverty rates that are unexpectedly low in light of modest per-capita food consumption. Less encouraging for the small island nation off India's southeast coast is its childhood malnutrition rate of 33 percent. A less positive exception is Pakistan. There, food consumption per capita is higher than in the Philippines and Thailand, where average incomes are higher. Nevertheless, infant mortality is inordinately high. This disturbing indicator is related in part to the sluggish GDP growth that India's western neighbor registered during the 1990s.

Something else can be discerned from Table 11.8, which is that low economic inequality – as expressed by the Gini coefficient – does not correlate with limited

---

[3] A similar concern could be raised about India, where food self-sufficiency improved somewhat during the 1980s and 1990s in spite of substantial consumption growth. In part for cultural reasons, consumption of animal products is limited at present. However, a sharp upswing in demand for these products, as might well be observed if there is a large and sustained rise in GDP per captia, would cause Indian imports of feed grains to increase, perhaps enough to drive up global prices.

Table 11.8 Recent food insecurity indicators in Asia

Country (ranked by column 1)	Food consumption per capita 1997–1999 (in cereal-equivalent tons) (1)	Infant mortality rate 2002 (per 1,000 live births) (2)	Child malnutrition 2002 (% children <5 years) (3)	Poverty rate late 1990s/early 2000s (population below $1 per day) (4)	Gini coefficient late 1990s (5)
Hong Kong	1.79	–	–	–	43.4
South Korea	1.04	5	–	<2	31.6
Malaysia	1.01	8	–	<2	49.2
China	0.91	30	10	16.6	44.7
Pakistan	0.74	76	–	13.4	33.0
Thailand	0.66	24	–	<2	43.2
Philippines	0.65	28	–	14.6	46.1
Vietnam	0.56	20	34	17.7	36.1
Nepal	0.53	62	48	37.7	36.7
Indonesia	0.51	32	25	7.5	34.3
Myanmar	0.49	77	–	–	–
India	0.48	65	–	34.7	32.5
Laos	0.48	87	40	26.3	37.0
Sri Lanka	0.44	16	33	6.6	34.4
Cambodia	0.43	96	45	34.1	40.4
Bangladesh	0.35	48	48	36.0	31.8
Mongolia	–	58	–	13.9	44.0

Source: Table 11.7 and World Bank (2004).

poverty. In general, countries with coefficients below 40 have high poverty rates, just as the incidence of poverty is low where the inequality measure is above 40. Clearly, economic equality is not a good thing if it is only a consequence of universal poverty. By the same token, inequality is not discouraging if this is the result of a dynamic economy in which the earnings of some people are being pulled up faster than those of others. The latter experience happens to have been shared by a number of Asian nations in recent years.

11.5 Summary

By the turn of the twenty-first century, Asia was moving to the center of the world economic stage, thanks to GDP increases that no other part of the world – developing or otherwise – has matched. Since diminished natural increase has coincided with economic expansion, there has been a corresponding improvement in average living standards. It is undeniable that hundreds of millions of Asians remain impoverished. Also, economic performance in some countries compares unfavorably to regional trends, just as growth has been interrupted at times in nations where GDP per capita is much higher today than what it was a generation ago. Nevertheless, the region's economic accomplishments are impressive – impressive enough that issues that formerly aroused debate and concern only in the OECD, such as food safety and environmental protection, are becoming progressively more important in South Korea, Taiwan, and other nations.

Agriculture, which is the focus of much of the environmental debate, has had an important role in Asia's development. The Green Revolution allowed famine to be averted in places like the Indian subcontinent, even when average incomes were not rising quickly. During the last two decades of the twentieth century, China's economic transformation began with the abandonment of doctrinaire communism in the countryside. To be sure, challenges remain, such as meeting the increased demand for livestock products resulting from higher living standards. Asia is addressing such challenges, however, which is helping the region to grow more prosperous and food secure.

Study Questions

1. What obstacles to sustained economic growth must still be overcome in China? In India?
2. What Asian countries have experienced the greatest relative declines in human fertility in recent decades?
3. Why does a policy of one-child-per-family no longer make economic sense for China?

4. Why is fertilizer used more intensively in Asia than in other parts of the world?
5. In what ways was the Green Revolution particularly well suited to conditions in Asia?
6. Analyze the declines in food self-sufficiency – defined as domestic consumption (in cereal equivalents) divided by domestic production – that a number of Asian countries experienced in the late twentieth century.

12

Latin America and the Caribbean

Settled right after the last Ice Age, when low sea levels allowed the ancestors of Native Americans to travel from Siberia to Alaska, the Western Hemisphere experienced agricultural revolutions not connected in any way with those of Eurasia and Africa. Domestication of corn, potatoes, and other New World crops gave rise to a series of civilizations in Mesoamerica and the Andes. During the 1520s and 1530s, the Spanish conquest of the Aztec and Incan empires ushered in three centuries of colonial rule, which came to an end on the American mainland early in the nineteenth century.

Having achieved independence long before most of Africa, Asia, and the Middle East, not to mention the Caribbean, Latin America has undergone prolonged economic stagnation at times and prospered in other periods. For example, GDP per capita declined by one-third in Mexico between 1800 and 1850 (Engerman and Sokoloff, 1997), which set the stage for the loss of the northern third of the country to the United States during the 1840s as well as a French invasion in the 1860s. However, there was an upswing in commerce during the last third of the nineteenth century and the first two decades or so of the twentieth, as the Industrial Revolution created new demands for primary commodities produced throughout the Western Hemisphere and as transoceanic shipping costs were driven down by the introduction of steam power, metal hulls, and other advances.

Globalization came to a halt during the 1930s, when protectionism in the United States and other leading nations caused international trade to collapse. Shaken by the Great Depression, Argentina, Brazil, Mexico, and other countries did not fully engage the global economy after the Second World War, but instead tried to develop by sheltering domestic industry from foreign competitors. This approach, called Import-Substituting Industrialization (ISI) (Chapter 7), eventually led to large fiscal and trade deficits and the Debt Crisis of the 1980s (Fraga, 2004). In some ways, the region is still coming to terms with that crisis, the protectionist approach to development that precipitated it, and the pronounced socioeconomic inequality that is in part a legacy of Iberian colonialism.

12.1 Trends in GDP Per Capita

Although Asia has experienced greater relative improvements during the last three or four decades, living standards south of the US border still compare favorably with those of many other places. As pointed out in Chapter 9, per-capita income in the region was $6,620 in 1999, as opposed to $3,740 in East and Southeast Asia, $2,110 in South Asia, $5,000 in the Middle East and North Africa, $5,980 in Eastern Europe and the Former Soviet Union, and $1,500 in Sub-Saharan Africa (Table 9.1).

Latin America and the Caribbean will cease being the most prosperous part of the developing world if it undergoes another period like the 1980s, when GDP expansion failed to match population increases in most countries (Table 12.1, column 1). Prior to the "lost decade," the region covered trade and fiscal deficits, which expanded as governments tried to stimulate industrialization by manipulating market forces, by borrowing heavily from international banks, which were eager to lend "petrodollars" deposited by oil exporters that had benefited from energy price increases. Export growth during the 1980s (column 2) continued to be anemic. Moreover, with foreign banks no longer willing to lend anywhere in the region after Mexico defaulted on its foreign debt in 1982, a number of governments printed money to cover the difference between public spending and tax revenues. This ignited inflation (column 3), which halted or reversed capital accumulation in all but a few nations (column 4).

Aside from Colombia, the only important exception during these years was Chile. Major steps toward free trade having been taken during the 1970s, annual export growth in the country was nearly 7 percent during the 1980s. Although high by Asian standards, Chilean inflation compared favorably to what neighboring lands were suffering. Thus, conditions for investment were relatively positive. Thanks to an economic revival after 1984, yearly increases in GDP per capita averaged 2.6 percent for the entire decade.

Triple-digit inflation, which during the 1980s plagued Argentina, Brazil, and other countries and threatened in several other nations, takes a sizable toll. Not all prices march up in unison, which means that businesses must continuously review the terms under which output is sold and inputs are purchased. At the same time, households must be on constant alert lest inflation erode the real value of their savings. Toward the end of the lost decade, the costs of this vigilance had risen high enough that the political will required for macroeconomic reform was mustered. Fiscal deficits were trimmed, which eased inflationary pressures. This was a key element of structural adjustment, which the International Monetary Fund (IMF) was urging for countries caught up in the debt crisis. Other elements included trade liberalization and deregulation of domestic markets, freer capital flows, the privatization of state-owned enterprises (SOEs), and stronger property rights (Fraga, 2004).

Although reform varied from nation to nation, structural adjustment was followed by economic growth. Between 1990 and 1997, GDP per capita for Latin America and the Caribbean as a whole rose by 2.0 percent per annum (Ocampo, 2004). Inflation declined sharply in Argentina, Brazil, Peru, and other nations, which

Table 12.1 Patterns of economic growth in Latin America and the Caribbean, 1980–2003

A. Average annual rate of growth 1980–1990 (%)					B. Average annual rate of growth 1990–1999 (%)					C. Average annual rate of growth 2000–2003 (%)			
Country (ranked by column 1)	GDP per capita	Exports	Inflation	Gross investment	Country (ranked by column 5)	GDP per capita	Exports	Inflation	Gross investment	Country (ranked by column 9)	GDP per capita	Economic Freedom Index* – 2002	Freedom status** 2001–2002
(1)	(2)	(3)	(4)		(5)	(6)	(7)	(8)		(9)	(10)	(11)
Chile	2.6	6.9	20.7	4.3	Chile	5.7	9.7	8.6	11.4	Dominican Republic	2.3	3.00	F
Colombia	1.5	7.5	24.8	0.6	Dominican Republic	3.8	7.5	9.8	7.4	Chile	2.0	1.85	F
Dominican Republic	0.9	4.5	21.6	4.3	Peru	3.7	9.0	28.7	9.0	Ecuador	2.0	3.45	PF
Jamaica	0.8	5.4	18.6	4.1	Argentina	3.6	8.7	6.2	9.1	Costa Rica	1.3	2.65	F
Brazil	0.7	7.5	284.0	0.2	Uruguay	3.0	7.0	36.0	8.9	Peru	1.3	2.75	F
Costa Rica	0.3	6.1	23.6	5.2	El Salvador	2.8	11.7	8.1	7.2	Panama	1.0	2.70	F
Uruguay	−0.2	4.3	61.3	−8.2	Panama	2.4	0.0	2.0	12.1	Brazil	0.8	3.10	PF
Honduras	−0.4	1.1	5.7	−0.7	Costa Rica	2.1	9.7	16.8	3.4	Colombia	0.8	2.85	PF
Ecuador	−0.5	5.4	36.4	−2.9	Bolivia	1.8	4.9	9.4	10.1	Honduras	0.8	3.15	PF
Paraguay	−0.5	12.2	24.4	−1.4	Guatemala	1.6	6.4	10.7	5.0	El Salvador	0.5	2.05	F
El Salvador	−0.9	−3.4	16.3	2.2	Brazil	1.5	4.9	264.3	3.1	Jamaica	0.5	2.90	F
Mexico	−1.0	7.0	71.5	−3.4	Colombia	1.4	5.2	20.5	7.5	Mexico	0.5	2.90	F
Venezuela	−1.5	2.8	19.3	−5.4	Nicaragua	1.4	10.3	38.5	12.6	Bolivia	0.3	2.70	F
Panama	−1.6	−0.9	1.9	−12.8	Mexico	0.9	14.3	19.3	3.9	Guatemala	0.3	2.80	PF
Guatemala	−1.7	−1.8	14.6	−2.1	Honduras	0.3	2.0	19.8	6.0	Nicaragua	0.0	3.15	PF
Haiti	−2.1	1.2	7.5	−3.4	Ecuador	0.1	4.4	33.7	1.1	Haiti	−1.8	3.80	NF
Argentina	−2.2	3.8	391.1	−8.3	Paraguay	−0.3	5.1	13.7	1.5	Paraguay	−1.8	3.10	PF
Bolivia	−2.2	1.0	327.2	−10.7	Venezuela	−0.5	5.6	47.6	2.9	Argentina	−2.8	2.50	PF
Peru	−2.5	−1.6	231.3	−5.0	Jamaica	−0.8	0.1	25.8	3.9	Uruguay	−4.0	2.55	F
Nicaragua	−4.6	−3.9	422.3	−4.5	Haiti	−0.8	2.4	23.3	1.7	Venezuela	−5.0	3.65	PF

Note: *The Economic Freedom Index ranges from 1.00 (completely free markets) to 5.00 (completely repressed markets). See Heritage Foundation (2002).
**Political freedom status is F (free), PF (partly free), and NF (not free). See Freedom House (2002).
Source: World Bank (2000/2001), pp. 278–279, 294–295; World Bank (2005).

stimulated investment (Table 12.1, columns 7 and 8). Trade revived as well, thanks in no small part to major reductions in tariff and non-tariff barriers.[1] Annual export growth during the last decade of the twentieth century (column 6) exceeded what it had been during the 1980s in every country other than Brazil, Colombia, Ecuador, Jamaica, and Paraguay. Of the nations listed in Table 12.1, Haiti, Jamaica, Paraguay, and Venezuela were the only four where human numbers rose faster than GDP between 1990 and 2000 (column 5).

Living standards stagnated after 1997, including from 2000 through 2003 (column 9). Reflecting on what many of them call the lost half-decade, those who had criticized structural adjustment in the late 1980s and early 1990s now feel vindicated. These critics concede that trade has increased because of diminished protectionism, greater fiscal and monetary discipline has resulted in lower inflation, and investment has occurred. However, they also complain that GDP expansion has been modest and uneven (Ocampo, 2004). Fraga (2004) disputes this, contending that nations with sustained macroeconomic stability because of sound fiscal and monetary policies also have enjoyed consistent economic expansion. Chile falls into this category. So does Mexico, at least after 1995. In contrast, an inflexible exchange rate throughout the 1990s created currency over-valuation in Argentina, with negative impacts on the balance of trade. Meanwhile, public-sector indebtedness mounted, which set the stage for a severe financial crisis in late 2001. Much richer than its neighbors from the late 1800s through the middle of the twentieth century (Engerman and Sokoloff, 1997), Argentina is still dealing with the wreckage.

As do many of his colleagues from the region, Fraga (2004) sees economic populism, which favors regulation and protection of the national economy and is not fastidious about the control of inflation, as the main impediment to economic progress in Latin America and the Caribbean. The populist threat is largely contained in Chile, where sound economic policies have been followed by each of the democratic governments that have held office since military rule ended in 1990. The same probably holds for Mexico, where a single party had maintained an iron grip on public institutions for more than seven decades before losing the presidency in 2000; no future government is likely to take the country out of the North American Free Trade Agreement (NAFTA), which has eliminated all but a few barriers to trade among Canada, Mexico, and the United States. A Brazilian, Fraga (2004) is optimistic about his nation. At the turn of the twenty-first century, Colombia, which is the third most populous country in the region (after Brazil and Mexico), was embroiled in a multi-sided civil war in which the very survival of the national state hung in the balance. Having received considerable support from the United States, it now appears to be making headway against the forces of anarchy (Córdoba, 2004); particularly if left- and right-wing terrorists and their drug-dealing

---

[1]Lora (2001, cited by Fraga, 2004) reports that the average tax on imports for the 12 largest Latin American nations fell from 49 percent in 1985 to less than 20 percent in 1994. Whereas non-tariff barriers applied to 38 percent of imports in the middle 1980s, this share was a little over 6 percent 10 years later.

associates are dealt a decisive defeat, there is reason to hope for application of the standard model (Chapter 8).

The prospects are more clouded for the other three countries that along with Brazil, Chile, Colombia, and Mexico produce 90 percent of the region's GDP. In Peru, sharp divisions between indigenous groups and the rest of the population, which date back to the Spanish conquest, regularly express themselves as populism. Faithful to a demagogic tradition that antedates Juan Domingo Perón's election to the presidency in 1946, Argentina's current leadership seems more comfortable at times whipping up public resentment over recent economic problems instead of actually solving them. In Venezuela, Hugo Chávez is cannily aggrandizing autocratic powers. Briefly imprisoned for leading a failed coup d'etat in the early 1990s, Chávez won the presidency in a landslide at the end of the decade. He also makes no secret of his affinity for populism, which is perpetuating the long economic slide of the region's leading producer and exporter of petroleum.[2]

Venezuela is representative, albeit extremely so, of most nations in the Western Hemisphere with pronounced declines in GDP per capita in recent years. Other than Uruguay, which is democratic and generally follows sound economic policies, few of these nations are fully free (Table 12.1, column 10). The failure of populist governments to follow sound macroeconomic policies and to let markets work are indicated by unfavorable indices of economic freedom (column 11) as well as the absence of economic progress.

12.2 Population Dynamics

Aside from decrying slow economic expansion after 1997, the critics of structural adjustment argue that economic inequality has remained the same or grown worse and that non-economic measures of human well-being have deteriorated (Ocampo, 2004). As do all other experts on Latin America and the Caribbean, Fraga (2004) admits that inequality, which is an age-old problem in the region, remains acute. However, he notes that Gini coefficients have declined in some nations – including Brazil, where there has long been a yawning gap between the privileged few and the rest of the population.

The same economist disagrees that structural adjustment has caused social conditions to deteriorate with no less vigor than he defends the causal linkage between sound macroeconomic, trade, and other policies and improved living standards. In particular, he points out that progress toward wider adult literacy, increased life expectancy at birth, diminished infant mortality, and greater school enrollment was made during the last decade of the twentieth century. Indeed, illiteracy fell, life spans went up, infant mortality became rarer, and the percentage of children

---

[2]Neither does he hide his admiration for Fidel Castro, one of the world's few remaining communist tyrants and the architect of prolonged economic stagnation in Cuba. The dimensions of this stagnation are not examined in this chapter because reliable economic data for the country are lacking.

attending school increased between 1990 and 2000 in each of the seven countries that account for nine-tenths of regional GDP – even in Venezuela, where GDP per capita was falling (Fraga, 2004). As in other parts of the world, non-economic indicators of well-being in Latin America and the Caribbean have converged toward indicators in affluent nations even in the absence of economic convergence, conventionally defined.

Reduced Human Fertility

Mortal threats to life are contained at least as well in Latin America and the Caribbean as in other non-affluent parts of the world. Life expectancy at birth in the region is 70 years – versus 68 or 69 years in East and Southeast Asia, the Middle East and North Africa, and Eastern Europe and the Former Soviet Union, 63 years in South Asia, and less than 50 years south of the Sahara (Table 9.2). As is to be expected, progress against mortal illness has been followed by lower human fertility.

Alhough a little less dramatic than the fertility decline that Asia experienced in the late twentieth century (Table 11.2), total fertility rates (TFRs) have fallen markedly since 1980 throughout the Western Hemisphere (Table 12.2, columns 1 and 2). Distinct from demographic conditions in Asia, where human fertility has reached the replacement level in nearly every country with a GDP per capita of $3,000 or more and where the average number of births per woman exceeds 2.1 nearly every place that has not achieved this standard of living, no similar threshold exists in Latin America and the Caribbean, at least not yet. As a rule, families are larger in the Western Hemisphere than in Asia. This is true of countries with living standards (column 3) well above Asian norms, such as Argentina, Colombia, and Venezuela. It is also true of Bolivia and Honduras, where GDP per capita is below $3,000. Brazil, Chile, and Uruguay are the only three countries with TFRs at the replacement level.

Fertility differences cannot really be ascribed to the historical predominance of Roman Catholicism in the region. TFRs will soon reach 2.1 births per woman in Argentina and Colombia, each of which has a constitution that establishes Catholicism as the state religion. In these countries and all their hemispheric neighbors, religious affiliation has not prevented women and their partners in recent decades from acting on the desire to have fewer children. A stronger sociocultural influence is ethnicity. In particular, large families are the rule among indigenous peoples, which have been marginalized since the Spanish conquest. This is reflected in high TFRs for Bolivia, Guatemala, and Paraguay, where these groups remain impoverished and comprise a majority of the population.

Notwithstanding sociocultural differences, factors that favor smaller families are all but irresistible throughout the region. Between 1980 and 2000, the TFR fell from 5.9 to 4.2 births per woman in Haiti, where grinding poverty is the norm. Even greater declines occurred in Honduras and Nicaragua, which are poorer than the rest of Central America. The same trend was evident in non-affluent nations with large indigenous populations – Ecuador and Peru as well as Bolivia, Guatemala,

Table 12.2 Social indicators for Latin America and the Caribbean in 2002

Country (ranked by column 3)	Total fertility rate		2002 Income per capita ($PPP)	Urban % of total population in 2002	Adult illiteracy rate in 2002 (% people 15 and over)		Female % of labor force		Contraceptive prevalence rate 1990–2002 (% women 15–49)
	1980	2002			Males	Females	1980	2002	
	(1)	(2)	(3)	(4)	(5)	(6)	(7)	(8)	(9)
Argentina	3.3	2.4	10,880	88	3	3	27.6	34.4	–
Chile	2.8	2.2	9,820	86	4	4	26.3	34.5	–
Mexico	4.7	2.4	8,970	75	7	11	26.9	33.8	65
Costa Rica	3.6	2.3	8,840	60	4	4	20.8	31.6	–
Uruguay	2.7	2.2	7,830	92	3	2	30.8	42.2	–
Brazil	3.9	2.1	7,770	82	14	13	28.4	35.5	77
Dominican Republic	4.2	2.6	6,640	67	16	16	24.7	31.4	65
Colombia	3.9	2.5	6,370	76	8	8	26.2	39.1	72
Panama	3.7	2.4	6,170	57	7	9	29.9	35.7	–
Venezuela	4.2	2.7	5,380	87	7	8	26.7	35.4	–
Peru	4.5	2.6	5,010	73	5	15	23.9	31.9	63
El Salvador	4.9	2.9	4,890	62	18	24	26.5	37.3	57
Paraguay	5.2	3.8	4,610	57	6	8	26.7	30.4	52
Guatemala	6.3	4.3	4,080	40	24	39	22.4	30.1	35
Jamaica	3.7	2.3	3,980	57	17	9	46.3	46.2	64
Ecuador	5.0	2.8	3,580	64	7	10	20.1	28.7	61
Honduras	6.5	4.0	2,600	55	25	25	25.2	32.6	53
Nicaragua	6.3	3.4	2,470	57	34	33	27.6	36.6	52
Bolivia	5.5	3.8	2,460	63	8	21	33.3	38.0	48
Haiti	5.9	4.2	1,610	37	48	52	44.6	42.8	23

Source: World Bank (2004).

and Paraguay. Everywhere else, women generally are bearing fewer than three children each.

Rapid movement toward replacement fertility is very probable. Aside from enjoying average incomes that are the envy of most developing nations, Latin America and the Caribbean are quite urbanized (Table 12.2, column 4), much more than Asia for example. Also, levels of educational attainment do not differ greatly between the sexes. True, male illiteracy (column 5) is appreciably lower than female illiteracy (column 6) in a handful of places: El Salvador and Guatemala in Central America as well as Bolivia and Peru in the Andes. Elsewhere, illiteracy rates are comparable. In spite of an increase since 1980, female participation in the labor force (columns 7 and 8) is a little below that of Asia. Nevertheless, contraceptive use (column 9) is, in general, very high. Indeed, choices about birth control made by women south of the US border are pretty much the same as choices made by females in affluent parts of the world.

Natural Increase

Latin America and the Caribbean are well beyond the stage in which population growth is spurred by diminished mortality. Haiti, where public sanitation and health care are appalling, is about the only place where substantial declines in the crude death rate (CDR) (Table 12.3, column 2) could be achieved, through investment in drinking water systems, hospitals and clinics, etc. The same sort of investment also could drive down the death rate by one or two per thousand in Bolivia, Guatemala, and other poor nations. Elsewhere, CDRs cannot fall any more and, if anything, will rise this century as populations age. Uruguay is an unusual case. In demographic terms more like an OECD member than a developing country, it has the region's second highest death rate, exceeded only by Haiti's. In addition, Uruguay's CDR is bound to rise as its population matures. The same thing is happening in Argentina.

With CDRs at or near their respective nadirs and with TFRs approaching the replacement level in all large countries as well as a number of smaller states, natural increase (Table 12.3, column 4) in Latin America and the Caribbean is mainly a consequence of demographic momentum, which is yet another demographic parallel with Asia. A gap between the birth rate (column 3) and the CDR persists for this reason in Argentina, Uruguay, and other places where the demographic transition is far advanced. In other countries, adolescents and young adults comprise a large segment of the population. Accordingly, differences between crude birth rates (CBRs) and CDRs are greater and natural increase will persist farther into the future. As in Asia, however, natural increase is diminishing, which means that current gaps between birth and death rates are smaller than average demographic expansion during the last two decades of the twentieth century (column 5).

Perhaps the most important demographic difference between Asia and the developing nations of the Western Hemisphere has to do with international migration. The attraction of remunerative employment in the United States and other affluent

Table 12.3 Population dynamics in Latin America and the Caribbean for 2002

Country (ranked by column 1)	Income per capita 2002 ($PPP)	CDR – 2002 (per 1,000 population)	CBR – 2002 (per 1,000 population)	Rate of natural increase (%)	Average annual population growth 1980–2000 (%)
	(1)	(2)	(3)	(4)	(5)
Argentina	10,880	8	19	1.1	1.2
Chile	9,820	5	17	1.1	1.5
Mexico	8,970	4	20	1.6	1.8
Costa Rica	8,840	4	20	1.5	2.5
Uruguay	7,830	10	16	0.6	0.6
Brazil	7,770	7	19	1.2	1.7
Dominican Republic	6,640	7	23	1.6	1.9
Colombia	6,370	6	21	1.6	2.0
Panama	6,170	5	20	1.5	1.9
Venezuela	5,380	5	23	1.9	2.3
Peru	5,010	6	22	1.6	2.0
El Salvador	4,890	6	26	2.0	1.5
Paraguay	4,610	5	30	2.5	2.6
Guatemala	4,080	7	33	2.7	2.6
Jamaica	3,980	6	20	1.4	0.9
Ecuador	3,580	6	23	1.7	2.2
Honduras	2,600	6	30	2.3	2.9
Nicaragua	2,470	5	29	2.4	2.8
Bolivia	2,460	8	29	2.1	2.3
Haiti	1,610	14	32	1.9	2.0

Source: World Bank (2002a), pp. 48–50; World Bank (2004).

settings has caused millions of people to relocate from the Caribbean, Central America, Mexico, and even the Andean Region. As a result, overall population growth – by definition, equal to natural increase plus the difference between immigration and emigration – is exceeded by the difference between birth and death rates in the latter places, just as the reverse holds in countries with net immigration. This state of affairs is unlikely to change very much as long as international differences in GDP per capita remain sizable.

12.3 Agricultural Development

In contrast to the general demographic congruence between Asia and Latin America and the Caribbean, agricultural differences between the two regions are considerable.

These differences derive largely from varying resource endowments. As noted in Chapter 3, the scope for extensifying farming is much greater in the New World – in South America, to be specific – than in China, India, and neighboring lands. But just as land is abundant in much of the region, labor is relatively scarce. Other than in Caribbean micro-states, such as Antigua and Barbados, rural population density exceeds 300 per square kilometer in just five small nations: Costa Rica, El Salvador, and Guatemala in Central America as well as Haiti and Jamaica in the Caribbean (Table 12.4, column 1). Of the 15 Asian countries listed in Table 11.4, only Cambodia and Mongolia are below this density level.

Even within the region, agriculture has developed in diverse ways. By no means do modes of production in the temperate, southern reaches of the American mainland seem exotic to farmers visiting from the United States. The crops are the same: corn, soybeans, and wheat in Argentina and Uruguay and soybeans in southern and central Brazil. Machinery is widely used, which is expected since labor is scarce. Rural areas in Argentina and Uruguay, for example, are more sparsely populated than the US countryside, where there are 37 people per square kilometer of arable land (Table 10.5). Also, commercial farming is complemented by agribusinesses that efficiently supply a wide range of inputs and marketing services. As noted in Chapter 7, Chilean agriculture has been growing as rapidly as the rest of the economy because local service providers are adept at adding value to the horticultural products in which the country holds a comparative advantage.

Considerable differences exist between agriculture in Latin America's southern cone and farming closer to the equator. In part, the differences have to do with commodities produced in the tropics and subtropics: sugar and bananas on lowland plantations and coffee at higher altitudes. But in other places, people raise crops that their distant ancestors domesticated, but which now are produced more efficiently far away in temperate latitudes. One indigenous food is corn, which was first cultivated thousands of years ago in Mesoamerica. Another is the potato, which is from the Andes. By and large, production of these traditional crops in their places of origin involves limited use of machinery, commercial fertilizer, and other modern inputs. Yields are minimal.

Rural hinterlands in the Andes and Mesoamerica are among the poorest parts of the hemisphere. Even after most Native Americans died from diseases brought over to the New World by Columbus and the Europeans who followed him, the two areas offered large, subjugated populations that Spaniards could put to work in mines and on large estates (Engerman and Sokoloff, 1997). In spite of land reforms in the middle 1900s, most rural Native Americans in places like Southern Mexico (e.g., the state of Chiapas, where there was an uprising in the middle 1990s), Guatemala, Peru, and Bolivia possess tiny holdings, called *minifundios*. Given the choice, many of these people would work outside agriculture. But given their lack of educational credentials and the economic isolation associated with a deficient road network, they are relegated to subsisting on their own limited harvests.

People from outside the region who visit the Andean or Mesoamerican countryside are often struck by the juxtaposition of commercial farms, which can be quite

Table 12.4 Agricultural inputs in Latin America and the Caribbean, 1979–2001

Country (ranked by income per capita 2002)	Rural population density (people per sq. km of arable land) 2001	Fertilizer use (kg per ha)		Irrigated land (% of cropland)		Agriculture machinery (tractors per sq. km arable land)	
	(1)	1979–1981 (2)	1999–2001 (3)	1979–1981 (4)	1999–2001 (5)	1979–1981 (6)	1999–2001 (7)
Argentina	9	4	25	5.2	4.5	0.6	0.9
Chile	108	34	242	31.1	82.7	0.9	2.7
Mexico	102	57	74	20.3	23.1	5.0	1.3
Costa Rica	697	265	710	12.1	20.6	2.1	3.1
Uruguay	20	56	85	5.4	13.5	2.4	2.6
Brazil	54	78	110	3.0	4.4	1.2	1.4
Dominican Republic	263	57	87	11.7	17.2	0.2	0.2
Colombia	420	81	240	7.7	20.2	0.8	0.8
Panama	230	69	58	5.0	5.1	1.2	0.9
Venezuela	122	70	101	10.1	16.9	1.3	1.9
Peru	191	38	72	32.3	28.4	0.4	0.4
El Salvador	370	138	119	4.6	5.0	0.6	0.5
Paraguay	77	4	23	3.4	2.2	0.5	0.6
Guatemala	516	73	145	5.0	6.8	0.3	0.3
Jamaica	648	123	109	10.1	8.8	2.0	1.8
Ecuador	285	47	118	24.8	29.0	0.4	0.9
Honduras	288	17	141	4.1	5.2	0.2	0.4
Nicaragua	117	42	16	6.2	4.5	0.2	0.2
Bolivia	110	2	2	6.6	4.2	0.2	0.2
Haiti	664	4	16	6.4	6.8	0.0	0.0

Source: World Bank (2004).

profitable, and impoverished *minifundios*. The same juxtaposition occurs in the Caribbean (although most of the region's labor now works in light manufacturing, tourism, and other non-agricultural sectors), northeastern Brazil, and other places. The mix of modern, commercial farming, often on large estates, and subsistence or near-subsistence production on small parcels within individual countries must be kept in mind when reviewing national-level data on the use of agricultural inputs. Similarly, pronounced variation from one part of Latin America to another must be remembered when the entire region is being examined.

Factor Use

With respect to some categories of agricultural input use, it is hard to distinguish between the northern- and southern-most reaches of the Western Hemisphere. In Canada, which has a comparative advantage in grain production, fertilizer applications averaged 55 kilograms per hectare between 1999 and 2001 (Table 10.5). This is between contemporaneous averages (Table 12.4, column 3) for Argentina (25 kilograms per hectare) and Uruguay (85 kilograms per hectare), which also compete successfully in international grain markets. Likewise, Argentina and Australia closely resemble each other in some ways. For example, the irrigated portion of arable land (column 5) is almost identical in Argentina (4.5 percent) and the antipodal continent (4.7 percent).

For Chile, which does not produce grain as efficiently as do its neighbors east of the Andes, a better benchmark from outside the region is New Zealand. Although dairy products and sheep are mainstays of agriculture in the latter nation, both specialize in horticultural products. Rates of fertilizer application are high in both countries. Differences in the irrigated portion of farmland – over four-fifths in Chile versus 9 percent or so in New Zealand – relate to very dry conditions along most of South America's Pacific Coast. These conditions oblige Chilean farmers, and their counterparts in coastal Peru and Ecuador, to divert water onto their fields from rivers and streams flowing out of nearby mountains. In contrast, rainfall is generous in New Zealand – in fact, excessive enough to preempt all farming in some places.

Differences between factor use in the southern cone and factor use in Australia, Canada, and New Zealand relate mainly to the high opportunity cost of labor in the latter nations. For example, ratios of tractors to land (Table 12.4, columns 6 and 7) are much higher among OECD members. However, farming is considerably more mechanized in Argentina, Chile, and Uruguay than in other parts of the hemisphere where GDP per capita is lower.

Aside from the three southern-most states in the region, all Latin American and Caribbean countries are tropical or subtropical. Some of the latter countries display similarities with Asian nations in the same latitudes. One of these is Colombia, where a high ratio of agricultural workers to arable land reflects a comparative advantage in the labor-intensive production of cut flowers. In this northern Andean nation, fertilizer use averaged 240 kilograms per hectare from 1999 through

2001 – an application rate comparable to that of Chile. Another country that resembles Asia is Costa Rica, which has a higher rural population density than any other nation listed in Table 12.4. Exceeding 700 kilograms per hectare, fertilizer applications in Costa Rica are elevated. By Asian standards, development of water resources for agriculture may seem limited. However, precipitation is abundant in this small state, which nevertheless irrigates a larger portion of its arable land than all but four other countries in the Western Hemisphere. Along with Chile, Ecuador, and Peru, Mexico has a higher irrigated percentage – mainly because of modest precipitation in much of the country, including the Sinaloa Valley and other settings where fruits and vegetables are produced for US markets.

Elsewhere in the American tropics and subtropics, resemblances with Asia are few. Even though Brazil has small-holder farming, along with the poverty that normally accompanies this activity, agriculture in Latin America's leading nation is eminently commercial and directly reflects the relative scarcity of labor. No more than one in every 20–25 hectares of arable land is irrigated, and crop production is highly mechanized. Widespread use of machinery and the development by national scientists of crop varieties that thrive in local conditions also have allowed Brazilians to extend soybean farming into the central part of the country, thereby increasing the competition that US exporters face in global markets.[3]

Other countries in the low latitudes have relative endowments of land and people closer to those of Asia, but have experienced less agricultural intensification. In some places, structural transformation helps to explain the discrepancy. For example, improved employment prospects outside of agriculture, including for rural households, are a reason why fertilizer applications in El Salvador, Panama, and Jamaica were lower between 1999 and 2001 than what these had been between 1979 and 1981. Even though the ratio of rural population to arable land is high in each of these three places, the option of non-agricultural work discourages spending on soil fertility.

The same three countries also have their share of *minifundios*, which are plentiful in nations characterized by severe poverty in the countryside, and levels of fertilizer use and irrigation that are modest by the standards of Asian nations with comparable ratios of rural population to arable land. Some of these nations are in or near the Andes: Paraguay, Bolivia, Peru, and Ecuador. Others are in Mesoamerica: Guatemala, Honduras, and Nicaragua, as well as southern Mexico. In each of these places, large segments of the rural population are experiencing material deprivation comparable to that of the poorest African and South Asian peasants. At an extreme, Haiti, the most miserable place in the Western Hemisphere, is Sub-Saharan in all important respects other than location.

---

[3] As more and more grasslands along the southeastern fringes of the Amazon Basin are converted to soybean fields, new routes to foreign markets are being developed. One option is to pave a road that runs north to Santarem, a port on the Amazon River located a few hundred kilometers upstream from the Atlantic Ocean. Environmentalists around the world are greatly concerned about the agricultural colonization and expanded logging that are bound to occur alongside the road as it is improved (Anonymous, 2004).

Intensification, Extensification, and Output Growth

Although yield growth has had a greater impact than expanded agricultural land use on production, the difference between the two in Latin America and the Caribbean has not been as great as in Asia (Chapter 11). From 1970 to 2000, yield increases (Table 12.5, column 3) greatly exceeded area expansion (column 2) for corn, rice, and wheat. In contrast, extensification was the main cause of output growth (column 4) for oil crops, mainly because of dramatic expansion (from a small initial base) in the area planted to soybeans in Brazil and other nations. As in other regions, extensification has been an important driver of increased fruit and vegetable output in Latin America and the Caribbean.

Where impoverished peasants with smallholdings and using few purchased inputs do a large share of the farming, trends in production per hectare have been disappointing. Cereal yields actually fell during the last two decades of the twentieth century in Haiti and Jamaica (Table 12.6, columns 1 and 2). The only three nations in Mesoamerica and the Caribbean to score more than a 33 percent gain during the same period were Costa Rica, the Dominican Republic, and Panama. Peru achieved this feat as well, although Ecuador to the north and Bolivia and Paraguay to the southeast did not.

Where commercial producers predominate and where they have found it in their interest to use more fertilizer and other purchased inputs, production per hectare has risen by 50 percent or more. This is obviously the case in the southern

Table 12.5 Trends in crop area, yield, and output in Latin America, 1970–2000

	2000 Production (million tons)	Average annual change, 1970–2000 (%)		
		Area	Yield	Production
	(1)	(2)	(3)	(4)
A. Crops				
Fruits	99	2.8	0.1	2.8
Maize	76	0.3	2.1	2.3
Roots and tubers	53	−0.1	0.4	0.2
Vegetables	32	1.3	1.8	3.3
Wheat	24	0.4	2.1	2.5
Rice	23	−0.1	2.3	2.2
Oil crops	16	3.1	2.4	5.7
Fibers	2	−3.8	2.8	−1.1
B. Animal products				
Total milk	60	–	–	2.9
Total meat	31	–	–	3.5
Total eggs	5	–	–	4.3
Total wool	0.2	–	–	−2.0

Source: Dixon and Gulliver with Gibbon (2001), pp. 269, 270.

Table 12.6 Agricultural and food output in Latin America and the Caribbean, 1979–2002

Country (ranked by income per capita 2002)	Cereal yield (kg per ha)		Arable and permanent cropland (1,000 ha)		Food Production Index (1989–1991 = 100)		Agricultural productivity (value added per worker – 1995 $)	
	1979–1981 (1)	2000–2002 (2)	1980 (3)	1999 (4)	1979–1981 (5)	2000–2002 (6)	1979–1981 (7)	1999–2001 (8)
Argentina	2,184	3,347	35,200	27,200	91.7	142.5	7,148	10,352
Chile	2,124	4,827	5,530	2,294	71.5	140.2	3,488	6,040
Mexico	2,164	2,847	23,330	27,300	85.3	135.7	1,482	1,802
Costa Rica	2,498	3,943	598	505	69.5	150.0	3,139	5,259
Uruguay	1,644	3,336	1,449	1,307	87.1	124.8	6,563	8,366
Brazil	1,496	2,861	71,120	65,200	69.5	153.2	2,049	4,779
Dominican Republic	3,024	4,262	1,420	1,571	85.2	107.8	2,129	3,351
Colombia	2,452	3,343	5,650	4,364	75.5	120.3	3,034	3,566
Panama	1,524	2,488	574	655	85.5	105.8	–	2,888
Venezuela Republic	1,904	3,360	3,755	3,490	80.2	135.0	3,935	5,304
Peru	1,946	3,241	3,400	4,210	77.3	175.0	1,299	1,835
El Salvador	1,702	2,181	725	810	88.9	111.7	1,925	1,712
Paraguay	1,535	1,976	1,920	2,285	60.8	141.0	2,641	3,389
Guatemala	1,578	1,775	1,750	1,905	68.0	136.2	2,143	2,115
Jamaica	1,667	1,172	265	274	93.6	125.9	1,123	1,534
Ecuador	1,633	2,132	2,620	3,001	77.4	153.8	3,839	3,313
Honduras	1,170	1,374	1,757	1,827	88.3	121.1	696	1,005
Nicaragua	1,475	1,689	1,246	2,746	117.8	154.3	1,549	1,729
Bolivia	1,183	1,730	3,370	2,205	71.5	151.6	693	750
Haiti	1,009	868	890	910	101.2	101.7	–	–

Source: World Bank (2004); FAO Production Yearbook (1983), pp. 48–50; FAO Production Yearbook (2000), pp. 5–8.

cone and Brazil. It is also true in Venezuela. During the 1980s and 1990s, cereal yields in Colombia only went up by 36 percent. However, this country does not produce grain (other than rice) as efficiently as its trading partners in temperate latitudes. Instead, Colombian agriculture specializes in coffee and, these days, cut flowers. The global dominance achieved by producers of the latter good is indicated both by the greenhouses that extend for kilometers out from the airports of Bogotá and other major cities and by the tags reporting country of origin in florists' shops throughout North America and Europe.

In a statistical analysis, Southgate (1994) has shown that rapid increases in crop yields tend to coincide with slow geographic expansion of agriculture in the Western Hemisphere. By the same token, sluggish growth in production per hectare often goes hand in hand with the rapid conversion of forests and other habitats to farmland. This finding is generally consistent with land-use trends reported in columns 3 and 4 of Table 12.6. To be sure, there are some nations, including Bolivia as well as smaller countries in Mesoamerica and the Caribbean, where virtually all soils that lend themselves to crop production were occupied by farmers long ago. In these nations, agricultural extensification during the 1980s and 1990s was modest, even if production per hectare was not going up very much. Conversely, extensification can easily occur in spite of appreciable yield growth if land that is well suited to farming is covered with trees and other natural vegetation. Expansion of the area planted to soybeans in Brazil is illustrative in this regard.

Although further increases in the geographic domain of agriculture are possible in Brazil and a few of its neighbors, intensification has been the primary response of Latin American and Caribbean farmers to demand growth. Where yields have risen appreciably, in Brazil, the southern cone, and other places, production (Table 12.6, columns 5 and 6) has gone up at least as much as human numbers. In contrast, output increases have been disappointing in poorer parts of the region, where there has been more population growth and where yields have been stagnant.

A similar pattern holds for trends in agricultural total factor productivity (TFP). As in other parts of the world, value added per agricultural laborer (Table 12.6, columns 7 and 8) correlates with GDP per capita as well as agriculture's mechanization and commercial orientation. Needless to say, this indicator of productivity is higher in the southern cone, Venezuela, and Costa Rica, not to mention the more prosperous parts of Brazil, Colombia, and Mexico. In contrast, value added per agricultural worker is low in impoverished settings in the Andes, Mesoamerica, and the Caribbean.

12.4 Dietary Change, Consumption Trends, and Food Security

As is the case in Asia, where as already emphasized demographic trends are much like those of the Western Hemisphere, population growth is having a smaller impact on food demand in Latin America and the Caribbean with each passing decade. The influence of improved living standards on per-capita consumption is not as great in the latter region, for the simple reason that GDP per capita has not been increasing very rapidly in most countries. However, the retail side of the food economy has been greatly affected by the proliferation of supermarkets and other large stores.

As Reardon and Berdegué (2002) have documented, large retail outlets, which accounted for 10–20 percent of all food sales in 1990, now account for three-fifths of the total. This growth has far-reaching implications for small growers, since large retailers prefer to do business with suppliers that can provide steady and sizable flows of high-quality produce. Moreover, there is little chance of a return to small shops and traditional markets. Employing managerial techniques nearly identical to those used by their counterparts in the OECD, supermarket chains keep track of inventories and manage deliveries with considerable efficiency. Also, they are better at controlling quality. This is an important selling point for wealthy, middle class, and even lower middle class households, which comprise the majority of consumers in Latin America and the Caribbean.

Just as people south of the US border buy food in stores much like those in a rich country, dietary composition is changing in some places much as it is in affluent settings. For example, per-capita consumption of animal products (Table 12.7, column 2) has been on a downward path in Costa Rica. The same thing is happening in Argentina, where consumption of meat, dairy goods, and other livestock products 25 years ago was elevated even by rich countries' standards. Elsewhere in the region, a positive correlation holds between how many animal products people eat and their earnings, exactly as in Asia and other parts of the developing world. During the 1980s and 1990s, rising consumption of these goods accounted for the entire increase in per-capita food intake (column 3) in Brazil, Chile, Mexico, and a number of other countries. In places like Haiti and Venezuela, where GDP per capita has been contracting for many years, diminished per-capita food intake mainly reflects reduced consumption of livestock products.

In light of the agricultural potential of Latin America and the Caribbean, which is not close to being fully exploited, increases in per-capita food intake do not always result in reduced food self-sufficiency (Table 12.7, column 4), defined here as in other regionally focused chapters as cereal-equivalent production divided by cereal-equivalent consumption. Mexico exhibits a pattern familiar in Asia, where growth in per-capita consumption mainly has to do with livestock products and usually coincides with declines in food self-sufficiency. Of course, Mexico deals with the latter decline exactly as various countries on the other side of the Pacific have done, by trading its non-agricultural exports (e.g., petroleum and tourism services) for imported food. A different approach makes sense in Brazil, Chile, and other nations, where food self-sufficiency has gone up as diets have improved. In the latter countries, a comparative advantage in agriculture is exploited to the benefit of living standards, generally, and how well people eat, specifically.

Table 12.7 Food consumption and production trends in Latin America and the Caribbean, 1979–1999

Country (ranked by income per capita 1999)	Years	Plant product consumption per capita (in cereal-equivalent tons) (1)	Animal product consumption per capita (in cereal-equivalent tons) (2)	Total food consumption per capita (in cereal-equivalent tons) (3)	Total food self-sufficiency* (4)
Argentina	1979–1981	0.25	2.30	2.55	1.26
	1989–1991	0.24	1.84	2.08	1.30
	1997–1999	0.25	1.81	2.07	1.41
Chile	1979–1981	0.26	0.69	0.95	0.69
	1989–1991	0.24	0.77	1.01	0.77
	1997–1999	0.26	0.99	1.25	0.99
Mexico	1979–1981	0.30	0.71	1.02	0.87
	1989–1991	0.30	0.74	1.05	0.79
	1997–1999	0.30	0.85	1.15	0.75
Costa Rica	1979–1981	0.25	0.81	1.06	1.19
	1989–1991	0.27	0.81	1.08	1.09
	1997–1999	0.27	0.76	1.02	1.12
Uruguay	1979–1981	0.21	1.98	2.20	1.98
	1989–1991	0.19	1.74	1.93	1.74
	1997–1999	0.21	2.06	2.27	2.06
Brazil	1979–1981	0.27	0.74	1.01	1.05
	1989–1991	0.27	0.91	1.18	1.02
	1997–1999	0.27	1.12	1.40	1.07
Dominican Republic	1979–1981	0.23	0.44	0.67	0.97
	1989–1991	0.23	0.49	0.72	0.88
	1997–1999	0.23	0.57	0.79	0.79

(Continued)

Table 12.7 (Continued)

Country (ranked by income per capita 1999)	Years	Plant product consumption per capita (in cereal-equivalent tons) (1)	Animal product consumption per capita (in cereal-equivalent tons) (2)	Total food consumption per capita (in cereal-equivalent tons) (3)	Total food self-sufficiency* (4)
Colombia	1979–1981	0.23	0.63	0.86	0.96
	1989–1991	0.23	0.67	0.91	0.97
	1997–1999	0.25	0.69	0.94	0.91
Panama	1979–1981	0.21	0.83	1.05	0.78
	1989–1991	0.21	0.81	1.02	0.84
	1997–1999	0.22	0.81	1.03	0.84
Venezuela	1979–1981	0.27	0.90	1.16	0.78
	1989–1991	0.24	0.70	0.94	0.87
	1997–1999	0.22	0.71	0.93	0.86
Peru	1979–1981	0.21	0.39	0.61	1.18
	1989–1991	0.2	0.35	0.55	1.36
	1997–1999	0.25	0.40	0.66	2.04
El Salvador	1979–1981	0.23	0.33	0.56	0.87
	1989–1991	0.25	0.29	0.54	0.85
	1997–1999	0.25	0.34	0.60	0.85
Paraguay	1979–1981	0.24	1.13	1.37	0.95
	1989–1991	0.23	1.04	1.27	1.24
	1997–1999	0.23	1.28	1.51	1.05

Guatemala	1979–1981	0.25	0.22	0.46	1.17
	1989–1991	0.26	0.24	0.51	1.11
	1997–1999	0.24	0.30	0.53	1.03
Jamaica	1979–1981	0.26	0.54	0.80	0.61
	1989–1991	0.25	0.61	0.86	0.58
	1997–1999	0.26	0.64	0.90	0.56
Ecuador	1979–1981	0.23	0.50	0.73	0.92
	1989–1991	0.25	0.49	0.74	0.97
	1997–1999	0.26	0.63	0.90	0.99
Honduras	1979–1981	0.22	0.38	0.60	1.22
	1989–1991	0.24	0.34	0.57	1.03
	1997–1999	0.23	0.37	0.60	0.83
Nicaragua	1979–1981	0.23	0.51	0.73	1.01
	1989–1991	0.23	0.31	0.54	0.95
	1997–1999	0.24	0.26	0.50	0.88
Bolivia	1979–1981	0.20	0.74	0.94	0.90
	1989–1991	0.21	0.71	0.92	0.99
	1997–1999	0.21	0.74	0.95	1.08
Haiti	1979–1981	0.22	0.21	0.43	0.84
	1989–1991	0.20	0.18	0.37	0.76
	1997–1999	0.21	0.19	0.39	0.67

Note: *Total food self-sufficiency equals total domestic food production divided by total domestic food consumption.
Source: Derived from FAOSTAT Agriculture Data, commodity balances and food supply data from 1979 to 1999 (www.fao.org).

Diets are meager and food self-sufficiency is low or declining in most, though not all, of the poorest countries in the Western Hemisphere. In Central America, El Salvador, Guatemala, and Honduras cope with this problem in part by exporting non-agricultural goods, including clothing and other products of assembly plants (called *maquilas* in Spanish) in which output is manufactured from imported materials by local workers using imported machinery. But these countries also export people, who remit home money earned in the United States so that relatives can buy food and other necessities, build houses, and start small businesses. Similarly, hundreds of thousands of Nicaraguans now live in Costa Rica, performing agricultural and other work disdained by people accustomed to higher standards of living. Migration and remittances are common in the Caribbean as well, as indicated by the risks that Haitians run to reach the shores of Florida and by their willingness to cut sugarcane and perform other menial work across the border in the Dominican Republic.

No less than in other parts of the world, frequently used measures of food security correlate with per-capita food consumption in Latin America and the Caribbean. There are some positive outliers in the region. The most celebrated case is Costa Rica, which has experienced less economic inequality than its neighbors in times past and has invested heavily in education and health care since disbanding its armed forces in 1949. The country's infant mortality rate (Table 12.8, column 2) actually compares favorably to those of OECD members, including the United States. Receiving much less international attention, though, is Chile. Its GDP per capita (Table 12.2, column 3) is a little above Costa Rica's and its Gini coefficient (Table 12.8, column 5) is much higher. Nevertheless, the incidence of extreme poverty (column 4) is identical in the two countries, as is the infant mortality rate. Whereas 5 percent of the Costa Rica's children are malnourished, only 1 percent of the children in Chile fall into this category.

Aside from Jamaica, there are no Sri Lankas among the nations listed in Table 12.8 – that is, places where GDP per capita is very low and yet economic inequality is limited and conventional indicators of food insecurity are low (Chapter 11). In contrast, a number of Latin American and Caribbean nations have indicators that are excessive in light of average incomes. A notorious example is Guatemala, which has a GDP per capita of $4,080 and an infant mortality rate of 36 per thousand live births and where a quarter of all children are malnourished and nearly one-third of the population is extremely poor. In contrast, the incidence of extreme poverty (Table 11.8, column 4) is about the same in India, which has a GDP per capita of just $2,650 (Table 11.2, column 3). In Indonesia, where GDP per capita equals $3,070, less than 10 percent of the population subsists on $1 a day or less (Table 11.8, column 4) and the infant mortality rate is 32 per thousand live births (column 2), which is lower than Guatemala's rate.

Obviously, substandard food consumption as well as poor indicators of food security, such as elevated infant mortality, go hand in hand with the hemisphere's lowest living standards in Haiti.

Table 12.8 Recent food insecurity indicators in Latin America and the Caribbean

Country (ranked by column 1)	Total food consumption per capita 1997–1999 (in cereal-equivalent tons) (1)	Infant mortality rate – 2002 (per 1,000 live births) (2)	Child malnutrition late 1990s/early 2000s (% children <5years) (3)	Poverty rate late 1990s/early 2000s (% population below $1 per day) (4)	Gini index late 1990s/early 2000s (5)
Argentina	2.07	18	5	3.3	52.2
Chile	1.25	10	1	2.0	57.1
Mexico	1.15	24	8	9.9	54.6
Costa Rica	1.02	9	5	2.0	46.5
Uruguay	2.27	14	–	2.0	44.6
Brazil	1.40	33	6	8.2	58.57
Dominican Republic	0.79	32	5	2.0	47.4
Colombia	0.94	19	7	8.2	57.6
Panama	1.03	19	8	7.2	56.4
Venezuela	0.93	19	4	15.0	49.1
Peru	0.66	30	7	18.1	49.8
El Salvador	0.60	33	12	31.1	53.2
Paraguay	1.51	26	–	14.7	56.8
Guatemala	0.53	36	24	16.0	48.3
Jamaica	0.90	17	4	2.0	37.9
Ecuador	0.90	25	14	17.7	43.7
Honduras	0.60	32	17	23.8	55.0
Nicaragua	0.50	32	10	45.1	55.1
Bolivia	0.95	56	8	14.4	44.7
Haiti	0.39	79	17	–	–

Source: Table 12.7 and World Bank (2004).

12.5 Summary

Long more prosperous than the rest of the developing world, Latin America, and the Caribbean is being overshadowed by Asia, which is far more populous and where living standards are rising at a much faster clip thanks to macroeconomic and other policies that stimulate economic expansion. In spite of differences in economic trajectories, the two regions resemble each other demographically. Death rates have fallen to very low levels and family size has plummeted. Human numbers will peak in a generation or two.

Thanks to a more generous endowment of natural resources relative to the region's population, agricultural development in Latin America and the Caribbean has differed substantially from what Asia has undergone. In wealthier parts of the region with low rural population densities, such as the temperate southern cone, crop production is mechanized and farm commodities are exported. Some settings in the tropics and subtropics have a comparative advantage in traditional crops, such as coffee, rice, and sugar, as well as new products, such as cut flowers. But throughout the Andes, northeastern Brazil, Mesoamerica, and some parts of the Caribbean, rural poverty is severe. Ironically, few people are more deprived than indigenous producers of corn and potatoes, who are the descendants of people who domesticated these same crops.

Food insecurity south of the US border, which is not as widespread as undernourishment in other developing regions, has almost nothing to do with limited agricultural output. Aside from Haiti, every Latin American and Caribbean nation can feed itself from domestic production and imports, paid for with non-agricultural exports and international migration and remittances. As in other parts of the world, reducing hunger depends primarily on the alleviation of poverty.

Study Questions

1. What did ISI in Latin America and the Caribbean have to do with the lost decade of the 1980s?
2. Does a linkage exist south of the United States between economic prosperity and political freedom?
3. Can a case be made that economic growth in Latin America and the Caribbean has had little positive effect on life expectancies and other non-economic measures of human well-being?
4. How do human fertility and population growth in the region compare with TFRs and demographic expansion in Asia?
5. Why are patterns of agricultural development and factor use so heterogeneous in Latin America and the Caribbean?
6. Compare and contrast the supply side of the food economy, particularly beyond the farm gate, in Latin America and the Caribbean with the same set of activities in OECD nations.
7. Are indicators of food insecurity south of the US border higher or lower than one would expect, based solely on GDP per capita in the region?

13

The Middle East and North Africa

The river valleys of the Middle East and North Africa having been cradles of civilization, the region also was where the world's three great monotheistic religions – Judaism, Christianity, and Islam – began. The youngest of these, Islam was dynamic and aggressive from the time of its founding in the seventh century and spread quickly through Arab lands, Persia (now Iran), and present-day Turkey. The religion eventually won adherents in southeastern Europe, across central and southern Asia, and south of the Sahara.

During the Middle Ages, the Islamic World eclipsed Christendom in astronomy, mathematics, philosophy, and knowledge of agriculture. In part, this was because of greater tolerance of religious minorities among those embracing the faith of Mohammed than among those who professed to follow Jesus. Islamic dominance did not stop immediately once Europe experienced a Renaissance and began exploring and colonizing the wider world. Arab occupation of the Iberian Peninsula, which lasted for centuries, did not come to a complete end until 1492, as Columbus was embarking on the first of his four voyages across the Atlantic Ocean. The Ottoman Turks, who had conquered the Arabs after converting to their religion, besieged Vienna as late as the 1680s.

The emerging ascendancy of Europe was entirely clear a little more than a century later, when the Ottomans opened permanent embassies in London, Vienna, Berlin, Saint Petersburg, and Paris – the first any Islamic land had seen fit to establish in any western capital (Lewis, 2002, p. 40). Overshadowed by Europe and its offshoots for more than 200 years now, the Middle East and North Africa continue to struggle with modernization. Representative government, separation of religion and state, women's rights, and tolerance of dissent are all very much the exception to the rule in the region. Likewise, the rule of law and other institutional underpinnings of capitalism are weak or non-existent (Kuran, 2004).

These deficiencies have not been fully compensated by an enormous endowment of oil and natural gas. The Middle East and North Africa have gained hundreds of billions of dollars from the sale of fossil fuels. Even nations in the region with meager natural resources have benefited as their workers have found employment in members of the Organization of Petroleum Exporting Countries (OPEC). Yet overall economic accomplishments have been disappointing. As the authors of the Arab Human Development Report (AHDR) point out, the aggregate GDP of all 22 Arab countries[1] from Morocco (across the Strait of Gibraltar from Spain) to Oman (across the Straits of Hormuz from Iran), which have a combined population of nearly 300 million, was $531 billion in 1999. The same year, Spain, which has slightly more than 40 million inhabitants and is by no means the richest place in Western Europe, had a GDP of $596 billion (UNDP, 2002a, p. 85).

13.1 Trends in GDP Per Capita

This lack of success would have been surprising from the perspective of the middle and late 1970s. In October 1973, Israel repulsed a surprise attack by Egypt and Syria, which had received abundant military assistance from the Soviet Union, during the Yom Kippur War. Arabian oil exporters retaliated against the United States and other western nations that had provided arms to the Jewish state (the only affluent democracy in the region) by cutting petroleum exports drastically. This caused oil prices to triple in late 1973 and early 1974. With higher prices sustained after the export embargo came to an end, Arab economies boomed. Between 1975 and 1980, GDP per capita rose from $1,845 to $2,300 thanks to annual growth of 5.6 percent (UNDP, 2002a, p. 88).

The oil bonanza was renewed in early 1979, when the overthrow of the Shah of Iran precipitated another price-spike. However, prosperity based exclusively on exporting fossil fuels at elevated values could not, and indeed did not, last. Price surges led to exploration and development in the North Sea, Alaska's North Slope, the Gulf of Mexico, and other locations outside OPEC. As new sources of petroleum came online, downward pressure was exerted on market values. The falling back of prices would not have caused growth to cease in oil-exporting countries had windfalls from the 1970s and early 1980s been harnessed effectively for economic expansion and diversification. However, oil earnings were swallowed up by hegemonic, bureaucratic states, which tended to favor consumption by favored elites over human capital formation and other kinds of investment favoring the general population. Growth consequently stalled, or even reversed, during the past two decades of the twentieth century.

To be sure, the expansion of government and the corresponding stunting of private economic activity in nations with abundant mineral wealth have not been an

---

[1] The governments of three of the 22 countries – Djibouti, Mauritania, and Sudan – identify themselves as Arab, even though non-Arab population are sizable in each place. In this book, we follow the convention of various international agencies of categorizing the three as Sub-Saharan.

exclusively Middle Eastern and North African phenomenon. The same pattern can be detected in Nigeria (Chapter 15) and Venezuela (Chapter 12), for example. Nevertheless, governmental hegemony has been especially pronounced in the Arab World. This probably relates to the weak attachment that many Arabs have for their respective nation-states, which in many cases were carved out of the long-decaying Ottoman Empire after the First World War by European negotiators who drew international boundaries in the region arbitrarily. Lacking the sort of legitimacy that governments have where people have a strong sense of national identification and citizenship, Arab states compensate by over-reaching. In the economic sphere, key sectors, such as natural resource industries and banking, are state-run and glaringly inefficient as a result. Also, private enterprise, relegated in many countries to traditional commerce, tends not to be internationally competitive since it has grown dependent on high tariffs and other protection from foreign competition provided by government (Yousef, 2004). Widespread protection is reflected by low levels of economic freedom in most Middle Eastern and North African nations (Table 13.1, column 11). In addition, corruption and rent-seeking abound as protection from imports is sought and won by privileged firms and individuals.

The most egregious over-reaching by states of doubtful legitimacy is in the security sphere. As noted in Chapter 9, military spending as a portion of GDP in the Middle East and North Africa is triple or quadruple the levels recorded in other parts of the world, including the United States. Frequently rationalized in terms of threats from Israel, generous budgets for the armed forces are more directly a response to the menace posed by co-religious neighbors, as exemplified by the Iran–Iraq War of the 1980s, Saddam Hussein's invasion of Kuwait, and Syria's long occupation of Lebanon. In addition, military spending is motivated by concerns about internal security. This security is threatened when an appreciable segment of the population harbors a loyalty to pan-Arabism or, more frequently these days, Islamic fundamentalism, either of which is antithetical to the nation-state. Internal security concerns also drive governments to operate extensive networks of secret police and informers, muzzle the press, and curb freedoms of assembly and association. Of course, the repression of civil liberties breeds public frustration and alienation, which in turn add to rulers' discomfort and ever more repression and military spending.

Now that the Soviet Union no longer exists, the Middle East and North Africa comprise the least free part of the world. Of the 19 countries listed in Table 13.1, all but six were categorized as not free (NF) by the Freedom House in 2002 (column 10). The only free (F) nation was Israel, where everyone – including Arab citizens who are Muslim or Christian – enjoys democracy and is protected by the rule of law. Turkey, which became a republic after the First World War, has regular elections and is endeavoring to harmonize its institutions with those of Europe in the hope of joining the European Union (EU). As a result, it rates a partly free (PF) designation from Freedom House. So does Lebanon, which was once the freest and most prosperous of all Arab countries and where Christians and Muslims not only coexisted but shared political power. After suffering protracted civil war during the

Table 13.1 Patterns of economic growth in the Middle East and North Africa, 1980 through 2003

A. Average annual rate of growth 1980–1990 (%)					B. Average annual rate of growth 1990–1999 (%)					C. Average annual rate of growth 2000–2003 (%)			
Country (ranked by column 1)	GDP per capita (1)	Exports (2)	Inflation (3)	Gross investment (4)	Country (ranked by column 5)	GDP per capita (5)	Exports (6)	Inflation (7)	Gross investment (8)	Country (ranked by column 9)	GDP per capita (9)	Economic Freedom** Index – 2002 (10)	Freedom Status** 2001–2002 (11)
Oman	4.4	—	−3.6	—	Lebanon	5.9	15.6	24.0	18.4	Iran	4.3	4.55	NF
Turkey	3.1	16.9	45.3	5.3	Tunisia	3.0	5.1	4.7	3.4	Tunisia	3.0	2.85	NF
Egypt	2.9	5.2	11.7	2.7	Syria	2.9	4.7	8.7	7.9	Morocco	2.5	3.05	PF
Morocco	2.0	6.8	7.2	2.5	Turkey	2.6	11.9	77.9	4.6	Oman	2.3	2.90	NF
Israel	1.7	5.5	101.5	2.2	Egypt	2.5	3.1	9.1	6.7	Algeria	2.3	3.10	NF
Tunisia	0.8	5.6	7.4	−1.8	Israel	2.1	9.1	10.6	5.5	Egypt	1.8	3.55	NF
Algeria	−0.2	4.1	8.0	−2.3	Iran	1.8	0.2	26.7	1.4	Turkey	1.8	3.35	PF
Iran	−0.9	—	—	—	Oman	1.0	—	—	—	Yemen	1.8	3.75	NF
Jordan	−1.2	5.9	7.0	7.3	Morocco	0.5	3.0	3.2	1.5	Jordan	1.3	2.70	PF
Syria	−1.8	3.6	15.3	−7.0	Jordan	0.4	7.4	3.2	3.4	Kuwait	1.3	2.70	PF
Kuwait	−3.1	−2.3	−2.4	−4.5	Algeria	−0.6	2.2	19.0	0.2	Bahrain	1.0	2.00	NF
Saudi Arabia	−5.2	—	−3.7	—	Yemen	−1.0	10.2	26.0	7.7	Lebanon	0.3	3.15	PF
Libya	−6.3	—	—	—	United Arab Emirates	−1.5	—	—	—	Saudi Arabia	0.3	3.00	NF
United Arab Emirates	−7.7	0.0	0.7	−8.7	Saudi Arabia	−1.8	—	1.4	—	Syria	0.0	4.10	NF
Lebanon*	—	—	—	—	Kuwait*	—	—	—	—	Israel	−0.5	2.65	F
Yemen*	—	—	—	—	Bahrain*	—	—	—	—	United Arab Emirates	−1.0	2.15	NF
Bahrain*	—	—	—	—	Qatar*	—	—	—	—	Iraq*	—	5.00	NF
Qatar*	—	—	—	—	Libya*	—	—	—	—	Libya*	—	4.75	NF
Iraq*	—	—	—	—	Iraq*	—	—	—	—	Qatar*	—	2.95	NF

Note: * No macroeconomic data are available for these countries, which are listed at the bottom of the table for convenience.

Note: ** The Economic Freedom Index ranges from 1.00 (completely free markets) to 5.00 (completely repressed markets). See Heritage Foundation (2002). Political Freedom Status is F (free), PF (partly free), and NF (not free). See Freedom House (2002).

Source: World Bank (2000/2001), pp. 278–279; 294–295; World Bank (2005).

1970s and 1980s, the country has been permitted only a modest measure of civil and political liberties by Syrian occupiers and their allies, the Iranian-backed Hezbollah. Also partially free are Jordan, Kuwait, and Morocco, where paternalistic monarchies operate through parliaments with limited power.

There are signs of reform in the region. National parliaments are gaining decision-making power and growing a little more assertive in Jordan and some of the small states bordering the Persian Gulf. Thanks to the invasion of Iraq by the United States and its allies in 2003, Saddam Hussein, who was one of the world's cruelest tyrants, was deposed; the elections Iraq held subsequently represented a major step toward democracy. Elsewhere, political freedoms were entirely absent as of early 2005. The civilian regimes of Egypt and Tunisia remain autocratic. Dictatorships in Algeria, Libya, and Syria are closely identified with a single party (e.g., the Baathists), the military, a strongman, or a combination of the three. By and large, royal rulers in Saudi Arabia and a number of Gulf states are authorities unto themselves. Finally, a theocratic autocracy in Iran keeps political and civil liberties firmly in check.

With oil prices falling off the lofty peaks attained shortly after the Iranian Revolution and then plummeting after 1985,[2] living standards eroded in the Middle East and North Africa during the 1980s. No more than six of the 19 countries listed in Table 13.1 reported a positive change in GDP per capita (column 1); of these, Oman was the sole exporter of fossil fuels. The other 13 nations either reported a decline or failed to furnish macroeconomic data, presumably in each case because living standards were deteriorating. Within this group, Jordan and Syria (where average incomes fell), Lebanon (where heavy fighting precluded documentation of an even worse decline), and Yemen (which is destitute by the standards of any of its regional neighbors) do not belong to OPEC. The other eight all have enormous energy resources. In Kuwait, Libya, Saudi Arabia, and the United Arab Emirates, GDP per capita was lower in 1990 than it had been in 1980. The same was probably true of Bahrain, Iraq, and Qatar. Clearly, the strategy of using oil-export revenues to buildup natural resource industries, banks, and other loss-leading enterprises owned and (mis)managed by the state (Bennett, 2003) yielded negative dividends when fossil fuels grew cheaper (Yousef, 2004).

The economic slide was arrested for OPEC members during the last decade of the twentieth century. After contracting by 2.7 percent per annum during the 1980s,

---

[2] During the first half of 1985, President Ronald Reagan, Defense Secretary Caspar Weinberger, and CIA Director William Casey lobbied the rulers of Saudi Arabia, where oil was being produced for about $1.50/barrel and sold for 20 times that amount, to increase output and, so, bring down prices. The former officials were motivated primarily by the US economic stake in cheaper petroleum, including accelerated growth and diminished trade deficits. However, lower prices for fossil fuels were also part of a broader strategy of the Reagan Administration to undermine the Soviet Union, which earned a large share of its foreign currency from sales of oil and natural gas. Interested in good relations with the United States and fiercely anti-communist, the Saudis ramped up output in 1985, from 2 million barrels/day to 9 million by the end of the year. As a result, the price of a barrel of petroleum sank from $30 to $12, much to the Soviets' detriment (Schweizer, 1994, pp. 216–220, 233, 242–243).

GDP per capita for the exporters held steady in the ensuing decade. Meanwhile, growth among the non-exporters, which had been positive though anemic (1.1 percent per annum) during the 1980s, picked up in the 1990s, to an annual rate of 2.1 percent. None of the half-dozen countries with the fastest increases in GDP per capita in the latter period (Table 13.6, column 5) belong to OPEC. Instead, economic expansion was achieved by promoting trade and encouraging investment.

With energy prices rising appreciably since the turn of the twenty-first century, GDP per capita improved in a number of OPEC members between 2000 and 2003 (Table 13.1, column 1). The most dramatic gains have been registered in nations where demographic growth has been relatively modest, such as Iran and Algeria. In contrast, population increases have nearly kept pace with economic expansion in Saudi Arabia, where GDP per capita approached $20,000 a quarter century ago but subsequently fell by more than 50 percent in real terms. Reserves of foreign currency, which were substantial during the 1980s, were gone by the turn of the century. In their place was a large national debt, which continued to grow since much of the population depended on subsidies provided by state agencies with little cash to spare (Singer, 2003).

13.2 Population Dynamics

Along with illustrating the pitfalls of oil-based development, Saudi Arabia stands out demographically. Far from being entirely dissipated, oil wealth has been used to improve health services and sanitation. Consequently, the incidence of disease has gone down and life spans have increased. While diminished mortality normally leads to a lowering of human fertility, including in many parts of the Middle East and North Africa, this reaction has been muted in the Arab World's leading oil producer.

Human Fertility

Many of the factors that normally coincide with lower total fertility rates (TFRs) are at work in the Middle East and North Africa. One of these is urbanization (Table 13.2, column 4), which outside of Egypt and Yemen has proceeded quite far. Also, average incomes (column 3) are by no means the lowest in the world, and not just in countries with abundant energy resources. Indeed, more than a few places where the vast majority of the population lives in cities and has little fear of extreme poverty bear more than a passing demographic resemblance to, say, the leading nations of Latin America. Especially in the Mahgreb (i.e., Morocco, Algeria, and Tunisia), Iran, and Turkey, half or more of all females aged 15 to 49 years use contraceptives (column 9) and human fertility, which was extraordinarily high as recently as a quarter century ago (column 1), has since fallen nearly to replacement levels (column 2).

But just as there are parts of Latin America and the Caribbean where TFRs are still well above 2.1 births per woman for cultural and other reasons (Chapter 12), human fertility remains elevated in spite of widespread urbanization and relatively

Table 13.2 Social indicators for the Middle East and North Africa in 2002

Country (ranked by income per capita 2002)	Total fertility rate		2002 Income per capita ($PPP)	Urban % of total population in 2002	Adult illiteracy rate in 2002 (% of people 15 and over)		Females % of labor force		Contraceptive prevalence rate 1990–2002 (% of women 15–49)
	1980	2002			Males	Females	1980	2002	
	(1)	(2)	(3)	(4)	(5)	(6)	(7)	(8)	(9)
Israel	3.2	2.7	19,530	92	3	7	33.7	41.7	–
Bahrain	5.2	2.3	17,170	93	8	16	10.9	21.6	62
Kuwait	5.3	2.5	16,240	96	15	19	13.1	32.1	–
Oman	9.9	4.0	13,340	–	18	35	6.2	18.9	24
Saudi Arabia	7.3	5.3	12,650	87	16	31	7.6	17.7	21
Tunisia	5.2	2.1	6,760	67	17	37	28.9	32.1	60
Iran	6.7	2.0	6,690	65	17	30	20.4	28.4	68
Turkey	4.3	2.2	6,390	67	–	–	35.5	38.1	64
Algeria	6.7	2.8	5,760	58	22	40	21.4	29.0	51
Lebanon	4.0	2.2	4,360	90	–	–	22.6	30.1	61
Jordan	6.8	3.5	4,220	79	–	–	14.7	25.6	47
Egypt	5.1	3.1	3,810	43	–	–	26.5	31.0	51
Morocco	5.4	2.8	3,810	77	37	62	33.5	34.9	50
Syria	7.4	3.4	3,620	52	9	26	23.5	27.6	43
Yemen	7.9	6.0	870	25	–	–	32.5	28.3	15
Iraq*	6.4	4.1	–	68	–	–	17.3	20.4	–
Libya*	7.3	3.3	–	88	8	29	18.6	24.0	45
Qatar*	5.6	2.5	–	93	–	–	6.7	16.4	43
United Arab Emirates*	5.4	3.0	–	88	24	19	5.1	15.9	–

Note: * 2002 income per capita is not reported for these countries, which are listed at the bottom of the table for convenience.
Source: World Bank (2004).

comfortable living standards. Clearly, Saudi Arabia is a case in point. That country's form of Islam is conservative. This militates against female empowerment, as is evident from pronounced gender discrepancies in literacy rates (Table 13.2, columns 5 and 6) and limited participation by women in the labor force (columns 7 and 8). The same pattern applies in four of Saudi Arabia's peninsular neighbors: Bahrain, Oman, Qatar, and the United Arab Emirates. In other countries with TFRs well above the replacement level, families are large for the same reasons that hold outside the Islamic World. Though not nearly as high as it was a generation ago, human fertility in Egypt, which is the most populous Arab state, still exceeds three births per woman largely because the country is largely rural and poor. As previously indicated, Yemen has the lowest living standards in the Middle East and North Africa. It is also much less urbanized than its neighbors. It is no coincidence, then, that Yemen has the highest TFR as well.

Access for females to schools and jobs ought to improve in the years to come in the Middle East and North Africa. As this happens, TFR declines, which every country (even poverty-stricken Yemen) has experienced since 1980, will continue. Replacement fertility is either being reached or is within sight in much of the region.

Natural Increase

Other than in Yemen, death rates (Table 13.3, column 2) have reached minimal levels in the Middle East and North Africa. These rates might even be going up slowly in Turkey and a few other places where the demographic transition is further advanced and, consequently, average age is gradually rising.

With crude death rates (CDRs) stable or nearly so, natural increase is being driven by changes in birth rates (Table 13.3, column 3). Typical of settings where TFRs greatly exceeded the replacement level in the recent past, demographic momentum keeps CBRs well above death rates in the Middle East and North Africa. Iran is the only nation in the region where the annual rate of natural increase (column 4) is close to 1.0 percent. In several countries, including Iraq, Jordan, Saudi Arabia, and Syria, the rate exceeds 2.0 percent per annum. Few nations anywhere in the world are experiencing more natural increase than Yemen, where the annual rate is 3.1 percent.

Needless to say, demographic momentum reflects changes in the age profile of a population – in particular, the size of the cohort of women of childbearing age and their male partners. Of course, this is the same cohort one normally finds in the labor force. Whether or not that labor force is fully employed, however, depends on job creation, which has flagged in the Middle East and North Africa since the 1980s precisely as millions of young people have left school and sought work (Gardner, 2003). Despairing of finding employment in their homelands, large numbers of these people have moved to Europe from Algeria, where the unemployment rate is 30 percent (Gardner, 2003), and its neighbors in the Mahgreb as well as from Turkey. This migration is swelling the Muslim populations of France, Germany, the Netherlands, and many other nations.

Table 13.3 Population dynamics in the Middle East and North Africa for 2002

Country (ranked by column 1)	Income per capita 2002 ($PPP) (1)	Crude death rate – 2002 (per 1,000 population) (2)	Crude birth rate – 2002 (per 1,000 population) (3)	Rate of natural increase (%) (4)	Average annual population growth 1980–2002 (%) (5)
Israel	19,530	6	20	1.4	2.4
Bahrain	17,170	4	21	1.7	3.4
Kuwait	16,240	3	20	1.7	2.5
Oman	13,340	3	26	2.3	3.9
Saudi Arabia	12,650	4	31	2.7	3.9
Tunisia	6,760	6	18	1.2	2.0
Iran	6,690	6	18	1.2	2.4
Turkey	6,390	7	22	1.5	2.0
Algeria	5,760	5	22	1.7	2.4
Lebanon	4,360	6	19	1.3	1.7
Jordan	4,220	4	28	2.4	3.9
Egypt	3,810	6	24	1.8	2.2
Morocco	3,810	6	21	1.5	1.9
Syria	3,620	4	29	2.4	3.1
Yemen	870	10	41	3.1	3.6
Iraq*	–	8	29	2.2	2.8
Libya*	–	4	27	2.2	2.7
Qatar*	–	4	14	1.0	4.6
United Arab Emirates*	–	4	17	1.4	5.3

Note: *2002 income per capita is not reported for these countries, which are listed at the bottom of the table for convenience.
Source: World Bank (2002a), pp. 48–50; World Bank (2004).

At the same time, oil exporters with small native populations – Kuwait and Saudi Arabia, for example – have been attracting foreign labor for decades. Some of the immigrants have been from poorer Arab lands, most notably Egypt and Yemen. Workers also have arrived from Bangladesh, India, Pakistan, and the Philippines. This sort of migration has slackened in recent years as GDP has contracted and unemployment among young men has increased around the Persian Gulf.

13.3 Agricultural Development

As discussed in the preceding chapter, agricultural revolution in the Western Hemisphere gave humankind corn, potatoes, and other widely used crops. But as far as the Middle East and North Africa are concerned, not to mention Europe and its overseas offshoots, the emergence of farming thousands of years ago in the valleys of the Nile, Euphrates, and Tigris Rivers was of fundamental and lasting importance. Various crops, including wheat, were domesticated, as were sheep and goats. The agricultural significance of the Middle East and North Africa also relates to the region's having served as a conduit for the westward diffusion of rice, sugar, and other species originally cultivated in Asia (Dixon and Gulliver with Gibbon, 2001, pp. 83–87).

Agriculture in Egypt and present-day Iraq began alongside rivers because of productive land in this setting – to be specific, land made fertile because of the deposition of soil eroded from the upper reaches of drainage basins (Ethiopian highlands, in the case of the Nile). Riparian agriculture also makes sense in the Middle East and North Africa because of sparse precipitation in much of the region. In Egypt, around the Persian Gulf, and in many other places, there is no rain-fed agriculture. Instead, the water supplied to farm fields must be diverted from rivers or extracted from underground.

In recent times, agricultural development has been greatly influenced in various nations by the presence of fossil-fuel resources. As mentioned in Chapter 9, exploitation of these resources has created "Dutch Disease" (Perkins *et al.*, 2001, pp. 643–651), as exports of oil and natural gas have strengthened national currencies and thereby discouraged the production of goods (other than fossil fuels) that can be exported or imported. Among these goods are farm commodities. A few petroleum-rich nations in the Middle East and North Africa have done nothing to alleviate the agricultural disincentives resulting from Dutch Disease. Others, though, have compensated farmers with subsidies, such as under-priced irrigation or low prices for energy used to pump water from rivers or underground aquifers. Where the latter route has been followed, abundance of one sub-surface resource (i.e., petroleum) has contributed to mounting scarcity of sub-surface deposits of water, which actually is the most prized and vital resource in the region.

Factor Use

Yet another reason for irrigation's spread in the Middle East and North Africa is that no part of the world outside of Asia has been more affected by the sort of agricultural intensification set in motion by the Green Revolution. As reported in Chapter 9, the ratio of rural population to arable land in the region is comparable to rural population densities in South, Southeast, and East Asia. In the Nile River Valley and a few other parts of Egypt where agriculture has developed, there are well over 1,000 rural people per cultivated hectare (Table 13.4, column 1). Where the countryside receives little rain and yet is this densely populated, abundant irrigation and fertilization are called for in order to raise yields. In a few countries, rates of fertilizer application (columns 2 and 3) have been reduced since the early 1980s. Kuwait is an unusual case in that rates were undoubtedly excessive 25 years ago and hence could be reduced without a huge sacrifice in output. Algeria and Libya are two prime examples of fossil-fuel producers that make little attempt to compensate for Dutch Disease; diminished fertilization in these two countries should

Table 13.4 Agricultural inputs in the Middle East and North Africa, 1979 through 2001

Country (ranked by income per capita in 2002)	Rural population density (people per square kilometer of arable land) 2001	Fertilizer use (kilograms per hectare of arable land)		Irrigated land (% of cropland)		Agricultural machinery (tractors per square kilometer of arable land)	
		1979–1981	1999–2001	1979–1981	1999–2001	1979–1981	1999–2001
	(1)	(2)	(3)	(4)	(5)	(6)	(7)
Israel	157	238	270	49.3	46.0	8.1	7.3
Bahrain	2,564	–	–	17.8	66.7	–	–
Kuwait	684	450	100	83.3	85.8	2.2	0.9
Oman	1,533	48	169	74.5	78.2	0.4	0.4
Saudi Arabia	79	23	104	28.9	42.8	0.1	0.3
Tunisia	118	21	39	4.8	7.7	0.8	1.2
Iran	160	43	91	35.5	44.2	0.6	1.6
Turkey	97	53	83	9.6	16.9	1.7	3.9
Algeria	170	28	13	3.4	6.8	0.7	1.2
Lebanon	257	166	311	28.3	32.0	1.4	4.5
Jordan	449	40	91	11.0	19.3	1.5	2.3
Egypt	1,306	286	440	100.0	100.0	1.5	3.1
Morocco	146	27	42	15.0	13.5	0.3	0.5
Syria	173	25	73	9.6	22.5	0.5	2.1
Yemen	923	9	10	19.9	30.2	0.3	0.4
Iraq*	134	17	67	32.1	60.8	0.4	1.1
Libya*	36	36	36	10.7	21.9	1.3	1.9
Qatar*	236	–	–	75.6	61.9	–	–
United Arab Emirates*	773	225	709	237.7	32.6	1.2	0.7

Note: * 2002 income per capita is not reported for these countries, which are listed at the bottom of the table for convenience.
Source: World Bank (2004).

be interpreted in terms of weak incentives for agriculture in the face of national currencies made strong by exports of oil and natural gas.

Likewise, the irrigated portion of cropland (Table 13.4, columns 4 and 5) in Algeria and Libya is modest relative to what other nations in the region have achieved. Aside from reflecting poor incentives for farming, limited irrigation is explained in part by the absence of major rivers in those two countries. Also, Algeria has upland areas with enough precipitation to allow for rain-fed farming complemented by limited, small-scale irrigation. Similar environments exist in Morocco, Tunisia, the Levant (Israel, Jordan, Lebanon, and Syria), and Turkey, as well as northern Iraq and Iran. Approximately 27 million upland farmers produce cereals and legumes and harvest fruit, olives, and other tree crops (Dixon and Gulliver with Gibbon, 2001, p. 88). Elsewhere in the Middle East and North Africa, irrigation has played a crucial role in developing the agricultural sector, generally, and in raising yields, specifically.

As in other parts of the world, agricultural mechanization depends on GDP per capita, which has a direct impact on the opportunity cost of rural labor. Israel, where living standards are comparable to those of many OECD members, has the highest ratio of tractors to arable land (Table 13.4, columns 6 and 7). Several countries in the region have experienced a considerable increase in machinery use since 1980, including Egypt, Iran, Syria, and Turkey. As a result, the tractor-to-land ratio in the Middle East and North Africa has risen slightly above that of Latin America and the Caribbean, and considerably exceeds ratios in Asia and Sub-Saharan Africa (Chapter 9).

Intensification, Extensification, and Trends in Output

The Middle East and North Africa stand out in the developing world for having experienced significant yield growth, resulting from increased fertilization and irrigation, as well as a great deal of mechanization, which has facilitated the sector's geographic expansion in several countries. The region's limited comparative advantage in agriculture is indicated by modest grain harvests (Table 13.5, column 1); increased production of wheat and other cereals, such as it is, has resulted mainly from intensification (columns 2–4). Different from other parts of the world, fruit and vegetables, raised in many countries for domestic markets, are the region's main crops. This category of output has risen more in recent decades because of extensification than because of yield growth.

Egyptian agriculture is noteworthy for its intensity, with crop yields being extremely high by global as well as regional standards. As already mentioned, all the country's limited farmland is irrigated. In addition, the natural fertility of fluvial soils in the Nile River Valley has been enhanced by fertilizer application rates that, like output per hectare, are elevated by global standards, not just regional norms. Fertilizer applications went from less than 300 kilograms to more than 400 kilograms per hectare during the past two decades of the twentieth century (Table 13.4, columns 2 and 3), which explains much of the 75 percent increase in cereal yields during the same period (Table 13.6, columns 1 and 2). This intensification

Table 13.5 Trends in crop area, yield, and output in the Middle East and North Africa, 1970–2000

	2000 production (million tons) (1)	Average annual change, 1970–2000 (%)		
		Area (2)	Yield (3)	Production (4)
A. Crops				
Vegetables	44	2.3	1.7	4.0
Fruits	30	3.0	1.0	4.1
Wheat	23	0.4	1.8	2.2
Rice	9	1.3	1.3	2.6
Maize	8	0.3	3.3	3.6
Barley	4	0.0	−0.7	−0.7
Pulses	2	0.8	0.0	0.8
Oil crops	1	0.8	1.8	2.7
B. Animal products				
Milk	17	–	–	3.4
Meat	6	–	–	4.5
Eggs	2	–	–	5.4

Source: Dixon and Gulliver with Gibbon (2001), pp. 94–95.

greatly exceeded relative growth in cropland between 1980 and 1999 (columns 3 and 4) and is the main reason why annual food production doubled during the 1980s and 1990s (columns 5 and 6).

Natural conditions for agriculture are considerably different in Turkey and Iran, which along with Egypt are the three most populous countries in the region. Compared to its neighbors to the south, Turkey has a favorable environment for rain-fed farming, as indicated by limited irrigation in the country.[3] Also, fertilizer application rates are relatively modest. Nevertheless, crop yields are moderately high, having risen during the 1980s and 1990s. Even though agricultural land use diminished during the same period, food production increased by nearly one-half. Iran, which exports petroleum, has made a substantial investment in irrigation. This has helped cause cereal yields to rise above 2,000 kilograms per hectare at the turn of the twenty-first century, after barely exceeding 1,000 kilograms two decades earlier. With farmed area only going up by one-fifth, the increase of nearly 150 percent in food production that occurred during the 1980s and 1990s was almost entirely a consequence of agricultural intensification.

---

[3]Irrigation will expand substantially in Turkey as the national government proceeds with its project to dam the upper courses of the Tigris and Euphrates Rivers. This project greatly concerns Iraq and Syria, through which these two waterways flow on their way to the Persian Gulf. Their worry is that water availability downstream from Turkey will diminish, while reservoirs are being filled as well as later because of the evaporation of a portion of the flow diverted for irrigation and other uses (Beaumont, 1998).

Table 13.6 Agricultural and food output in the Middle East and North Africa, 1979 through 2002

Country (ranked by income per capita 2002)	Cereal yield (kilogram per hectare)		Arable and permanent cropland (1,000 hectare)		Food production Index (1989–1991 = 100)		Agricultural productivity (value added per worker – 1995$)	
	1979–1981 (1)	2000–2002 (2)	1980 (3)	1999 (4)	1979–1981 (5)	2000–2002 (6)	1979–1981 (7)	1999–2001 (8)
Israel	1,783	2,749	413	440	85.0	115.3	–	–
Bahrain	–	–	2	6	125.8	89.9	–	–
Kuwait	3,157	2,318	1	7	81.0	229.0	–	–
Oman	941	2,320	41	77	62.1	163.1	–	–
Saudi Arabia	693	3,790	1,105	3,785	26.7	98.5	2,152	15,326
Tunisia	756	1,334	4,700	5,100	66.3	115.0	1,743	3,150
Iran	1,151	2,018	15,950	19,265	61.2	154.8	2,165	3,698
Turkey	1,868	2,220	28,479	26,672	75.8	114.6	1,872	1,850
Algeria	661	1,040	7,509	8,215	68.8	136.2	1,357	1,939
Lebanon	1,329	2,554	348	308	60.6	108.9	–	–
Jordan	542	1,514	1,380	387	57.4	147.4	1,141	1,120
Egypt	4,029	7,254	2,855	3,300	68.5	158.2	721	1,299
Morocco	974	777	7,719	9,445	55.8	103.6	1,146	1,537
Syria	1,060	1,803	5,684	5,502	93.6	163.6	2,206	2,618
Yemen	1,007	1,031	2,790	1,668	74.8	142.6	–	406
Iraq*	797	646	5,450	5,540	77.3	77.5	–	–
Libya*	369	634	2,080	2,150	78.7	134.1	–	–
Qatar*	2,516	3,683	2	21	38.7	153.6	–	–
United Arab Emirates*	2,326	485	13	134	42.7	549.9	–	–

Note: *2002 Income per capita is not reported for these countries, which are listed at the bottom of the table for convenience.
Source: World Bank (2004); FAO Production Yearbook (1983), pp. 50–52; (2000), pp. 3–5; 8–10.

There has been less agricultural development in the four North African nations west of Egypt. Although rainfall is very sparse in Libya, that country has not made a large investment in irrigation. Likewise, fertilizer use is modest. As a result, cereal yields are among the lowest in the region. Starting from a modest base 25 years ago, food production has increased, although Libya does the economically sensible thing of exporting fossil fuels and using a portion of the proceeds to pay for edible imports. Algeria is in much the same position, although it has more land suited to rain-fed farming. Neither of the other two Mahgreb nations has oil or natural gas. While food production nearly doubled in Tunisia during the last two decades of the twentieth century, through a combination of intensification and extensification, Moroccan output was little changed in the late 1990s from what it had been at the beginning of the decade – this after substantial increases during the 1980s.

Mirroring other sufferings the country has undergone, Iraqi agriculture deteriorated significantly under the despotism of Saddam Hussein. Although an increase was reported in the irrigated portion of arable land, irrigation systems in fact fell into serious disrepair. Output per hectare consequently registered a sharp decline, in spite of a reported increase in fertilizer use (which may or may not have happened). With Iraq's cereal yields among the lowest in the Middle East and North Africa, there was absolutely no change in food production during the 1980s and 1990s, even though demographic expansion was driving up demand.

More encouraging agricultural developments have occurred in two of the region's smallest states. Israel and Lebanon, each of which has less than a half million hectares of farmland, have developed good irrigation systems. They also register high levels of fertilizer use and small-scale mechanization. Having come up with drip irrigation, a technique that saves water, Israel has been able to make its arid lands bloom, thereby allowing it to export a wide range of horticultural products to Europe. Lebanon has entered this market as well, which is the main reason why value added per agricultural worker has risen far above levels in those neighboring countries for which data are available (Table 13.6, columns 7 and 8). The two larger Levantine nations, Syria and Jordan, have relatively low levels of irrigation and fertilizer use as well as limited mechanization. Each has moderate cereal yields, but a good record of food production in recent decades.

In the Arabian Peninsula, Yemen stands out. Lacking energy resources, the country, as already reported, has the lowest living standards in the Middle East and North Africa, which has hampered agricultural development. Irrigation, fertilizer use, and mechanization are all limited and cereal yields, which were slightly above 1,000 kilograms per hectare 25 years ago, have stagnated. In contrast, every other peninsular nation is richly endowed with oil and natural gas. As data on agricultural land use in Kuwait, Oman, Qatar, and the United Arab Emirates indicate, farming in these countries is unimportant, consisting largely of niche operations that supply fresh fruit and vegetables to local consumers. The largest country on the peninsula, Saudi Arabia attempted a major expansion of farming after oil prices rose in the early 1970s. A portion of the earnings generated by petroleum exports was used to underwrite generous subsidies for fertilizer, agricultural machinery, and above all else irrigation.

Cereal yields multiplied and, as can be seen in Table 13.6, food production around 1990 was nearly four times what it had been 10 years earlier.

Table 13.6's food production indices for the kingdom also indicate that this effort faltered during the 1990s, with output levels in the early twenty-first century little different from what these had been 10 years beforehand. The best interpretation of this decline is that it amounts to a concession to economic reality. Like Libya, Saudi Arabia holds no comparative advantage in farming. Rather than trying to be agriculturally self-sufficient, it is better off producing and exporting fossil fuels, which it does at a very low cost, and using the proceeds to buy imported food. By the same token, capital formation in agriculture, such as the construction of irrigation systems, carries an appreciable opportunity cost, since the funds required could instead be channeled to more productive investment, such as building up human capital.

Not least among the economic demerits associated with subsidizing inefficient irrigation are its adverse environmental impacts. If water is applied to farm fields without proper drainage, land may become water logged. Another problem is soil salinization, which occurs as irrigation water evaporates (as it does quickly in arid places close to the equator) and leaves the minerals it contains on the ground. Furthermore, where groundwater is made available cheaply to farmers and other users, aquifers are being depleted, as extraction outpaces recharge. This problem is arising not only in Saudi Arabia, but in Jordan, Libya, and Yemen as well (Dixon and Gulliver with Gibbon, 2001, p. 91). The long-term costs of aquifer depletion in countries as dry as these are sure to dwarf whatever advantages are derived in the short term from irrigated agriculture that is both uneconomical and unsustainable.

13.4 Dietary Change, Consumption Trends, and Food Security

The Middle East and North Africa possess abundant energy resources and have living standards that are the envy of many lands to the south and east. Due mainly to the scarcity of water, the region's comparative advantage in agriculture is limited and food imports are substantial. Outside of a few pockets of poverty, such as Yemen, food insecurity is not a great concern.

Applying Rask's (1991) methodology to FAO data, as in other regionally focused chapters, one finds the usual linkage between average income and per capita food consumption. For one thing, animal products comprise a large part of the diet (in terms of cereal equivalents) where GDP per capita is high (Table 13.7, columns 2 and 3). Such is the case in Israel as well as Kuwait and the United Arab Emirates, in the Persian Gulf. Conversely, consumption of meat, eggs, and dairy goods is limited in poorer nations, such as Morocco (and presumably Yemen, for which data of the sort reported in Table 13.7 are unavailable). In poorer settings, consumption of plant products (column 1) is comparable to consumption of animal products, even in terms of cereal equivalents.

Table 13.7 Food consumption and production trends in the Middle East and North Africa, 1979 through 1999

Country (ranked by income per capita 1999)	Years	Plant product consumption per capita (in cereal equivalent tons) (1)	Animal product consumption per capita (in cereal equivalent tons) (2)	Total food consumption per capita (in cereal equivalent tons) (3)	Total food self-sufficiency (4)
Israel	1979–1981	0.29	1.02	1.31	0.68
	1989–1991	0.32	1.02	1.33	0.73
	1997–1999	0.33	1.10	1.44	0.66
Bahrain	1979–1981	–	–	–	–
	1989–1991	–	–	–	–
	1997–1999	–	–	–	–
Kuwait	1979–1981	0.26	1.20	1.46	0.44
	1989–1991	0.22	0.96	1.18	0.34
	1997–1999	0.28	1.24	1.51	0.43
Oman	1979–1981	–	–	–	–
	1989–1991	–	–	–	–
	1997–1999	–	–	–	–
Saudi Arabia	1979–1981	0.28	0.67	0.95	0.30
	1989–1991	0.28	0.65	0.94	0.41
	1997–1999	0.29	0.65	0.94	0.43
Tunisia	1979–1981	0.30	0.38	0.68	0.75
	1989–1991	0.34	0.46	0.79	0.75
	1997–1999	0.35	0.50	0.85	0.60
Iran	1979–1981	0.28	0.50	0.79	0.72
	1989–1991	0.30	0.43	0.74	0.74
	1997–1999	0.31	0.45	0.77	0.82
Turkey	1979–1981	0.33	0.55	0.88	1.02
	1989–1991	0.37	0.57	0.94	0.99
	1997–1999	0.36	0.53	0.89	0.99
Algeria	1979–1981	0.28	0.34	0.62	0.51
	1989–1991	0.30	0.45	0.75	0.54
	1997–1999	0.31	0.42	0.73	0.55

(Continued)

Table 13.7 (Continued)

Country (ranked by income per capita 1999)	Years	Plant product consumption per capita (in cereal equivalent tons) (1)	Animal product consumption per capita (in cereal equivalent tons) (2)	Total food consumption per capita (in cereal equivalent tons) (3)	Total food self-sufficiency (4)
Lebanon	1979–1981	0.26	0.67	0.93	0.44
	1989–1991	0.32	0.53	0.84	0.48
	1997–1999	0.32	0.55	0.88	0.55
Jordan	1979–1981	0.27	0.46	0.73	0.33
	1989–1991	0.29	0.49	0.78	0.37
	1997–1999	0.29	0.45	0.74	0.43
Egypt	1979–1981	0.31	0.34	0.65	0.66
	1989–1991	0.34	0.38	0.72	0.67
	1997–1999	0.36	0.46	0.82	0.75
Morocco	1979–1981	0.30	0.34	0.64	0.77
	1989–1991	0.33	0.41	0.73	0.88
	1997–1999	0.33	0.37	0.70	0.76
Syria	1979–1981	0.29	0.58	0.87	0.85
	1989–1991	0.32	0.49	0.82	0.80
	1997–1999	0.34	0.51	0.84	0.86
Yemen	1979–1981	–	–	–	–
	1989–1991	–	–	–	–
	1997–1999	–	–	–	–
Iraq	1979–1981	0.30	0.39	0.69	0.52
	1989–1991	0.33	0.30	0.65	0.49
	1997–1999	0.27	0.13	0.40	0.52
Libya	1979–1981	0.33	1.05	1.39	0.72
	1989–1991	0.33	0.65	0.98	0.56
	1997–1999	0.34	0.59	0.92	0.67
Qatar	1979–1981	–	–	–	–
	1989–1991	–	–	–	–
	1997–1999	–	–	–	–
United Arab Emirates	1979–1981	0.30	1.42	1.72	0.26
	1989–1991	0.26	1.26	1.52	0.35
	1997–1999	0.27	1.34	1.60	0.40

Note: *Total food self-sufficiency equals total domestic food production divided by total domestic food consumption.
Source: Derived from FAOSTAT Agriculture Data, commodity balances, and food supply data from 1979 to 1999 (www.fao.org).

Since average incomes have grown slowly in many countries and contracted in a number of others, neither per capita food intake, generally, nor average consumption of animal products has increased dramatically in recent years. About the only notable exception is Egypt, where GDP per capita has increased steadily during the past 25 years. Well below the regional average around 1980, average food consumption in the country climbed by more than one-fifth during the last two decades of the twentieth century, with the intake of animal products accounting for most of the increase. In contrast, the human diet, generally, and consumption of meat, eggs, and dairy products, specifically, have deteriorated noticeably in oil-based economies suffering a sharp contraction since the 1970s. Libya is one of these countries; Iraq is another.

Egyptian experience is distinctive in that the ratio of domestic food production to food consumption (Table 13.7, column 4) has gone up even as the denominator of this indicator of self-sufficiency has been increasing. Similar change, on a more modest scale, has occurred in Algeria. In five other countries, the ratio of production to consumption rose: Iran, Jordan, Lebanon, Saudi Arabia, and the United Arab Emirates. But in each of these, the increase in the ratio was modest, per capita food intake failed to go up (or actually diminished), or both. Turkey is the only nation in the region where self-sufficiency is consistently achieved, although distortions of agricultural trade there are not great (Chapter 10) and edible imports and exports are sizable. Elsewhere in the Middle East and North Africa, importing one-third to two-thirds of the domestic food supply is not at all unusual.

Based on the region's experience, it cannot be argued that dependence on food imports leads to food insecurity. To the contrary, the tie between economic well-being, broadly defined, and food security that holds for other parts of the world also applies to the Middle East and North Africa. As in other places, per capita food consumption (Table 13.8, column 1), which is a good indicator of general well-being, is inversely related to standard indicators of food insecurity, such as infant mortality (column 2) and the percentage of children aged 5 years or less who are so underweight as to be categorized as malnourished (column 3).

Information is limited about how many people in the Middle East and North Africa are living below the international threshold of extreme poverty (i.e., daily earnings of $1). But data provided by a number of countries indicate that the incidence of poverty (Table 13.8, column 4) is not high in the region. This and other indicators of food insecurity compare favorably with indicators in other parts of the developing world for various reasons. GDP per capita is not especially low. Also, part of the wealth created by fossil-fuel development has been used to pay for hospitals and other public services and to lift people out of extreme poverty. Moreover, as the authors of the AHDR contend, the Muslim tradition of charity has beneficial effects (UNDP, 2002a, p. 91).

One might be tempted to offer similar explanations for the last indicator reported in Table 13.8, which is the Gini coefficient of income inequality (column 5). Among the countries for which estimates of this index are available, inequality is intermediate to low, definitely comparing favorably with what one finds in Sub-Saharan Africa

Table 13.8 Recent food insecurity indicators in the Middle East and North Africa

Country (ranked by column 1)	Total food consumption per capita 1997–1999 (in cereal equivalent tons) (1)	Infant mortality–rate 2002 (per 1,000 live births) (2)	Child malnutrition late 1990s/early 2000s (% of children <5 years) (3)	Poverty rate late 1990s/early 2000s (% of population below $1 per day) (4)	Gini Index late 1990s/early 2000s (5)
United Arab Emirates	1.60	8	7	–	–
Kuwait	1.51	9	2	–	–
Israel	1.44	6	–	–	35.5
Saudi Arabia	0.94	23	–	–	–
Libya	0.92	–	5	–	–
Turkey	0.89	35	8	2.0	40.0
Lebanon	0.88	28	3	–	–
Tunisia	0.85	21	4	2.0	39.8
Syria	0.84	23	7	–	–
Egypt	0.82	33	4	3.1	34.4
Iran	0.77	34	11	2.0	43.0
Jordan	0.74	27	5	2.0	36.4
Algeria	0.73	39	6	2.0	35.3
Morocco	0.70	39	9	2.0	39.5
Iraq	0.40	102	16	–	–
Qatar	–	11	6	–	–
Bahrain	–	13	9	–	–
Oman	–	9	18	–	–
Yemen	–	83	46	15.7	33.4

Source: Table 13.7 and World Bank (2004).

and Latin America and the Caribbean (Chapter 9). In Yemen's case, modest inequality helps to explain why less than one-sixth of the population is living on less than a dollar a day, even though GDP per capita is under $1,000. However, Gini estimates are unavailable for much of the Middle East and North Africa – presumably because accurate estimates would reveal that income inequality was pronounced in Saddam-era Iraq (where infant mortality was appalling) and continues to be so in Kuwait, Libya, Saudi Arabia, and other non-reporting nations.

13.5 Summary

Like every other part of the developing world, the Middle East and North Africa are wrestling with modernization, which among other things involves the establishment of the rule of law and representative government as well as a market economy. This task, which sometimes is made difficult by the region's Islamic heritage, is not obviated by a generous endowment of energy resources. Indeed, revenues from international sales of oil and natural gas have provided various governments with the means to interfere with private enterprise and to suppress the rights of individual citizens. The results of this interference and suppression include poor economic performance since the petroleum boom of the 1970s came to an end.

Demographic change, in contrast, has been dramatic. True, large families are still the rule in the Arabian Peninsula and a few other places. Elsewhere, however, fertility rates have fallen to the replacement level in Iran, Turkey, and several other countries. Since these rates were very high a generation ago, demographic momentum is strong. As long as this is the case, human numbers will continue to increase rapidly. Since GDP and employment are expanding slowly in the region, some of this growth will spill over into Europe, which as reported in Chapter 10 is on the brink of demographic contraction.

Much of the Middle East and North Africa is arid or nearly so. Accordingly, agricultural comparative advantage in many places does not extend far beyond raising fruit and vegetables for local markets. Moreover, crop and livestock production has been discouraged in many nations because of Dutch Disease, and offering subsidies to ameliorate the effects of currencies strengthened by fossil-fuel exports generally has been unwise. The worst consequence of subsidizing agriculture has been accelerated mining of water – undoubtedly the scarcest resource in the region – from underground aquifers.

If the presence of energy resources has lowered incentives for domestic agricultural production, these also provide the financial wherewithal for food imports. Except in Turkey, agricultural imports exceed agricultural exports throughout the region. The absence of food self-sufficiency does not create hunger. To the contrary, none of the standard indicators of food insecurity, such as infant mortality and the poverty rate, are elevated. As the Middle East and North Africa define and follow a path toward freer markets and freer societies, hunger seems destined to continue diminishing.

Compared to what one finds in Asia, Latin America and the Caribbean, and Sub-Saharan Africa, poverty is not particularly widespread in the Middle East and North Africa. Thus, the terrorism originating in the region cannot be blamed simply on widespread destitution. Instead, it is to be hoped that, as political and economic freedoms are nurtured and living standards improve, advocates of violence will win fewer recruits.

Study Questions

1. Has oil wealth created a lasting economic boom in the Middle East and North Africa?
2. How has the lack of political freedom in the Middle East and North Africa affected the region's economic performance?
3. Is it accurate to characterize the Middle East and North Africa as a region where human fertility is uniformly high?
4. Describe the impacts of Dutch Disease on Middle Eastern and North African agriculture.
5. How has the presence of one subterranean resource – oil and natural gas – contributed to depletion of another resource, namely water?
6. Should Middle Eastern and North African nations strive for food self-sufficiency?
7. Are indicators of food security in the region above or below what one would expect, in light of average incomes?

14

Eastern Europe and the Former Soviet Union

Considerable though the challenge of modernization has been in the Middle East and North Africa, no part of the world has gone through a more profound transformation in recent years than Eastern Europe and the Former Soviet Union. Entirely dominated by steel, mining, and other traditional industries, which were rigidly locked into energy-intensive production technology from the 1930s (Kotkin, 2001, p. 17), the region's economies were strictly controlled by bureaucrats for decades and proved incapable of harnessing advances in electronic and information technology that created unprecedented prosperity elsewhere in the world in the late twentieth century. Stagnation was setting in during the 1970s and living standards deteriorated noticeably in a number of countries, including the USSR, for several years before the Berlin Wall fell in 1989 (Aslund, 2002, p. 41).

The events of 1989 kindled great expectations behind the Iron Curtain (as Winston Churchill had described the boundary between free and communist Europe at the outset of the Cold War). So did the defeat in 1991 of an attempted coup d'etat by communist military officers in Moscow. But rather than being marked by spreading affluence, the closing years of the twentieth century were a time of mounting hardship, with GDP per capita plummeting in fledgling nations carved out of the USSR as well as in many former Soviet satellites.

In some ways, the transition from central planning to a market economy in Eastern Europe and the Former Soviet Union was more difficult than the recovery of West Germany (which was spared occupation by the Red Army) and Japan after the Second World War. True, factories and infrastructure in the defeated axis powers were in a shambles in the late 1940s. However, a *tabula rasa* had been created by the deracination of totalitarian institutions, not to mention fascist leaders. Under US supervision, the rule of law and other elements of an institutional framework favorable to economic progress were put firmly in place. At the same time, development was bolstered by substantial foreign aid from the United States, through the Marshall Plan. The situation was vastly different four decades later in the Former Soviet Union and its erstwhile satellites, which suffered neither defeat in war nor occupation by a conquering army. Especially in Russia and other parts of the old USSR, the communist party elite, the *nomenklatura*, did not go away, but instead drew on old connections to secure privileged positions in the new regime and to grab natural resource industries and other valuable properties for themselves. Often resorting to bribery and strong-arm tactics, the same group was unenthusiastic about establishing the rule of law, which helps to explain the slowness with which private property rights and judicial mechanisms for contract enforcement were created. Moreover, the old Soviet bloc did not benefit from massive investment by overseas descendants of Russian emigrants, as China benefited after Mao's demise from remittances by descendants of its emigrants (Chapter 11). Other capital inflows, including foreign aid, were similarly limited.

Even if an institutional *tabula rasa* had existed in Russia and neighboring lands in 1989, the post-communist transition was never going to be easy. A way had to be found to root out deeply ingrained cultural habits required for survival in a self-styled workers' paradise with pervasive central planning – habits such as avoiding even the appearance of entrepreneurial initiative. Furthermore, economic waste of epic dimensions presented an enormous challenge. Somewhat tentatively, McKinnon (1991) observed as the USSR was expiring that "industries producing finished goods might well exhibit negative value added at world market prices" (p. 165). Subsequent analysis (Reed, 2004, pp. 224–225) confirmed that, indeed, the Soviet economy was fundamentally an engine of value destruction (as opposed to value creation), in the straightforward sense that GDP was less than the value of petroleum, bauxite, and other resources being consumed. Thoroughly unsustainable, this system inevitably had to be dismantled. Just as inevitably, though, dismantlement of an entire economic regime, even a woefully inefficient one, involved great dislocation and considerable belt tightening.

To summarize, the damage wrought under communism was enormous, in terms of economics and in every other respect. Many countries and peoples subjected to this system have yet to recover fully.

14.1 Patterns of Economic Growth since the Fall of Communism

As long as development was gauged in terms of the physical output of traditional industries, the economic shortcomings of communism were not immediately obvious. Through the 1960s and even into the 1970s, Soviet leaders could point with pride to steady increases in the production of iron and steel, tractors, and so forth – increases that at times matched the growth of manufactured output in the OECD – even though a great deal of communist production lay unused and rusting away at factories or in the parking lots of collective farms and other "customers." However, a service-dominated economy, one that developed and adopted new technology at a very rapid pace, proved incompatible with central planning. The ground lost during the 1970s and 1980s by the USSR and its allies (which remained locked into traditional manufacturing) to the capitalist West (which adapted readily to innovative structural change) is difficult to determine, for the simple reason that communist officials never measured or reported GDP per capita accurately. Nevertheless, the gap in living standards between the Soviet bloc and the OECD was widening during these two decades because of economic stagnation or worse in the former setting.

Far less ambiguous has been the erosion of living standards since the transition away from inefficient, centrally planned systems began. During the 1990s, GDP fell by 15–20 percent in Eastern Europe. Farther to the east, in the old Soviet heartland, the contraction was more severe, generally in the range of 30–50 percent (Aslund, 2002, p. 118). As this variance in the degree of economic decline suggests, some places handled the transition much better than others. One of the positive standouts was Poland, where opposition to communism was galvanized in 1979 by the triumphal return of John Paul II less than a year after he became Pope. *Solidarity*, a free labor union, was soon organizing openly. Subsequently, during the 1980s, organized dissent grew in Czechoslovakia (which divided amicably into the Czech Republic and Slovakia in 1993) and Hungary. Not coincidentally, reform-minded cadres were ready in each of these countries to usher in the institutional changes needed to put market economies on a sound footing soon after the Berlin Wall came down. In case these cadres needed reinforcement, the clear signal that reform would be rewarded with membership in the European Union (EU) and the North Atlantic Treaty Organization (NATO) added to the impetus for institutional change.

As indicated in Table 14.1 (column 1), the Czech Republic, Hungary, Poland, and Slovakia are among the minority of nations in Eastern Europe and the Former Soviet Union where GDP increased faster than human numbers during the last decade of the twentieth century. Albania, where a prior Maoist regime had created the worst economic conditions in all of Europe, also experienced improved living standards. So did three former parts of Yugoslavia, which broke up violently during the 1990s: Croatia, Macedonia, and Slovenia (which because of proximity and other reasons had always enjoyed the closest ties with Western Europe).[1] In general, countries where GDP per capita was on an upward trend soon after the end of communism had solid growth in exports (column 2) and, thanks to some success in the struggle against inflation (column 3), relatively high rates of investment (column 4). Much of this investment was undertaken by western firms, which were attracted by market-friendly reform. In more than a few instances, foreign corporations modernized inefficient state companies after acquiring a full or partial ownership stake.

Much more troubled were countries where communists' stranglehold on power was longer, tighter, or (more often than not) both. Throughout the old USSR, the

---

[1] Bosnia-Herzegovina and the remnant of Yugoslavia (Serbia and Montenegro) suffered prolonged armed conflict during the 1990s, which took a severe economic toll and impeded data collection.

Table 14.1 Patterns of economic growth in Eastern European and Former Soviet countries, 1990–2003

Country (ranked by column 1)	Average annual rate of growth 1990–1999 (%)				Country by (ranked column 5)	B. Average annual rate of growth 2000–2003 (%)				
	GDP per capita	Exports	Inflation	Gross investment		GDP per capita	Exports	Inflation	Economic Freedom Index* – 2002	Political freedom status* 2001–2002
	(1)	(2)	(3)	(4)		(5)	(6)	(7)	(8)	(9)
Poland	4.5	10.8	24.5	11.9	Turkmenistan	17.0	29.0	9.5	4.40	NF
Slovenia	2.5	−0.5	23.5	10.2	Armenia	11.3	22.8	2.3	2.70	PF
Albania	2.0	13.6	51.5	22.4	Kazakhstan	11.3	14.0	10.3	3.60	NF
Slovak Republic	1.7	12.0	10.8	4.6	Azerbaijan	9.8	12.7	5.5	3.50	PF
Hungary	1.3	8.2	20.7	8.4	Tajikistan	9.0	12.0	20.5	3.85	NF
Macedonia	1.2	1.2	13.5	6.7	Ukraine	8.0	9.8	11.3	3.85	NF
Czech Republic	1.0	9.0	13.7	6.3	Latvia	8.0	8.2	2.3	2.50	F
Croatia	0.4	—	131.2	—	Russia	7.5	7.0	21.0	3.70	PF
Estonia	−0.4	10.2	62.7	−1.8	Estonia	7.3	8.8	4.3	1.80	F
Romania	−0.8	6.1	105.5	−11.8	Lithuania	7.0	13.3	0.0	2.35	F
Bulgaria	−2.0	0.3	111.8	−0.9	Georgia	6.5	13.0	4.8	3.40	PF
Lithuania	−3.8	2.9	90.4	8.8	Belarus	5.8	10.0	84.7	4.35	NF
Latvia	−3.8	0.7	58.8	−4.4	Bulgaria	5.8	10.5	5.0	3.40	F
Armenia	−3.9	−21.5	269.5	−29.5	Albania	5.8	17.8	5.3	3.30	PF
Uzbekistan	−4.0	—	356.7	—	Moldova	5.8	16.3	15.8	3.35	PF
Belarus	−4.2	−11.1	449.9	−10.0	Croatia	4.8	7.8	3.8	3.40	F
Kazakhstan	−5.3	4.3	255.7	−11.7	Romania	4.8	17.3	32.8	3.70	F
Russia	−6.0	2.3	189.6	−13.3	Hungary	3.8	11.7	9.0	2.40	F
Turkmenistan	−6.4	—	622.8	—	Slovak Republic	3.5	12.0	4.8	2.90	F
Kyrgyz Republic	−8.2	6.7	157.3	12.6	Kyrgyz Republic	3.3	4.5	10.0	3.60	NF
Azerbaijan	−10.2	12.6	249.5	14.7	Slovenia	3.0	7.0	7.3	2.90	F
Georgia	−10.3	9.8	513.0	51.2	Czech Republic	3.0	9.5	2.8	2.40	F
Ukraine	−10.4	−3.6	440.0	−24.8	Uzbekistan	3.0	−0.3	37.8	4.35	NF
Moldova	−11.3	4.8	1142.5	−20.0	Poland	2.8	10.3	2.8	2.70	F
Tajikistan	−11.6	—	300.0	—	Macedonia	0.8	3.3	4.3	3.25	PF

Note: *The Economic Freedom Index ranges from 1.00 (completely free markets) to 5.00 (completely repressed markets). See Heritage Foundation (2002). Political freedom status is F (free), PF (partly free) and NF (not free). See Freedom House (2000, 2001).
Source: World Bank (2000/2001), pp. 278–279, 294–295; World Bank (2005).

dissident-reform tradition was weak at best. As a result, little prevented former communist *apparatchiks* from seizing control as old regimes crumbled. This group routinely ignored civil and political liberties and showed little interest in establishing open, competitive capitalist economies. In places like Belarus and Ukraine, *apparatchiks* actively subverted reform.[2] Under these circumstances, large-scale, communist-era enterprises either were never privatized or, when these were, ended up in the hands of favored *apparatchiks* thanks to officially sanctioned theft. Rent seeking and corruption proliferated in tightly controlled regimes characterized by predatory regulation. Furthermore, organized crime expanded, sometimes to meet demands for the enforcement of individual property rights and private contracts – a service that weak public institutions failed to provide.

No part of the Former Soviet Union was able to avoid significant reductions in GDP per capita during the 1990s – reductions, it must be added, that happened in spite of modest population growth or demographic contraction in the same period (see next section). The smallest decline occurred in the Baltic Republics – Estonia, Latvia, and Lithuania – that had been swallowed up by the USSR at the beginning of the Second World War.[3] The economic performance of these three small countries was comparable to that of Bulgaria and Romania, which did not make as swift a transition from communism as countries like Poland and Slovenia (Table 14.1, column 1). Elsewhere, inflation raged (column 3), which often led to negative rates of capital formation (column 4). With few exceptions, export growth was either anemic or negative (column 2). Annual rates of decline in GDP per capita ranged from 4 percent or so in Armenia, Belarus, and Uzbekistan to more than 10 percent in Azerbaijan, Georgia, Moldova, Tajikistan, and Ukraine. The yearly decline in Russia averaged 6 percent (column 1).

With living standards falling precipitously in the Former Soviet Union (and places in Eastern Europe, like Bulgaria and Romania, which bore a close resemblance to the former communist hegemon) and the old *nomenklatura* benefiting from various forms of corruption and theft, income distribution grew more skewed. In contrast, income inequality remained modest as GDP per capita rose in Central Europe and those parts of the old Yugoslavia that moved decisively away from communism. This accomplishment is explained by economic progress and broad access to education, as well as high living standards and modern institutions in many of these countries prior to the Second World War.

Since the turn of the twenty-first century, the fortunes of Eastern Europe and the Former Soviet Union have started to recover. Aside from Macedonia, each and every country has experienced an appreciable gain in GDP per capita (Table 14.1, column 5).

---

[2] In Ukraine, the opponents of reform, whom Russia's Vladimir Putin had supported, were dealt a defeat in late 2004, when the country's supreme court threw out the fraudulent results of a rigged election. After prevailing in a new election, which was fair and free, Victor Yushchenko, a democrat, was sworn in as president in early 2005.
[3] The Soviets' absorption of the Baltic states was part of a broader accord between Adolf Hitler and Joseph Stalin, which resulted in the partition of Poland between the western two-thirds (which the Germans conquered in 1939) and the eastern third (which became part of Ukraine in the USSR).

In Central Asian nations, such as Kazakhstan and Turkmenistan, as well as Russia and Ukraine, the nadir reached during the 1990s was so deep that a rebound was all but inevitable, even though current conditions remain worse in these places than conditions were for ordinary citizens in 1990. Armenia is a special case, benefiting as it has from a construction boom and other investment financed by a large infusion of funds from Armenian-Americans in California. The most potent driving force for recovery, however, has been a major increase in the prices of fossil fuels, which Russia, Azerbaijan, and Central Asia export. After hovering between $14 and $18 throughout the 1990s, the value of a barrel of oil shot up, averaging nearly $30 between 2000 and 2002 and rising above $50 in 2005.[4]

The handsome gains flowing to fossil-fuel exporters notwithstanding, sustained improvement has been noteworthy in the Baltics and Eastern Europe, which have little oil and natural gas to speak of aside from Romania's deposits. A basic reason for sustained growth of GDP per capita in the latter settings is indicated in the last two columns of Table 14.1. By and large, civil and political liberties that were introduced as communism collapsed took hold, as indicated by a free (F) ranking by the Freedom House (column 9). For the most part, such a ranking was accompanied by favorable indices of economic freedom (column 8), reflecting openness to trade, the containment of inflation, and the like. The Baltic countries, for example, increased their exports to Scandinavia. They also attracted substantial investment from the West, as did various Central European nations. Within this entire group, Poland was the sole exception to the general trend, which was for GDP per capita to grow during the 1990s and then accelerate upward after the turn of the century.

The Case of Russia

As indicated above, Russia, at the heart of the old USSR, underwent the same rebound after 2000 that happened in other parts of the region. A change in political direction occurred at the turn of the century as well, with Vladimir Putin replacing Boris Yeltsin as president.

Showing great courage as he faced down the attempted coup of August 1991, Yeltsin initially attempted a radical break from communism. His prime minister, Yegor Gaidar, tried to implement widespread deregulation of prices along with rapid though partial privatization of state enterprises. However, these reforms did not succeed. From 1993 onward, Yeltsin was served by a succession of risk averse, bureaucratic prime ministers, and inflation accelerated, unemployment rose, and poverty became more acute (Desai, 2005). Furthermore, the president's own leadership was undermined as time went on by his alcoholism, ill health, and unstable personality.

---

[4] As reported in a footnote in the preceding chapter, oil prices fell dramatically in 1985, as Saudi Arabia responded positively to US requests to boost output. Along with benefiting the US economy, this turn of events cut deeply into the USSR's hard currency earnings, and thus hastened the end of the communist regime (Schweizer, 1994, pp. 216–220, 233, 242–243).

During the 1990s, weak leadership at the top levels of the Russian government was symptomatic of state weakness, in general, which in turn led to an expansion of organized crime. For example, private security companies emerged to protect employers and to enforce contracts. Over time, much of this racketeering was taken over by official police forces administered by government officials (Aslund, 2002, pp. 358–359). Moreover, this official extortion had an especially severe impact on small businesses, which had none of the official protection enjoyed by powerful "oligarchs" (Safavian et al., 2001).

The most controversial move of the Yeltsin administration was the infamous shares-for-loans scheme. With no effective tax system in place to cover public-sector expenses and with destabilizing fiscal deficits on the rise, former members of the *nomenklatura* who had recently gone into private banking were asked to lend the government money, accepting as collateral recently created shares of stock in state-owned natural resource industries, some of which were large and potentially profitable. The government defaulted soon afterward, which left key energy and mineral enterprises in the hands of a few oligarchs (Guriev and Rachinsky, 2005). In return for being singled out to benefit from one of the most spectacular political thefts the world has ever seen, the oligarchs bankrolled Yeltsin's successful campaign for reelection in 1996.[5]

Although hand picked by Yeltsin, Putin represents a considerable departure from his predecessor. A former officer of the Soviet security and intelligence service, the Committee for State Security of the Soviet Union (KGB), Russia's current president is somewhat austere, an athlete (with a black belt in karate), and, what is rare for his countrymen, a teetotaler. With Putin's ascendancy to high office coinciding with elevated prices for oil and natural gas, the first few years of his administration were marked by brisk economic expansion, which added to the president's popularity. The expansion was reinforced by reforms introduced by Putin. One of these was a "flat" income tax, in which the government takes 13 percent of everyone's earnings. A major virtue of this arrangement over the former scheme, which featured steeply graduated rates, is that tax avoidance and other types of maneuvering have been greatly curtailed. Another is that tax revenues have increased, with a corresponding decline in fiscal deficits and the inflation these deficits precipitate. The banking sector also has benefited from Putin-era reforms, such as more stringent capital-adequacy requirements for banks that lower the risk of failure for these institutions.

---

[5] Although Russia's oligarchs often compare themselves to the US "robber barons" of the late nineteenth century, this is a self-serving and misleading conceit. While figures such as Carnegie, Harriman, Morgan, and Rockefeller were able for a time to shape the law to serve their interests, they were in the final analysis entrepreneurs of enormous dynamism, who created new industries that generated wealth, jobs, and innovation. They mobilized capital on a grand scale, thereby expanding financial markets in the United States. Moreover, they kept their wealth in the country, sustaining investment in their companies. The Russian oligarchs, on the other hand, are not entrepreneurs, but instead former communists who cleverly appropriated state properties. They neither mobilized capital nor played any role in developing financial markets. On top of everything else, the oligarchs transferred most of their ill-gotten, windfall gains abroad rather than investing these in Russia.

Contrary to the claim by former German Chancellor Gerhard Schroeder that Putin is a "dyed-in-the-wool democrat" (Stelzer, 2005, p. 27), the Russian leader has autocratic tendencies. He is pragmatic enough not to press disputes that the country, which has much less military power than the Soviet Union had even in its waning days, cannot win. For example, a US military presence has been tolerated in Central Asia and Georgia after September 11, 2001 for the sake of defending against Islamic terrorism, which threatens Russia no less than the United States. However, Russia under Putin has not shied away from bullying the "near abroad" of Belarus, Kazakhstan, and Ukraine (up until the presidential inauguration in early 2005 of Victor Yushchenko, who is democratic and pro-Western).

The most worrying exercise of Putin's autocratic impulses has been within Russia's borders. The reining in of the oligarchs, who dominated and corrupted Yeltsin's administration, met initially with approval, among Russians and foreigners alike. However, serious concerns have been aroused about Putin's commitment to the rule of law. For example, freedom of the press looks shaky in light of moves against the country's leading media magnate. Likewise, vigorous prosecution of the Russian oil conglomerate, Yukos, for tax delinquency and the jailing of its chief executive drove the company into bankruptcy. As far as many entrepreneurs are concerned, the whole episode has undermined the legal guarantees required for investment.

Since 2003, the Putin presidency has lost its luster. No end to the horrible conflict in Chechnya is in sight. Incidental to that conflict in 2004 was the occupation by terrorists of a schoolhouse full of hundreds of children and their teachers in Beslan; the government botched the ensuing siege and there was great loss of life. In economics, Russia can ill afford to shake investors' confidence, as has happened because of the Yukos affair. As already mentioned, the country does not benefit as China does from a community of emigrants and their descendants who are willing to invest tens of billions in the ancestral homeland.[6] If Russia is to attract more direct foreign investment, which as of 2000 had amounted to less than $7 billion (EBRD, 2002, p. 193), or even to discourage capital flight by its own citizens, then the rule of law must be solidified. Whatever political and financial ends for the government have been served by its takeover of Yukos (Stelzer, 2005), the damage done to the investment environment has been serious.

14.2 Demographic Trends

A difficult economic transition is not the only legacy of communism in Eastern Europe and the Former Soviet Union. The environmental aftermath of Marxist

---

[6] Since 1978, China has attracted close to $400 billion in direct foreign investment and the current annual flow is $40 billion (Broadman, 2003). Much of this investment has been made by ethnic Chinese living elsewhere in East and Southeast Asia.

totalitarianism also has been alarming. Especially in the old USSR, air and water pollution are widespread and dealing with toxic and nuclear wastes will be very costly. On top of this, the lack of a politically conscious and influential middle class has meant that the political constituency demanding a clean environment is weak. This is a sharp contrast with the situation in Western Europe, where "green movements" are powerful.

In addition, communism created lasting demographic impacts. Under Stalin, Russia and neighboring countries depopulated, in part due to the death of millions of combatants and civilians during the Second World War and also because the earlier collectivization of agriculture was accomplished with enormous loss of life (Chapter 7). Moreover, the years after communism fell were marked not just by a lowering of GDP per capita, but also a diminishing of life spans. In Russia, male life expectancy at birth was 64.2 years in 1989 – unchanged from what it had been a decade earlier (Powell, 2002) – but then proceeded to fall by 6.6 years during the next 5 years, reaching 57.6 years in 1994. Female life expectancy declined by 3.3 years during the same period. Similar changes occurred in the Baltic Republics, Belarus, and Ukraine (Brainerd and Cutler, 2005).

Life spans have been curtailed in Russia and neighboring countries for various reasons. The incidence of tuberculosis has been rising, as has the number of people infected with HIV/AIDS. Many people smoke heavily, which of course leads to cardiovascular disease and other illnesses. Also, industrial accidents and suicide take a heavy toll among working-age men (Powell, 2002). In a statistical analysis of the causes of increased mortality in the Former Soviet Union, Brainerd and Cutler (2005) conclude that deterioration of the communist-era health-care system has not had a major impact on life expectancy. Of much greater importance has been widespread alcoholism, which is related to homicides, suicides, and accidents. In addition, increased mortality has resulted from stress "associated with a poor outlook for the future" (p. 108). The latter pessimism could turn out to be self-fulfilling because, if the trend toward higher death rates is not reversed or is not compensated for by an upswing in human fertility, then Russia's demographic contraction, which is approaching 1 million per annum, will continue. This trend will have far-reaching implications for the national economy and the country's standing in the world – not to mention relations with potentially hostile neighbors to the south where human numbers are still increasing.

Human Fertility

In a region as extensive and diverse as Eastern Europe and the Former Soviet Union, total fertility rates (TFRs) vary. A generalization that holds pretty well, however, is that the number of births per woman in any particular setting resembles patterns of human fertility in adjacent areas with similar ethnic, religious, or other characteristics. In places like Bulgaria, Poland, Russia, and Ukraine, populations are much like those of Western Europe. For one thing, they are largely urban (Table 14.2, column 4). Also, women face few barriers to education or employment, female

Table 14.2 Social indicators for Eastern European and Former Soviet countries in 2002

Country (ranked by income per capita in 2002)	Total fertility rate		2002 income per capita ($PPP)	Urban % of total population in 2002	Adult illiteracy rate in 2002 (% people 15 and over)		Female % of labor force		Contraceptive prevalence rate 1990–2002 (% women 15–49)
	1980	2002			Males	Females	1980	2002	
	(1)	(2)	(3)	(4)	(5)	(6)	(7)	(8)	(9)
Slovenia	2.1	1.2	18,540	49	0	0	45.8	46.6	—
Czech Republic	2.1	1.2	15,780	75	—	—	47.1	47.2	69
Hungary	1.9	1.3	13,400	65	1	1	43.3	44.8	73
Slovak Republic	2.3	1.3	12,840	58	0	0	45.3	47.7	—
Estonia	2.0	1.3	12,260	69	0	0	50.6	48.9	—
Poland	2.3	1.3	10,560	63	—	—	45.3	46.5	—
Lithuania	2.0	1.3	10,320	69	1	0	49.7	48.0	—
Croatia	—	1.5	10,240	59	3	1	40.2	44.4	—
Latvia	1.9	1.2	9,210	60	0	0	50.8	50.5	—
Russian Federation	1.9	1.3	8,230	73	1	0	49.4	49.2	34
Bulgaria	2.1	1.3	7,130	68	2	1	45.3	48.0	—
Romania	2.4	1.3	6,560	55	3	1	45.8	44.5	61
Macedonia	2.5	1.8	6,470	60	—	—	36.1	42.0	—
Kazakhstan	2.9	1.8	5,870	56	1	0	47.6	47.1	63
Belarus	2.0	1.3	5,520	70	0	0	49.9	48.9	—
Ukraine	2.0	1.2	4,870	68	1	0	50.2	48.8	70
Albania	3.6	2.2	4,830	44	23	8	38.8	41.5	—
Turkmenistan	4.9	2.7	4,250	45	—	—	47.0	45.9	62
Azerbaijan	3.2	2.1	3,210	52	—	—	47.5	44.7	55
Armenia	2.3	1.2	3,120	67	2	1	47.9	48.6	61
Georgia	2.3	1.1	2,260	57	—	—	49.3	46.8	41
Uzbekistan	4.8	2.4	1,670	37	1	0	48.0	46.9	56
Kyrgyz Republic	4.1	2.4	1,620	34	—	—	47.5	47.2	60
Moldova	2.4	1.4	1,470	42	2	0	50.3	48.4	74
Tajikistan	5.6	2.9	980	28	1	0	46.9	45.2	—

Source: World Bank (2004).

illiteracy rates being very low and indistinguishable from male rates (columns 5 and 6) and women participating about as much as men in the labor force (columns 7 and 8). Even though GDP per capita (column 3) is generally lower than average incomes in Western Europe, TFR trends where ethnicity and culture are unmistakably European differ little from trends to the west of the old Iron Curtain. At or close to the replacement level 25 years ago (column 1), fertility rates have fallen since then. Indeed, few countries in the world, including members of the OECD, have rates as low as Armenia, the Czech Republic, Latvia, Slovenia, and Ukraine, where 1.2 births per woman are the norm. In Georgia, women only deliver 1.1 babies on average.[7]

Once predominantly Christian, countries such as these are now mainly secular. This characteristic, like advanced urbanization, is shared with Western Europe. However, secularism is also a consequence of the active suppression of religion under communism. The same sort of suppression was also directed against Islam, but with less effect. This probably helps to explain why fertility decline, though it definitely has occurred in the southern reaches of the old USSR, has been less extreme in Azerbaijan (in the Caucasus) and Turkmenistan, Uzbekistan, and other Central Asian republics. TFRs in these nations differ little from rates in co-religious, Middle Eastern nations (Chapter 13). As in other parts of the world, larger families coincide in this setting with low average incomes and limited urbanization. Such is the case with the Kyrgyz Republic and Tajikistan, for example.

Natural Increase

Since TFRs in much of the old USSR and all of its European satellites were close to the replacement level in 1980 and since human fertility continued to decline during the ensuing quarter century, there is little demographic momentum to speak of in Eastern Europe and the Former Soviet Union. Death and birth rates (Table 14.3, columns 2 and 3) are close to one another throughout the region, along with being comparable to rates in the OECD. Where crude death rates (CDRs) are relatively high – as these are in Albania and Macedonia in southeastern Europe as well as in much of Central Asia – births are not keeping pace, which means that natural increase (column 4) is negative.

Demographic contraction poses a significant economic challenge, since it is normally accompanied by an aging of the population. In the OECD, people and their elected representatives are wrestling with the funding of pensions for the elderly, who are becoming more numerous. Eastern Europe and the Former Soviet Union face similar issues, although without the high living standards enjoyed by their

---

[7] Extraordinarily low TFRs such as these also reflect the high frequency with which women have had abortions, both under communism and more recently (Powell, 2002). Multiple abortions tend to scar the reproductive tract, thereby causing infertility.

Table 14.3 Population dynamics in Eastern European and Former Soviet countries for 2002

Country (ranked by income per capita in 2002)	Income per capita 2002 ($PPP) (1)	CDR – 2002 (per 1,000 population) (2)	Crude birth rate – 2002 (per 1,000 population) (3)	Rate of natural increase (%) (4)	Average annual population growth – 1980–2002 (%) (5)
Slovenia	18,540	9	10	0.1	0.2
Czech Republic	15,780	9	11	0.2	0.0
Hungary	13,400	10	13	0.4	−0.2
Slovak Republic	12,840	11	10	−0.1	0.3
Estonia	12,260	9	14	0.5	−0.3
Poland	10,560	9	9	0.0	0.4
Lithuania	10,320	9	12	0.2	0.1
Croatia	10,240	10	12	0.2	−0.1
Latvia	9,210	8	14	0.6	−0.3
Russian Federation	8,230	10	15	0.6	0.2
Bulgaria	7,130	9	14	0.6	−0.4
Romania	6,560	10	13	0.3	0.0
Macedonia	6,470	14	9	−0.5	0.4
Kazakhstan	5,870	15	12	−0.3	0.0
Belarus	5,520	9	14	0.5	0.2
Ukraine	4,870	9	15	0.7	−0.1
Albania	4,830	17	6	−1.1	0.8
Turkmenistan	4,250	22	8	−1.5	2.3
Azerbaijan	3,210	16	7	−1.0	1.3
Armenia	3,120	9	8	−0.1	0.0
Georgia	2,260	8	10	0.2	0.1
Uzbekistan	1,670	20	6	−1.4	2.1
Kyrgyz Republic	1,620	20	7	−1.3	1.5
Moldova	1,470	11	13	0.2	0.3
Tajikistan	980	23	7	−1.6	2.1

Source: World Bank (2002a), pp. 48–50; World Bank (2004).

neighbors to the west. Put very simply, this region is growing old without the advantage of having grown rich first.

14.3 The Agricultural Sector

Along with GDP contraction, diminished life spans, and rock-bottom fertility rates, agricultural difficulties have arisen in Eastern Europe and the Former Soviet Union during communism's decline and fall. As with other aspects of the post-communist transition, some countries have succeeded in substituting capitalist farming for state-run systems, which predominated in all but a few of the region's countries for decades. Other nations, however, are still struggling with reform or have yet to undertake it.

Reform in formerly communist nations can be measured using a 10-point scale developed by the World Bank. Applying this scale to farm ownership, liberalization of agricultural markets, and related parameters in Eastern Europe and the Former Soviet Union, the European Bank for Reconstruction and Development (EBRD) has identified three sorts of countries (Table 14.4, column 1). Considerable progress has occurred in the first group, made up almost entirely of Central European and Baltic nations that generally have made a successful transition from communism (see above). From Hungary to Estonia, property rights of private farmers who had continued to operate during the communist era were strengthened, there was restitution of land confiscated when communism arrived after the Second World War (to original owners or their heirs), or both (column 2). Private ownership of rural land is now the rule in these countries (column 3) and market distortions, as measured by producer subsidy equivalents (PSEs), are diminishing (columns 5 and 6).

Countries in the second group, where reform is at an intermediate stage according to the EBRD, are in southeastern Europe (aside from Bulgaria, which is in the advanced group) as well as some of the old USSR. Private farming had survived under communism in three former Yugoslav republics: Bosnia-Herzegovina, Croatia, and Macedonia. Elsewhere, restitution has been impractical – in the old USSR because countless farmers had been liquidated in the rural collectivization ordered by Stalin. Rather than trying to divide large collective farms among individually owned parcels, these farms, which have remained intact, have been converted into private enterprises. Quite often this has involved the distribution of shares among workers. However, most of these people have been reluctant to assume the risks of ownership, which are compounded in many countries by limited services in the countryside and poorly organized channels for the marketing of inputs and outputs. Accordingly, they have tended to resell their allotted shares to former officials and managers, who are now rural entrepreneurs.

Similar things have occurred in the third group of countries, where reform is barely under way or has not even been attempted. With the exception of the remnants of Yugoslavia (i.e., Serbia and Montenegro), every member of this last group was part of the USSR. An important feature of agriculture in places like Russia and Belarus is that Soviet-era collectives not only have stayed intact, but have remained in the hands of the state as well.

The place of agriculture in Russia contrasts sharply with the sector's role in China. As explained in Chapter 11, the latter nation's reforms began in the countryside. The limited reintroduction of farmland ownership in the late 1970s, after the collective approach had been followed for a little less than two decades, led to a sharp increase in output, created a base for economic activity in villages and small towns (in part because rural households in China had not lost their knowledge of individual farming) and pointed the way for future changes in economic

Table 14.4 Indicators of agricultural reform in post-communist countries in Eastern Europe and the Former Soviet Union in 2000

	World Bank reform index, 2001	Privatization strategy	% Rural land in individual use circa 2000	Net agricultural trade (US $ million) 2000	Producer-equivalent support (PSE %)	
					1993	2000
	(1)	(2)	(3)	(4)	(5)	(6)
A. Advanced						
Hungary	9.2	Both	51	1,153	22.0	20.0
Czech Republic	9.2	Restitution	26	−676	27.0	16.0
Slovenia	9.2	Individual pre-1990	na	−361	28.0	43.0
Estonia	9.0	Restitution	61	−199	−32.0	10.0
Latvia	9.0	Restitution	91	−297	−40.0	18.0
Slovak Republic	8.2	Restitution	9	−382	30.0	3.0
Poland	8.0	Individual pre-1990	84	−533	12.0	7.0
Lithuania	8.0	Restitution	87	−106	−37.0	9.0
Bulgaria	8.0	Restitution	56	139	−4.0	2.0
B. Intermediate						
Albania	7.6	Distribution	94	−221	na	na
Armenia	7.4	Distribution	90	−184	na	na
Romania	7.4	Both	85	−593	16.0	11.0
Croatia	7.0	Individual pre-1990	66	−288	na	na
Azerbaijan	6.6	Distribution	5	−16	na	na
Georgia	6.6	Distribution	44	−69	na	na
Macedonia	6.6	Individual pre-1990	80	na	na	na
Bosnia-Herzegovina	6.2	Individual pre-1990	94	na	na	na
Kyrgyz Republic	6.2	Distribution	37	na	na	na
Moldova	6.0	Distribution	20	179	na	na
Ukraine	6.0	Distribution	17	442	na	na
C. Rudimentary						
Russian Federation	5.8	Distribution	13	5,689	−24.0	3.0
Kazakhstan	5.8	Distribution	24	138	na	na
Tajikistan	4.8	Distribution	9	na	na	na
Fr. Yugoslavia	4.8	Individual pre-1990	na	na	na	na
Uzbekistan	3.4	Distribution	14	na	na	na
Turkmenistan	2.0	Distribution	8	na	na	na
Belarus	1.8	Distribution	14	−603	na	na

Source: EBRD (2002), pp. 78–84 (for explanation of terms in column headings see text).

policy. Russian reformers, on the other hand, have been reluctant to deal with the difficult legacy of collectivization, other than to strengthen individual property rights in small, peri-urban holdings. Instead of agricultural reform contributing to general economic progress, as has been the case in China, the slow dismantlement of communist-style farming, which dates back to the 1930s, has held back reform in the Russian economy as a whole. A more positive move is that price distortions in the agricultural sector have been reduced, mainly because continued subsidization is all but impossible given the meager financial resources of the national government.

Factor Use

Recent trends in the utilization of farm inputs reflect the transition away from state-run agriculture, imperfect though it might have been in many parts of the region. Also influential has been sluggish demand resulting from demographic stagnation and falling standards of living.

Aside from Slovenia, Tajikistan, and a few other places, Eastern Europe and the Former Soviet Union feature low ratios of rural population to arable land (Table 14.5, column 1). Under these circumstances, yield-enhancing measures, such as heavy fertilization and widespread irrigation, are not a high priority. In the late communist era, however, inputs often were applied with little concern given to efficiencies and opportunity costs. For example, fertilizer applications (column 2) were frequently excessive. While applications remain heavy in a number of small countries with limited endowments of arable land, fertilizer use fell during the 1990s as subsidies for this input were eliminated. In Russia, application rates have reached very low levels (column 3).

Since precipitation is fairly generous in much of the region, many countries have invested little in irrigation. Again, the main exceptions are nations with limited land resources for crop production, particularly in the Caucasus and Central Asia (Table 14.5, columns 4 and 5). After the Second World War, Soviet authorities supported a large expansion of irrigated agriculture in Central Asia, where rainfall is not abundant, mainly for the sake of producing cotton for international as well as domestic markets. With large volumes of water diverted from rivers in the area, deserts spread and the Aral Sea dried up; another lasting environmental consequence of Soviet-era irrigation has been farmland salinization (Dixon and Gulliver with Gibbon, 2001, p. 130). Irrigation investment has yielded better returns in Armenia, Azerbaijan, and Georgia – the warmest parts of the old USSR, which these days export sizable quantities of fruits and vegetables (grown in large part on small, private farms) to Moscow and other northern cities. Romania, where the irrigated percentage of farmland increased during the 1980s and 1990s, engages in similar trade.

Just as the elimination of communist-era subsidies has resulted in lower levels of fertilization, diminished subsidization has driven down ratios of farm machinery

Table 14.5 Agricultural inputs in Eastern European and Former Soviet countries, 1979–2001

Country (ranked by income per capita in 2002)	Rural population density (per square km of arable land) 2001 (1)	Fertilizer use (kg per ha of arable land) 1979–1981 (2)	Fertilizer use (kg per ha of arable land) 1999–2001 (3)	Irrigated land (% of cropland) 1979–1981 (4)	Irrigated land (% of cropland) 1999–2001 (5)	Agricultural machinery (tractors per sq. km of arable land) 1979–1981 (6)	Agricultural machinery (tractors per sq. km of arable land) 1999–2001 (7)
Slovenia	581	–	438	–	1.3	–	65.6
Czech Republic	85	–	108	–	0.7	–	2.9
Hungary	78	291	84	3.6	4.6	1.1	2.3
Slovak Republic	–	–	61	–	11.2	–	1.6
Estonia	62	–	39	–	0.4	–	5.7
Poland	104	239	111	0.7	0.7	4.1	9.3
Lithuania	37	–	53	–	0.2	–	3.5
Croatia	128	–	145	–	0.2	–	0.2
Latvia	51	–	31	–	1.1	–	3.0
Russian Federation	32	–	12	–	3.6	–	0.6
Bulgaria	59	233	33	28.3	17.4	1.6	0.6
Romania	107	145	31	21.9	31.2	1.5	1.7
Macedonia	146	–	67	–	9.0	–	9.5
Kazakhstan	31	–	20	–	10.8	–	0.2
Belarus	49	–	129	–	2.1	–	1.2
Ukraine	48	–	14	–	7.2	–	1.0
Albania	309	156	28	53.0	48.6	1.7	1.4
Turkmenistan	148	–	60	–	100.1	–	2.9
Azerbaijan	230	–	6	–	74.8	–	1.8
Armenia	204	–	12	–	51.3	–	3.7
Georgia	286	–	52	–	44.2	–	2.2
Uzbekistan	352	–	164	–	88.6	–	3.8
Kyrgyz Republic	232	–	16	–	74.2	–	1.9
Moldova	–	–	3	–	14.1	–	2.3
Tajikistan	485	–	11	–	68.3	–	3.2

Source: World Bank (2004).

Table 14.6 Trends in crop area, yield, and output in Eastern Europe and the Former Soviet Union, 1970–2000

	2000 Production (million tons) (1)	Average annual change, 1970–2000 (%)		
		Area (2)	Yield (3)	Production (4)
A. Crops				
Wheat	111	−1.2	0.9	−0.4
Roots and tubers	101	−1.2	−0.3	−1.5
Vegetables	71	−0.3	1.2	1.5
Barley	40	−0.4	0.0	−0.3
Other grains	31	−1.8	0.8	−0.9
Fruits	30	−0.1	0.0	0.1
Maize	27	−0.5	0.1	−0.5
Oil crops	7	0.8	0.3	1.0
Pulses	5	−2.5	0.3	−2.2
B. Animal products				
Milk	103	–	–	−0.5
Meat	18	–	–	−0.2
Eggs	5	–	–	1.1

Source: Dixon and Gulliver with Gibbon (2001), pp. 138–139.

to arable land (Table 14.5, columns 6 and 7). Nevertheless, agriculture is more mechanized in Eastern Europe and the Former Soviet Union than in any other part of the world, save nations belonging to the OECD. This is sensible, of course, given that rural population densities are low in most of the region.

Agricultural Yields and Land Use and Trends in Output

Diminished use of fertilizer, machinery, and other non-land inputs to crop production has had a dampening effect on output per hectare. During the last three decades of the twentieth century, yields of barley, roots and tubers, and a number of other crops either rose slowly or fell (Table 14.6, column 3). Since cultivated areas have declined as well (column 2), output levels in 2000 were generally lower than what these had been 30 years earlier (columns 1 and 4). These reductions coincided with level or downward trends in human numbers and GDP per capita. Furthermore, food consumption has been affected by the removal of subsidies at the retail level.

As a rule, relative declines in food output have been modest in Eastern Europe. Part of the reason for this outcome has been the success of agricultural reform in countries such as Hungary and Poland. Larger declines have occurred in those

Table 14.7 Agricultural and food output in Eastern European and Former Soviet countries, 1979–2002

Country (ranked by income per capita 2002)	Cereal yield (kg per ha)		Arable and permanent cropland (1,000 ha)		Food production index (1989–1991 = 100)		Agricultural productivity (value added per worker – 1995 $)	
	1979–1981 (1)	2000–2002 (2)	1980 (3)	1999 (4)	1979–1981 (5)	2000–2002 (6)	1979–1981 (7)	1999–2001 (8)
Slovenia	–	5,241	–	202	–	100.9	–	36,188
Czech Republic	–	4,252	–	3,332	–	78.0	–	6,331
Hungary	4,519	4,156	5,333	5,039	90.7	79.5	3,390	5,469
Slovak Republic	–	–	–	1,594	–	–	–	–
Estonia	–	2,058	–	1,135	–	39.8	–	3,634
Poland	2,345	2,948	14,901	14,401	87.9	86.0	–	1,642
Lithuania	–	2,680	–	2,996	–	64.7	–	3,323
Croatia	–	4,630	–	1,590	–	68.5	–	9,383
Latvia	–	2,220	–	1,880	–	42.4	–	2,619
Russian Federation	–	1,823	–	126,820	–	66.6	–	3,574
Bulgaria	3,853	2,940	4,181	4,511	105.5	68.2	2,754	8,270
Romania	2,854	2,492	10,497	9,845	113.0	87.1	1,397	3,598
Macedonia	–	2,583	–	635	–	89.5	–	4,209
Kazakhstan	–	1,102	–	30,135	–	73.5	–	1,706
Belarus	–	2,118	–	6,306	–	62.1	–	2,894
Ukraine	–	2,474	–	33,615	–	52.4	–	1,494
Albania	2,500	3,157	750	699	–	–	1,184	1,826
Turkmenistan	–	2,179	–	1,695	–	131.6	–	642
Azerbaijan	–	2,520	–	1,983	–	83.7	–	977
Armenia	–	1,747	–	560	–	79.3	–	2,765
Georgia	–	1,772	–	1,063	–	74.9	–	–
Uzbekistan	–	3,239	–	4,850	–	122.3	–	1,426
Kyrgyz Republic	–	2,756	–	1,435	–	132.5	–	1,824
Moldova	–	2,350	–	2,181	–	51.1	–	941
Tajikistan	–	1,303	–	860	–	60.5	–	702

Source: World Bank (2004); FAO Production Yearbook (1983), pp. 52–55; FAO Production Yearbook (2000), pp. 8–12.

parts of the Former Soviet Union with a Slavic heritage, even including the three Baltic Republics. Demand reduction is a part of the explanation. So is the lack of agricultural reform in places like Belarus, Kazakhstan, and Russia. In contrast, minimal or absent reform has not prevented output from going up in response to demographically related demand growth in Uzbekistan and Turkmenistan.

Finally, value added per agricultural worker (Table 14.7, columns 7 and 8) is generally low. The major exceptions are in those Eastern European nations where a successful transition from communism has occurred. In addition, irrigated production of fruit and vegetables for foreign markets has enhanced value added per worker in the Armenian countryside.

14.4 Dietary Change, Consumption Trends, and Food Security

Just as the old USSR and its satellites constantly strove to match western output of manufactured items such as steel and tractors, communist leaders used to point with pride to levels of nourishment in Eastern Europe and the Former Soviet Union that were comparable to those of developed nations on the other side of the Iron Curtain. So that everyone could eat well, rural producers were heavily subsidized and prices paid by consumers were kept artificially low. When domestic harvests fell short, as happened during the 1970s, hard currency was expended on grain from North America and other non-communist producers (Chapter 4).

As communism declined and fell, insulating the food economy from market forces grew untenable. Subsidies for production and consumption were withdrawn and the hard currency needed for imports was in very short supply (Rask and Rask, 2004). The full dimensions of the resulting decline in per-capita food intake are difficult to discern since data for the 1980s are not available for the Former Soviet Union and much of Eastern Europe. If trends in Bulgaria, Hungary, Poland, and Romania (Table 14.8, column 4) are any guide, however, the contraction throughout the region during communism's waning decade was appreciable. Contraction having continued during the 1990s, current food consumption per capita, though superior to the amounts eaten by most Africans, Asians, and Latin Americans, is well below levels in the OECD.

Consumption of plant products has not changed very much (Table 14.8, column 1), although noticeable increases in Estonia, the Kyrgyz Republic, and a few other places reflect the substitution of these items for meat, eggs, and dairy goods. Instead, most of the adjustment in diets has involved the latter commodities, demand for which is fairly income elastic. In several countries where per-capita consumption of livestock products was still elevated when the Berlin Wall was pulled down, dramatic reductions occurred during the next several years. Among these countries were the Czech Republic, Poland, Russia, and Kazakhstan (column 2). A sizable decline also occurred in nations like Tajikistan and Turkmenistan,

Table 14.8 Food consumption and production trends in Eastern European and Former Soviet countries, 1979–1999

Country (ranked by income per capita 1999)	Years	Plant product consumption per capita (in cereal-equivalent tons) (1)	Animal product consumption per capita (in cereal-equivalent tons) (2)	Total food consumption per capita (in cereal-equivalent tons) (3)	Total food self-sufficiency* (4)
Slovenia	1992–1993	0.24	1.18	1.42	0.99
	1998–1999	0.24	1.39	1.63	0.94
Czech Republic	1993–1994	0.25	1.43	1.67	1.09
	1998–1999	0.28	1.26	1.54	1.04
Hungary	1979–1981	0.26	1.30	1.56	1.19
	1989–1991	0.27	1.34	1.61	1.29
	1997–1999	0.27	1.04	1.32	1.16
Slovak Republic	1993–1994	0.25	1.18	1.42	0.99
	1998–1999	0.27	1.07	1.34	0.97
Estonia	1992–1993	0.18	1.51	1.69	1.23
	1998–1999	0.27	1.11	1.38	0.91
Poland	1979–1981	0.27	1.43	1.71	0.92
	1989–1991	0.26	1.39	1.66	0.94
	1997–1999	0.29	1.10	1.38	0.97
Lithuania	1992–1993	0.24	1.47	1.70	1.24
	1998–1999	0.27	1.03	1.30	1.10
Croatia	1992–1993	0.22	0.67	0.88	0.88
	1998–1999	0.25	0.57	0.81	0.92
Latvia	1992–1993	0.23	1.52	1.74	1.22
	1998–1999	0.25	0.81	1.06	0.94
Russian Federation	1992–1993	0.26	1.26	1.52	0.81
	1998–1999	0.26	0.98	1.23	0.70

Country	Period			
Bulgaria	1979–1981	0.33	1.11	1.44	1.09
	1989–1991	0.30	1.33	1.63	1.02
	1997–1999	0.25	0.99	1.24	0.94
Romania	1979–1981	0.28	1.00	1.28	0.94
	1989–1991	0.27	0.98	1.25	0.84
	1997–1999	0.29	0.85	1.14	0.81
Kazakhstan	1992–1993	0.28	1.35	1.62	1.07
	1998–1999	0.20	0.98	1.19	1.06
Belarus	1992–1993	0.26	1.52	1.77	1.10
	1998–1999	0.27	1.31	1.58	1.00
Ukraine	1992–1993	0.30	1.11	1.41	1.09
	1998–1999	0.26	0.77	1.03	1.05
Turkmenistan	1992–1993	0.25	0.86	1.10	0.72
	1998–1999	0.26	0.76	1.02	0.92
Azerbaijan	1992–1993	0.22	0.51	0.72	0.76
	1998–1999	0.21	0.46	0.68	0.77
Armenia	1992–1993	0.17	0.45	0.61	0.77
	1998–1999	0.22	0.50	0.71	0.60
Georgia	1992–1993	0.20	0.43	0.63	0.76
	1997–1999	0.24	0.48	0.72	0.77
Uzbekistan	1992–1993	0.26	0.68	0.93	0.84
	1998–1999	0.29	0.73	1.02	0.81
Kyrgyz Republic	1992–1993	0.21	1.02	1.24	0.97
	1998–1999	0.27	0.94	1.20	0.99
Tajikistan	1992–1993	0.24	0.44	0.67	0.68
	1998–1999	0.21	0.24	0.45	0.57

Note: * Total food self-sufficiency equals total domestic food production divided by total domestic food consumption.
Source: Derived from FAOSTAT Agriculture Data, commodity balances and food supply data from 1979 to 1999 (www.fao.org).

where consumption of livestock products was more limited at the end of the communist era. In contrast, meat, eggs, and dairy goods were eaten in greater quantities in Slovenia, which is the wealthiest nation in the entire region, as well as Armenia, Georgia, and Uzbekistan. The same four countries were the only exceptions to the general pattern of average food intake being lower in the late 1990s than it had been at the beginning of the decade (column 3).

Sizable though declines in per-capita food consumption have been, these declines generally have been smaller than reductions in domestic output. Of course, a downward adjustment in the ratio of domestic production to consumption (Table 14.8, column 4) is not necessarily to be regretted, especially if GDP per capita is rising because of increased specialization and trade as has occurred in Slovenia. On the other hand, seeing that the ratio has remained above 1.00 after a successful transition away from communism, including in the agricultural sector, convinces one that a nation has a comparative advantage in food production. This is the case with the Czech Republic, Hungary, and Lithuania, for example.

Agricultural comparative advantage clearly has been neutralized in some parts of the old USSR because of halting reform (or worse) of state-run systems. One such setting is Ukraine, which was a major exporter when the international grain trade began in the middle 1800s (Morgan, 1979, pp. 27–28). Although domestic output still exceeds consumption, the gap between the two would be much greater if a more decisive break were made with communist agriculture. The same lesson holds for Russia, where the failure of reform in the countryside caused the self-sufficiency ratio to drop from 0.81 in the early 1990s to 0.70 at the end of the decade.

As living standards have deteriorated and food consumption has diminished, food insecurity, which formerly was all but unknown in Eastern Europe and the Former Soviet Union, has become a limited problem in the region. While infant survival is excellent in a number of countries, the infant mortality rate, which is inversely related to per-capita food intake, is higher elsewhere, particularly in the Caucasus and Central Asia (Table 14.9, column 2). Likewise, a sizable cohort of children is severely underweight in the southern reaches of the old USSR; even in Russia, 6 percent of children are malnourished by this standard measure (column 3). Extreme poverty is also now detectable in the region. One out of every 50 Eastern Europeans is living on a dollar a day or less (column 4). In the old USSR, the poverty rate is higher, equal to 6 percent in Russia and exceeding 20 percent in Moldova and Uzbekistan. For people suffering this deprivation, it is cold comfort that much larger segments of the population are impoverished in Africa, Asia, and Latin America.

Much of Eastern Europe and the Former Soviet Union have avoided major change in another indicator of food insecurity, which is income inequality. A sharp break with communism has not prevented nations such as the Czech Republic and Hungary from maintaining low Gini coefficients. The reason for this is that educational opportunities long have been widely available, thereby equipping much of the population for success in new market economies. Gini indices in the 20s

Table 14.9 Recent food insecurity indicators in Eastern European and Former Soviet countries

Country (ranked by income per capita 2002)	Total food consumption per capita 1998–1999 (in cereal-equivalent tons) (1)	Infant mortality rate – 2002 (per 1,000 live births) (2)	Child malnutrition (late 1990s/early 2000s, % children <5 years) (3)	Poverty rate (late 1990s/early 2000s, % population below $1/day) (4)	Gini index (late 1990s/early 2000s) (5)
Slovenia	1.63	4	–	2.0	28.4
Belarus	1.58	17	–	2.0	30.4
Czech Republic	1.54	4	–	2.0	25.4
Estonia	1.38	10	–	2.0	37.2
Poland	1.38	8	–	2.0	31.6
Slovak Republic	1.34	8	–	2.0	25.8
Hungary	1.32	8	–	2.0	24.4
Lithuania	1.30	8	–	2.0	31.9
Bulgaria	1.24	14	–	4.7	31.9
Russian Federation	1.23	18	6	6.1	45.6
Kyrgyz Republic	1.20	52	6	2.0	29.0
Kazakhstan	1.19	76	4	2.0	31.3
Romania	1.14	19	3	2.1	30.3
Latvia	1.06	17	–	2.0	32.4
Ukraine	1.03	16	3	2.9	29.0
Turkmenistan	1.02	70	12	12.1	40.8
Uzbekistan	1.02	55	19	21.8	26.8
Croatia	0.81	7	1	2.0	29.0
Georgia	0.72	24	3	2.7	36.9
Armenia	0.71	30	3	12.8	37.9
Azerbaijan	0.68	76	17	3.7	36.5
Tajikistan	0.45	90	–	10.3	34.7
Macedonia	–	22	6	2.0	28.2
Albania	–	22	14	2.0	28.2
Moldova	–	27	–	22.0	36.2

Source: Table 14.8 and World Bank (2004).

and low 30s are also still observed in countries that have not pursued reform vigorously, including Belarus and Ukraine. The reason, as Aslund (2002) puts it, is that the state-controlled system, which remains intact, is able to "keep income differentiation at bay" (p. 312). In contrast, Gini coefficients have risen above the middle 30s and even into the 40s in Russia and much of the Caucasus and Central Asia, mainly because incomplete reforms have created economic distortions that breed inequality. Primarily because of changes in this last group of nations, the Gini coefficient for the region as a whole rose from 24 in the late 1980s to 33 (comparable to what one finds in much of Western Europe) in the middle 1990s (Aslund, 2002, p. 311).

Russia is an interesting and special case. Aside from having the greatest income inequality in the entire region, it has experienced the fastest increase ever recorded in a national Gini coefficient, from the middle 20s two decades ago to 48.7 in the late 1990s (Aslund, 2002, p. 311). Part of this increase was a byproduct of hyperinflation during the early 1990s, which greatly benefited creditors by wiping out their debts. Among these creditors were people who had managed to take out loans carrying interest rates of 15–20 percent at a time when prices in general were going up by as much as 2,500 percent per annum. Another group of creditors were business proprietors who were chronically tardy paying wages, which were not adjusted for price increases. At about the same time, a handful of oligarchs reaped enormous gains thanks to the shares-for-loans scheme mentioned earlier in this chapter, precisely as earnings were plummeting for the vast majority of their countrymen. This peculation, along with slowness in settling wage arrears and other debts in the midst of very high inflation, has created economic inequality of Latin American dimensions at the very heart of the old Soviet Union.

14.5 Summary

While it had certain accomplishments, including the elimination of hunger and the widening of educational opportunities, communism created enormous harm in Eastern Europe and the old USSR, as it has done in all countries where it has been imposed. Aside from atrocious loss of life during the collectivization of Soviet agriculture, economic output contracted severely once centrally planned systems, which were mechanisms for value destruction as opposed to engines of value creation, wound down and finally were dismantled. The costs of dismantlement were hidden as long as communist authorities, who were never keen on the publication of honest data, remained in charge. These costs only became apparent, in the form of sizable reductions in GDP per capita, after Marxist states were terminated. During the 1990s, the only nations to avoid these reductions were those that made a decisive break with central planning. Growth has continued in the same nations, in spite of higher energy prices that have created a turnaround of sorts in places like Russia.

Where national economies have deteriorated, because the post-communist transition has been poorly handled, demographic contraction is setting in, at least in nations with a Slavic heritage. Life spans have been reduced and human fertility, which is held down because of female economic empowerment, has fallen to levels unknown anywhere else in the world, including Western Europe. Deaths already exceed births in a number of countries, including where reforms were successful during the last decade of the twentieth century, and natural increase will soon swing negative in many more.

As with other parts of the economy, agriculture generally has prospered where the transition from communism has been successful. Such is the case in Central Europe, where private farming is now reestablished. In contrast, the sector has languished where the communist past has not been dealt with effectively. With production subsidies cut drastically by governments strapped for cash, yields have grown slowly or not at all and farmed areas have diminished. Output of most commodities was lower at the turn of the twenty-first century than what it had been three decades earlier. In Ukraine and elsewhere, agriculture's comparative advantage is largely cancelled out because the countryside remains mired in the Marxist past.

Consumption of food, especially livestock products, has fallen as well, both because retail-level subsidies are largely gone and because of declines in population and living standards. Practically unheard of between the Second World War and the waning days of the USSR, food insecurity is now a problem, as indicated by the presence of severely underweight children and people struggling to survive on no more than a dollar a day. By the same token, income inequality has risen in various places, most notably Russia.

By no means are the prospects entirely dim for Russia and other countries that have suffered under communism and have had a difficult time ridding themselves of that system. Many of these nations have made progress toward democracy, as indicated by a succession of multiparty elections for example. Imperfect though it certainly is, this progress compares favorably with the experience of China, where the communist party is clinging steadfastly to its political monopoly. While economic growth to date in the colossus of East Asia far outstrips that of the old USSR, one can envision a day when Russia and its immediate neighbors will have harmonized their institutions with those of the West, thereby allowing them to take a more prominent place in the global economy.

Study Questions

1. In what sense was the old USSR economically unsustainable?
2. Compare and contrast trends in GDP per capita in countries that have made a decisive break with communism and countries that have not done so.
3. Who has gotten rich in post-communist Russia and how have they done so?

4. Why has life expectancy been going down in Russia?
5. Is demographic contraction avoidable in Eastern Europe and the Former Soviet Union?
6. Compare and contrast the role of agriculture in China's economic development since the late 1970s and the place of agriculture in the Russian economy since the fall of communism.
7. Why has use of fertilizer, machinery, and other agricultural inputs declined in much of Eastern Europe and the Former Soviet Union?
8. Has food security improved or grown worse in the region since communism came to an end?

15

Sub-Saharan Africa

The challenges of development vary considerably from one part of the world to another. Age-old problems of socioeconomic inequality have yet to be resolved in Latin America and the Caribbean, for example. One encounters in the Middle East and North Africa incontrovertible proof that a generous endowment of natural resources does not really compensate for failing to establish institutions conducive to broad-based economic progress. In Eastern Europe and the Former Soviet Union, some nations are still endeavoring to clear away the wreckage of Marxist central planning.

Notwithstanding difficulties yet to be overcome, prospects for each of these regions are far from discouraging. High inflation, of the sort experienced during the 1980s, no longer threatens in Latin America and the Caribbean. People in the Middle East and North Africa are starting to speak seriously of representative government and the rule of law. Likewise, the worst of the post-communist transition appears to be over in the old USSR. At the same time, the outlook for Asia continues to be positive.

Grounds for optimism are much shakier in Sub-Saharan Africa, which has the lowest GDP per capita in the world and where impediments to development are legion. Geography is part of the problem. Since much of the continent is of ancient geological origin, most soils are highly weathered and, hence, not very fertile. Other than in settings with torrential rainfall, such as the Congo River Basin, precipitation that is sufficient for non-irrigated agriculture is more the exception than the rule, with droughts a frequent occurrence. In addition, tropical diseases are a chronic threat to people and domesticated livestock. Likewise, crops are under constant attack from insects and other pests.

Even the region's great geographic expanse is not really a blessing. Whereas Asia (the only continent larger than Africa) has no more than 20 independent states and China, India, and other leading nations are both populous and extensive, Africa has more than 40 separate countries, many of them small.[1] As the cases of South Korea

---

[1] In West Africa, only Mali, Niger, and Nigeria have extensive territories. East Africa has just two large countries: Sudan and Ethiopia. The Democratic Republic of the Congo, in the central part of the continent, is sizable, as are Angola, Mozambique, and South Africa, farther south. Of these nine nations, only Congo, Ethiopia, Nigeria, and South Africa have large populations. The other five are sparsely inhabited.

and Taiwan demonstrate, limited national territory is not necessarily a problem. Ready access to overseas markets, not to mention policies conducive to commerce, more than makes up for acute land scarcity. However, one in every three African nations lacks such access for the plain and simple reason that it is landlocked. Just as Bolivia is poorer than the rest of South America and Afghans and the Nepalese have lower average incomes than other Asians, international trade is inhibited and living standards are below regional norms in most African countries that do not have their own seaports.

Another consequence of the continent's ample territorial dimensions is that, by and large, rural population densities are low. Where the countryside is inhabited by relatively few people, almost all of whom are destitute, the costs of roads and other infrastructure required for trade and development are far beyond local financing capacities. All too often, then, capital formation never happens. When investment occurs, usually thanks to foreign aid, development activities can easily prove difficult to sustain, as exemplified by decaying roads, crumbling schools and clinics, and other monuments to previous donor largesse that dot the landscape.

While the Sub-Saharan environment is daunting, the difficulties of development have been compounded by the misdeeds of rulers. Colonized long after the Western Hemisphere and several decades after many parts of Asia, Africa was carved up by European powers during the second half of the nineteenth century. The imperial era created a lasting linguistic legacy – Anglophone, Francophone, and Lusophone. However, structures required for effective governance were not put in place. Also, colonial boundaries, which later became national frontiers, were drawn with little or no regard for ethnic identifications and cleavages. The result once independence was achieved was country after country characterized by an absence of statecraft in the face of dysfunctional ethnic and regional groupings (Bates, 2000; Herbst, 2000; Salih, 2001).

All too often, this inauspicious start was followed by civil strife. In Burundi and Rwanda – a pair of tiny states to the northeast of the Democratic Republic of the Congo (DRC)[2] with the highest population densities in the entire continent – Hutus and Tutsis have fought often, most recently during a massacre of Tutsis by militant Hutus in Rwanda in 1994. Perpetual clan warfare has put an end to national government in Somalia. In large countries, population clusters in remote areas that are hostile to ascendant groups in capital cities have rebelled from time to time (Herbst, 2000, pp. 145–152). This was true of Nigeria in the 1960s and of Angola, Ethiopia, and Mozambique from the 1970s through the 1980s. An attempt is being made to end a war that has dragged on for decades between Arabs in northern Sudan and Dinka, Nuer, and other Africans in the south, although government-backed gangsters recently started perpetrating genocide in the western state of Darfur. The fighting that began in the DRC after the death of its long-time dictator, Mobutu Sese Seko (who during the early 1960s had prevailed in the civil war that erupted at the

---

[2]The DRC was formerly known as Zaire.

time of independence from Belgium), has persisted until very recently. Thus far, millions of Congolese have lost their lives.

During the Cold War, it was common for one faction to be allied with the United States and other Western nations and for its opponents to be supported by the USSR and its satellites. These days, the stakes involved may relate to mineral wealth – diamonds, for instance, in Liberia and Sierra Leone, which in the 1990s experienced social breakdown similar to that of Somalia. By and large, the outside world has had no direct interest other than the avoidance of state failure, which as the case of Taliban-controlled Afghanistan made clear can have serious global repercussions if ignored.

Aside from its enormous and tragic human toll, warfare has crippled a number of African economies. Even where armed conflict has been avoided, policies that discourage trade and investment frequently have been followed, thereby compounding difficulties related to geography (Collier and Gunning, 1999). That Sub-Saharan Africa's engagement of global commerce has been limited is indicated by small GDPs in the region. South Africa is the only economy of appreciable size. Its GDP is four times that of Nigeria, which is Africa's most populous nation and leading oil producer and has the region's second-largest economy. A scattering of other countries export fossil fuels, copper, and other natural resources, with very few rewards spilling over to local populations. It is telling that GDP for the entire region is comparable to the annual economic output of Belgium and that median national GDP is approximately $2 billion, which is about the same as the annual output of a city of 60,000 people in Western Europe or the United States (World Bank, 2000a, p. 7).

In 1958, when the Gold Coast ceased being a British colony and renamed itself Ghana, there were just two self-ruling states between the Arab nations of North Africa and South Africa, at the continent's antipodal tip. One was Liberia, which had been founded in the 1800s by emancipated slaves from the United States. The other was Ethiopia, which had been governed by the ancestors of Emperor Haile Sellassie for centuries. By the middle 1970s, when the Portuguese (who were the first Europeans with colonies south of the Sahara) completed their withdrawal from Angola, Mozambique, and other possessions, virtually the entire continent was independent, although white minorities ruled in South Africa, Namibia, and Rhodesia (now Zimbabwe). Three, four, or five decades is precious little time to develop the habits of modern statecraft (e.g., reaching compromises among groups with varied interests) almost from scratch, particularly in light of difficult natural conditions and ethnic fragmentation. It is hardly remarkable that African governments are weak, propped up by foreign aid totaling 10–20 percent of GDP in many countries and as much as 25–30 percent of GDP in a number of others (World Bank, 2000b, Table 21).

Yet there are glimmers of hope in the region. Nelson Mandela adroitly directed the transition from apartheid to majority rule in South Africa. Uganda has made notable progress in the fight against HIV/AIDS. Long dominated by military strongmen, Nigeria now has a democratically elected government. The challenges of getting economies on track and alleviating poverty, one has to think, are not insurmountable.

15.1 Trends in GDP Per Capita

Although its average income is much lower than GDP per capita in Latin America and the Caribbean, Africa went through economic travails during the 1980s similar to those experienced on the other side of the Atlantic Ocean (Chapter 12).

Some of the causes of poor economic performance were similar as well. Early in the decade, recession in North America and Europe cut into demand in those places for primary commodities imported from Africa. In addition, currency overvaluation arose after the middle 1980s in the Ivory Coast, Mali, Senegal, and other Francophone countries that maintained fixed exchange rates with the French franc, which appreciated relative to other European currencies as well as the US dollar. The result was that exports, which grew slowly throughout the region (Table 15.1, column 2), declined markedly in many of France's former colonies. Among the consequences of sluggish or negative export trends was a decline in tariff collections, which have been the main source of public revenues in states ill equipped to levy and collect taxes on domestic economic activity. Rather than cutting public spending in response to this decline, some governments monetized fiscal deficits, thereby igniting inflation (column 3). This had a dampening effect on gross investment (column 4), which is discouraged in Africa even in the best of times because the rule of law is weak.

The 1980s were not the best of times south of the Sahara, and not just because of disappointing exports and macroeconomic instability. Price controls were widespread. Also, key sectors were dominated by parastatal monopolies, which along with overstaffed bureaucracies and insolvent government-owned banks marginalized the private sector while simultaneously running up fiscal deficits (Collier and Gunning, 1999; World Bank, 2000a, pp. 18–28). The burden was especially severe for rural producers, who lacked the political clout of urban elites and laborers. State-run marketing boards, which were authorized to purchase all commodity output, paid low prices to farmers, just as had happened during the early years of the Soviet Union (Chapter 7). Another element of the anti-rural bias in public policy was minimal spending on education and public health in the countryside and on agricultural research and development.

Adding to Africa's economic misery during the 1980s were prolonged drought and the Cold War. Precipitation was well below average for several years in Botswana, Kenya, Mozambique, Tanzania, and Zimbabwe. The most spectacular drought lasted from 1982 to 1988 and took millions of lives in Ethiopia, Somalia, and Sudan, in spite of international aid efforts. Simultaneously, the USSR supported military dictatorships in Ethiopia and Somalia, as well as Angola, Benin, Congo, Guinea, Madagascar, and Mozambique. Political repression and economic autarky in these places had predictably dismal effects, as did the insurrections that broke out in Angola,

Ethiopia, Mozambique, and Somalia. Economic performance was also affected for the worse by authoritarianism in oil-rich Nigeria and mineral-rich Zaire (now the DRC), which were allied with the West, and by apartheid in Namibia and South Africa.

Weak export demand, debilitating macroeconomic and exchange-rate policies, governmental repression of private producers (especially in rural areas), drought, war, and repression all contributed to economic stagnation or worse in Sub-Saharan Africa. Of the 37 nations listed in Table 15.1, only three recorded sizable improvements in GDP per capita during the 1980s. Botswana (with great mineral wealth), Mauritius (an archipelago in the Indian Ocean), and Swaziland (a tiny state surrounded by South Africa) all have small populations and economies. In 20 of the 37, human numbers grew faster than economic output. There were anomalies. For example, GDP per capita declined in the DRC even though exports grew at an annual rate of 10 percent. GDP per capita also went down in Namibia in spite of strong capital inflows. Each of these results reflected the presence of large natural resource industries with few ties to the rest of the national economy.

As happened in Latin America and the Caribbean, the last decade of the twentieth century brought a measure of relief south of the Sahara. After recessionary conditions during much of the 1980s, the global economy has expanded more recently, which has allowed exports to recover among Africa's commodity producers. In addition, the disappearance of the USSR in 1991 ended the Cold War, which in turn caused proxy conflicts to wind down and obliged a number of former Soviet allies to abandon Marxist economic practices. Pro-market reform was also pursued in other African nations, often as part of structural adjustments recommended by the International Monetary Fund (IMF) and World Bank. State-owned enterprises were privatized and prices were deregulated. Barriers to efficient international trade were lowered as well. For example, much-needed devaluation occurred in Francophone countries that formerly had maintained fixed exchange rates with the French franc.

Among nations where GDP grew appreciably faster than human numbers during and after the 1990s because governmental control of the economy was scaled back and civil warfare came to an end were Eritrea (which achieved independence from Ethiopia), Ethiopia, and Mozambique (Table 15.1, column 5). The greatest improvement in living standards was registered in Uganda, which along with Mozambique was a willing convert to market-friendly policies. Exports grew faster in the former nation than in any other part of the region (column 6) and, with inflation barely in the double digits (column 7), investment took place (column 8).

However, per-capita GDP shrank during the waning years of the twentieth century in nearly half the nations listed in Table 15.1, with the worst contraction coinciding with armed conflict. The most atrocious violence occurred in Burundi and Rwanda (see above). In the DRC, Liberia, and Sierra Leone, rival bands squabbled over diamond deposits and other natural resources, the exploitation of which financed bloody insurrections and enriched rebel leaders. The same thing happened in Angola, where investment by oil firms in the Cabinda enclave did not begin to offset the damage

Table 15.1 Patterns of economic growth in Sub-Saharan African countries, 1980

	A. Average annual rate of growth 1980–1990 (%)				B. Average annual rate of growth	
Country (ranked by column 1)	GDP per capita	Exports	Inflation	Gross investment	Country (ranked by column 5)	GDP per capita
	(1)	(2)	(3)	(4)		(5)
Botswana	8.2	10.6	13.6	–	Uganda	4.2
Mauritius	5.3	10.4	9.5	10.2	Mozambique	4.1
Swaziland	3.8	14.4	13.8	–	Mauritius	3.9
Burundi	1.6	3.4	4.4	4.5	Eritrea	2.5
Lesotho	1.6	4.1	13.8	6.9	Lesotho	2.2
Chad	1.4	6.5	2.9	19.0	Ethiopia	2.0
Burkina Faso	1.1	−0.4	3.3	8.6	Benin	1.9
Angola	1.0	3.7	5.9	−6.8	Botswana	1.9
Kenya	0.8	4.3	9.1	0.8	Sudan	1.7
Uganda	0.7	1.8	104.0	9.6	Ghana	1.6
Cameroon	0.5	5.9	5.6	−2.7	Guinea	1.6
Republic of the Congo	0.5	5.1	0.5	11.9	Burkina Faso	1.4
Mali	0.4	4.8	4.5	5.4	Malawi	1.4
Senegal	0.3	3.7	6.5	3.9	Mauritania	1.3
Benin	0.1	−2.4	1.3	−6.2	Ivory Coast	1.1
Mozambique	0.1	−6.8	38.3	−2.5	Mali	0.8
Zimbabwe	0.1	4.3	11.6	1.3	Namibia	0.8
Ghana	−0.3	2.5	42.1	4.5	Senegal	0.6
Guinea	−0.5	–	–	–	Swaziland	0.2
Rwanda	−0.5	3.4	4.0	3.7	Tanzania	0.2
Ethiopia	−0.8	2.4	4.6	3.5	Zimbabwe	0.2
Mauritania	−0.9	3.6	8.4	−4.1	Gabon	−0.1
Malawi	−1.0	2.5	14.6	−2.8	South Africa	−0.1
South Africa	−1.0	1.9	14.9	−4.8	Central African Republic	−0.3
Sudan	−1.2	0.9	31.3	–	Nigeria	−0.4
Tanzania	−1.2	−1.5	25.1	–	Kenya	−0.5
Namibia	−1.4	−0.1	13.9	11.9	Chad	−0.6
Nigeria	−1.4	−0.3	16.7	−8.6	Togo	−0.8
Togo	−1.4	0.1	4.8	2.9	Niger	−0.9
Sierra Leone	−1.5	2.1	64.0	−6.5	Madagascar	−1.2
Democratic Republic of the Congo	−1.7	9.6	62.9	–	Cameroon	−1.4
Madagascar	−1.8	−1.7	17.1	4.9	Zambia	−1.7
Central African Republic	−2.0	−1.2	7.9	1.8	Republic of the Congo	−1.9
Zambia	−2.2	−3.4	42.2	−2.7	Angola	−2.4
Gabon	−2.7	3.0	1.9	−4.6	Rwanda	−3.5
Ivory Coast	−2.9	1.9	2.8	−28.8	Burundi	−5.1
Niger	−3.2	−2.9	1.9	−5.9	Sierra Leone	−7.2
					Democratic Republic of the Congo	−8.3

Source: World Bank (2001b), pp. 278–279 and 294–295; World Bank (2005). The Economic Freedom Index See Heritage Foundation (2002).
Political Freedom Status is F (free), PF (partly free), and NF (not free). See Freedom House (2002).

through 2003

1990–1999 (%)			C. Average annual rate of growth 2000–2003 (%)			
Exports	Inflation	Gross investment	Country (ranked by column 9)	GDP per capita	Economic freedom index –2002	Freedom status 2001–2002
(6)	(7)	(8)		(9)	(10)	(11)
16.3	13.7	9.9	Mozambique	5.0	3.05	PF
13.4	36.4	13.1	Botswana	4.5	2.90	F
–	–	–	Chad	4.5	3.60	NF
0.5	9.7	–	Tanzania	4.3	3.40	PF
11.3	9.6	2.3	Sudan	4.0	–	NF
9.3	7.4	13.4	Mauritius	3.8	3.00	F
1.9	9.4	5.3	Mali	3.5	2.90	F
2.5	10.0	−1.3	Angola	3.3	–	NF
–	–	–	Rwanda	3.3	3.40	NF
10.8	27.2	4.2	Sierra Leone	3.3	–	PF
4.7	6.2	2.4	Uganda	2.8	3.00	PF
0.4	6.2	4.8	Ghana	2.5	3.40	F
4.9	33.5	−7.5	Nigeria	2.5	3.60	PF
1.6	6.1	6.8	Cameroon	2.5	3.25	NF
4.7	8.0	17.6	Zamiba	2.3	3.25	PF
9.6	8.5	−0.8	Senegal	2.3	3.20	PF
4.3	9.8	2.5	Burkina Faso	2.0	3.20	PF
2.6	5.2	3.1	Republic of the Congo	1.8	3.75	PF
–	–	–	Mauritania	1.8	3.30	PF
9.5	23.2	−1.7	Lesotho	1.8	3.40	PF
11.0	23.8	−0.7	South Africa	1.3	2.90	F
–	–	–	Ethiopia	1.0	3.55	PF
5.3	10.2	3.0	Guinea	0.8	3.30	NF
6.7	4.9	−1.7	Namibia	0.5	2.90	F
2.5	34.8	5.8	Swaziland	0.5	3.1	NF
0.4	14.8	4.9	Niger	0.3	3.50	PF
5.0	7.6	4.4	Gabon	−0.3	3.25	PF
1.5	8.3	11.6	Burundi	−0.5	–	NF
1.7	6.4	5.4	Madagascar	−0.8	3.1	PF
3.6	20.6	0.9	Togo	−1.0	3.60	PF
2.7	5.5	0.0	Kenya	−1.0	3.20	NF
1.8	56.9	11.3	Malawi	−1.5	3.50	NF
4.3	7.1	4.7	Eritrea	−2.3	–	NF
8.2	813.8	12.9	Central African Republic	−2.5	3.1	PF
−6.0	16.3	2.1	Democratic Republic of the Congo	−2.8	–	NF
2.4	11.7	−12.4	Ivory Coast	−4.0	2.90	PF
−12.2	31.2	−10.3	Zimbabwe	−8.0	4.3	NF
−5.5	1423.0	−3.5				

ranges from 1.00 (completely free markets) to 5.00 (completely repressed markets).

done in the rest of the country by warfare and triple-digit inflation (the worst in the region outside the DRC). None of these places has come close to duplicating the success of Botswana, a lightly populated and landlocked nation just north of South Africa. Perhaps because practically everyone in the country is from the same ethnic group, which means there is no prospect of a split along ethnic lines, democratic governance has been sustained for decades and living standards have risen steadily, thanks mainly to large-scale diamond mining.

The linkage between political freedom, which has been spreading since the abolition of apartheid in South Africa in 1994,[3] and economic progress applies elsewhere in the region. Among African nations where annual growth in GDP per capita has exceeded 2.5 percent since the turn of the twenty-first century (Table 15.1, column 9), three are characterized by Freedom House as not free (column 11): Rwanda (where a rebound of some sort was inevitable after the 1990s), Angola (which benefited from sharply higher oil prices), and Chad. The remaining nine countries, though, are either free or partly so. In contrast, civil and political liberties are severely curtailed in four of the seven countries where available data indicate that GDP has failed to keep pace with population. Particularly instructive in this regard is Zimbabwe, where octogenarian Robert Mugabe – the only leader the country has known since white-minority rule came to an end in 1980 – is systematically dismantling speech and press freedoms and debauching electoral processes (Lloyd, 2002) and, in so doing, is driving the national economy into the ground.[4]

Less clear cut in Sub-Saharan Africa is the relationship between deregulation and other elements of economic freedom, on the one hand, and trends in GDP per capita. But even allowing for rebounds in places where market forces are kept under tight rein, such as Rwanda and Ethiopia, indices of economic freedom (Table 15.1, column 10) are generally better in countries where living standards have improved markedly than in those experiencing stagnation or worse. Perhaps the lack of sharper differences between good and poor performers reflects the difficulties of overcoming difficult geography and ancient ethnic animosities as well as a recent history marked by recurring violence and autocratic and kleptocratic rule. Another reason for limited evidence of the impacts of market-friendly policies is that the adoption of such policies has remained tentative throughout the region, as indicated by the general absence of superior indices of economic freedom in Table 15.1. An entirely new cause of poor economic performance emerged during the late twentieth century: the HIV/AIDS epidemic.

---

[3] Democratization in South Africa was soon followed by clean elections in neighboring Mozambique and Namibia. In West Africa, Mali, Benin, Senegal, Ghana, and finally Nigeria, which still has acute ethnic and religious divisions, subsequently elected democratic governments. In December 2002, the corrupt autocrat, Daniel arap Moi, was finally voted out of office in Kenya (Holmquist, 2003). Research by Ndulu and O'Connell (1999) indicates a positive relationship in Sub-Saharan Africa between good economic performance and democracy from the late 1960s to the 1990s.

[4] A special target of Mugabe has been white-owned commercial farms, which used to employ a quarter million black Zimbabweans and account for most of the country's agricultural exports. After violent invasions by the dictator's followers, these farms have been dismantled and subdivided (Lloyd, 2002).

15.2 Demographic Trends

Progress in modern times against mortal illness, which set demographic transition in motion and is the main reason why human numbers rose from one billion to six billion in less than 200 years (Chapter 2), is taken for granted in most of the world, affluent and developing alike. By the same token, the chance of serious reversals in the constant struggle against disease is rarely contemplated. To be sure, there have been recent setbacks – in the old USSR, for example. However, the shortening of life expectancies in places like Russia is largely related to the decline and fall of communism and therefore does not portend similar downturns elsewhere.

Epidemiological troubles in Sub-Saharan Africa are far more unsettling. Communicable diseases, such as malaria, were never brought fully under control in the region, and in fact are now resurging. Far worse is the spread of HIV/AIDS, which could precipitate demographic collapse. Such a collapse, which no part of the world has experienced since millions of indigenous people in the Western Hemisphere succumbed to illnesses introduced by Europeans after 1492 (Chapter 2), would shake each and every part of human existence, not least agriculture and the economy, to its very core.

The impacts of HIV/AIDS, which first appeared in Central Africa around 1980, have been projected for English-speaking nations in the southern part of the continent. With 20 percent or more of the adult population infected, life-spans are expected to plummet (Table 15.2, column 1) and death rates to rise (column 2). With human fertility projected to diminish (column 3), crude birth rates (CBRs) (column 4) are on a downward track. Around 2020, the annual decline in human numbers could approach ½ percent in Botswana (column 5). Demographic contraction is also quite possible in South Africa and neighboring lands. Events could follow a similar course in Lusophone countries in Southern Africa (i.e., Angola and Mozambique) as well as the central and eastern parts of the continent.

Human Fertility

The projections contained in Table 15.2 also indicate that, due to high death rates, total fertility rates (TFRs) will not have to fall to what under normal circumstances is the replacement level for natural increase to turn negative in Anglophone Southern Africa. Except for South Africa, where that level will probably be reached during the second decade of the twenty-first century, there will be 2.6 births per woman in Botswana and Zimbabwe, yet each of these countries will be losing population. Namibia will avoid a decline, but only barely, because its fertility rate will still be above 3.0.

Just as South Africa and its English-speaking neighbors are suffering more from HIV/AIDS than the rest of the continent, the former countries also have experienced significant reductions in human fertility. Sizable declines also have occurred in Ghana and the Ivory Coast as well as Kenya, where the TFR was 4.4 births per woman in 2000 as opposed to 7.8 in 1980 (Table 15.3, columns 1 and 2). In most

Table 15.2 Historical and forecasted population parameters for HIV/AIDS high-incidence countries in Southern Africa for selected 5-year intervals, 1975 through 2020

Country income (ranked by per capita)	Life expectancy at birth (years) (1)	Crude death rate (per 1,000 population) (2)	Total fertility rate (3)	Crude birth rate (per 1,000) (4)	Natural increase (per 1,000) (5)
1. South Africa					
(a) 1975–1980	55.6	11.6	5.0	35.3	23.7
(b) 1995–2000	58.2	10.0	2.9	24.6	14.6
(c) 2005–2010	41.5	22.9	2.4	21.0	−1.9
(d) 2015–2020	43.4	22.3	2.1	19.4	−2.9
2. Botswana					
(a) 1975–1980	60.0	9.3	6.4	46.1	36.8
(b) 1995–2000	56.3	10.2	4.0	32.4	22.2
(c) 2005–2010	32.2	30.7	3.3	27.7	−3.0
(d) 2015–2020	32.9	29.7	2.6	24.8	−4.9
3. Namibia					
(a) 1975–1980	52.9	14.3	6.5	43.4	29.1
(b) 1995–2000	54.5	12.3	5.1	38.0	25.7
(c) 2005–2010	39.4	21.5	3.9	29.5	8.0
(d) 2015–2020	41.0	20.6	3.1	27.2	6.6
4. Swaziland					
(a) 1975–1980	50.0	16.2	6.8	48.3	32.1
(b) 1995–2000	47.2	15.7	5.1	38.0	22.3
(c) 2005–2010	30.0	30.7	3.9	31.4	0.7
(d) 2015–2020	31.5	29.9	3.0	28.5	−1.4
5. Lesotho					
(a) 1975–1980	50.5	16.4	5.7	42.6	26.2
(b) 1995–2000	46.9	16.7	4.3	32.6	15.9
(c) 2005–2010	31.5	30.2	3.4	29.7	−0.5
(d) 2015–2020	33.4	29.0	2.8	27.3	−1.7
6. Zimbabwe					
(a) 1975–1980	58.0	11.1	7.3	46.5	35.4
(b) 1995–2000	40.8	19.5	4.5	34.8	15.3
(c) 2005–2010	31.2	29.6	3.4	30.7	1.1
(d) 2015–2020	33.0	29.0	2.6	28.1	−0.9

Source: UNPD (2003).

other places, families with five or six children continue to be normal, with Niger (a landlocked nation to the north of Nigeria) having the world's highest TFR (7.2 births per woman) in 2000.

The reasons for elevated fertility are readily discerned. Aside from the insular nation of Mauritius, Gabon (which has a small population and abundant oil and

Table 15.3 Social indicators for Sub-Saharan African countries in 2000

Country (ranked by income per capita)	Total fertility rate		2000 income per capita ($PPP)	Urban % of total population in 2000	Adult illiteracy rate in 2000 (% people 15 and over)		Female % of labor force		Contraceptive prevalence rate 1990–2000 (% women 15–49)
	1980	2000			Males	Females	1980	2000	
	(1)	(2)	(3)	(4)	(5)	(6)	(7)	(8)	(9)
Mauritius	2.7	2.0	9,940	41	12	19	25.7	32.6	75
South Africa	4.6	2.9	9,160	55	14	15	35.1	37.8	62
Botswana	6.1	4.0	7,170	50	25	20	50.1	45.3	–
Namibia	5.9	5.0	6,410	31	17	19	40.1	40.9	29
Gabon	4.5	4.2	5,360	81	–	–	45.0	44.7	33
Swaziland	6.2	4.4	4,600	26	19	21	33.5	37.7	–
Lesotho	5.5	4.4	2,590	28	28	6	37.9	36.9	23
Zimbabwe	6.4	3.8	2,550	35	7	15	44.4	44.5	54
Guinea	6.1	5.2	1,930	33	–	–	47.1	47.2	6
Ghana	6.5	4.2	1,910	38	20	37	51.0	50.5	22
Mauritania	6.4	5.7	1,630	58	49	70	45.0	43.6	–
Cameroon	6.4	4.8	1,590	49	18	31	36.8	38.0	19
Sudan	6.1	4.6	1,520	36	31	54	26.9	29.5	10
Ivory Coast	7.4	4.8	1,500	46	46	61	32.3	33.4	15
Senegal	6.8	5.1	1,480	47	53	72	42.2	42.6	11
Togo	6.8	5.0	1,410	33	28	58	39.3	40.0	24
Uganda	7.2	6.2	1,210	14	22	43	47.9	47.6	15
Angola	6.9	6.6	1,180	34	–	–	47.0	46.3	–
Central African Republic	5.8	4.7	1,160	41	40	65	–	–	15
Kenya	7.8	4.4	1,010	33	11	24	46.0	46.1	39
Benin	7.0	5.5	980	42	48	76	47.0	48.3	16
Burkina Faso	7.5	6.5	970	19	66	86	47.6	46.5	12

(Continued)

Table 15.3 (Continued)

Country (ranked by income per capita)	Total fertility rate		2000 income per capita ($PPP)	Urban % of total population in 2000	Adult illiteracy rate in 2000 (% people 15 and over)		Female % of labor force		Contraceptive prevalence rate 1990–2000 (% women 15–49)
	1980	2000			Males	Females	1980	2000	
	(1)	(2)	(3)	(4)	(5)	(6)	(7)	(8)	(9)
Eritrea	7.5	5.4	960	19	33	55	47.4	47.4	8
Rwanda	8.3	5.9	930	6	26	40	49.1	48.8	21
Chad	6.9	6.4	870	24	48	66	43.4	44.7	4
Madagascar	6.6	5.4	820	30	26	40	45.2	44.7	19
Mozambique	6.5	5.1	800	40	40	71	49.0	48.4	6
Nigeria	6.9	5.3	800	44	28	44	36.2	36.5	15
Mali	7.1	6.3	780	30	51	66	46.7	46.2	7
Zambia	7.0	5.3	750	45	15	29	45.4	44.8	26
Niger	8.0	7.2	740	21	76	92	44.6	44.3	8
Ethiopia	6.6	5.6	660	18	53	69	42.3	40.9	8
Malawi	7.7	6.3	600	15	26	53	50.6	48.6	31
Burundi	6.8	6.0	580	9	44	60	50.2	48.7	–
Republic of the Congo	6.3	6.0	570	63	13	26	42.4	43.5	–
Tanzania	6.7	5.3	520	28	16	33	49.8	49.1	25
Sierra Leone	6.5	5.8	480	37	–	–	35.5	36.8	–
Democratic Republic of the Congo	6.6	6.1	–	30	27	50	44.5	43.4	–

Source: World Bank (2004).

other resources), and most of Anglophone Southern Africa, income per capita is below $2,000 – well below that threshold in more than 20 countries (Table 15.3, column 3). Whereas at least half the population of the eight nations listed in Table 15.3 with an average income of at least $2,500 lives in cities, urbanization (column 4) is not very advanced in the rest of the region. To the contrary, 70 or even 80 percent of the population is rural in a number of countries. Yet another factor related to family size is educational access for women. Where GDP per capita is above $2,500, differences between male and female illiteracy rates (columns 5 and 6) have been reduced sharply. Indeed, differences have all but disappeared in Namibia, South Africa, and Swaziland and rates are actually lower for women than for men in Botswana and Lesotho. In contrast, illiteracy is widespread, especially for women, in the rest of the region.

Elevated human fertility, which is observed where most of the population is impoverished and rural and where the majority of women can neither read nor write, has often been reinforced by political interests, religious traditions, and public policy. In countries with ethnic divisions, political leaders worry about slower natural increase within their own constituencies, and hence encourage procreation. In Francophone nations inhabited by Muslims, Roman Catholics, or both, colonial-era edicts against birth control, which were motivated by a desire to foster the peopling of sparsely inhabited settings, continued to be applied after independence. To this day, no French-speaking nation in Africa actively promotes the use of contraceptives – as happens routinely in Anglophone countries, where higher living standards and other conditions correlate with the wider acceptance of birth control (Becker, 2002).

Large families have predominated south of the Sahara even though labor force participation by African women – usually a good indicator of female economic empowerment – was uniformly high 25 years ago and has changed little since then (Table 15.3, columns 7 and 8). While similarly high participation by women in the labor force helps to explain why TFRs are very low in Eastern Europe and the Former Soviet Union, human fertility in Sub-Saharan Africa is little affected by the fact that 40 percent or more of all the region's workers are female. The reason for this is that most of the jobs that women hold are agricultural or in the service sector and, with few exceptions, the work is menial and low paid. Accordingly, contraceptive prevalence (column 9) is low, as can be seen in Burkina Faso (in the interior of West Africa), Ethiopia, Mozambique, Tanzania, and many other places where grinding rural poverty and large families are the norm, with children laboring alongside their mothers to feed the family.

Finally, human fertility has been affected somewhat by economic crisis and HIV/AIDS. People are marrying at a later age, in the hope of finding work first. Likewise, spousal separation, often resulting from employment-related migration by men, is becoming more common. Behavioral changes such as these obviously diminish reproduction. Nevertheless, the global trend toward lower human fertility is weaker south of the Sahara than anywhere else. Whereas TFRs are approaching or have fallen below the replacement level in many other parts of the world, families with three or four children are small by African standards.

Natural Increase

With fertility rates greater than or equal to five births per woman in most of Sub-Saharan Africa – including in three of the region's four most populous nations (the DRC, Ethiopia, and Nigeria) – the demographic transition in this, the poorest part of the developing world, is not very far advanced. Except in Mauritius and South Africa, CBRs have yet to fall below 30 per thousand; indeed, birth rates above 40 per thousand are far from unusual (Table 15.4, column 3). HIV/AIDS and other diseases are preventing death rates from falling below 10 per thousand on the continent (column 2), as has occurred in the rest of the developing world. But in spite of the toll from mortal illness, births far outnumber deaths and the annual rate of natural increase has yet to fall below 2 percent (column 4), as has happened elsewhere.

As with fertility trends, the only exceptional cases (aside from insular Mauritius) are in Anglophone Southern Africa, which remains more prosperous than the rest of the continent. In South Africa and nearby countries, human numbers are growing more slowly not just because CBRs are declining but also because death rates are climbing. The suddenness of the deceleration is indicated by the wide discrepancy between rates of natural increase at the turn of the twenty-first century (Table 15.4, column 4) and average annual population growth during the preceding two decades (column 5).[5] The cause of impending demographic collapse in Southern Africa is indicated in column 6 of Table 15.4, in which the incidence of HIV/AIDS infection among adults is reported.

Large numbers of people are similarly infected elsewhere in the region: 20 percent of the adult population in Zambia, 14 percent in Kenya, and a little over 11 percent in Burundi and Rwanda. In addition, the disease is starting to take hold in West Africa. More than one out of every 20 adults in Nigeria is now HIV positive or suffering from AIDS. One in every 10 is similarly afflicted in the Ivory Coast. Almost alone in Sub-Saharan Africa, Uganda has significantly lowered the incidence of the disease through its ABC program, which features (A) abstinence education (with an emphasis on building up the self-esteem of pre-adolescent and adolescent girls, many of whom are being pressured by older males to engage in sexual intercourse), (B) the exhortation of couples to be faithful, and (C) education about condoms. Sadly, other countries have been slow to adopt the Ugandan program and the alternative approach, which stresses the use of condoms, has proven ineffective (Green, 2005). Thus, demographic collapse, instead of being confined to Southern Africa, is likely to spread.

This collapse will be an unmitigated disaster, as is starting to be apparent in the southern part of the continent. Labor scarcity is bound to rise, as are health-care expenditures. This will in turn lead to diminished productivity and profits, not to mention lower tax revenues (needed to pay for health care, among other things) and

---

[5] Population growth in excess of natural increase also has occurred in South Africa and some of its mineral-rich neighbors because of immigration by men from other nations seeking jobs in mines.

Table 15.4 Population dynamics in Sub-Saharan African countries for 2000

Country (ranked by column 1)	Income per capita 2000 ($PPP) (1)	Crude death rate – 2000 (per 1,000 population) (2)	Crude birth rate – 2000 (per 1,000 population) (3)	Rate of natural increase (%) (4)	Average annual population growth 1980–2000 (%) (5)	Percentage of adults with HIV/AIDS (late 1990s) (6)
Mauritius	9,940	7	17	1.0	1.0	0.0
South Africa	9,160	16	26	1.0	2.2	19.9
Botswana	7,170	20	32	1.2	2.8	35.8
Namibia	6,410	17	36	1.9	2.9	19.5
Gabon	5,360	16	36	2.0	2.9	4.2
Swaziland	4,600	15	36	2.1	3.1	25.3
Lesotho	2,590	17	33	1.6	2.0	23.6
Zimbabwe	2,550	18	30	1.2	2.9	25.1
Guinea	1,930	17	39	2.2	2.5	1.5
Ghana	1,910	11	30	1.9	2.9	3.6
Mauritania	1,630	15	42	2.7	2.7	0.5
Cameroon	1,590	14	37	2.3	2.7	7.3
Sudan	1,520	11	34	2.3	2.4	–
Ivory Coast	1,500	17	37	2.0	3.3	10.8
Senegal	1,480	13	37	2.4	2.7	1.8
Togo	1,410	15	37	2.2	2.9	6.0
Uganda	1,210	19	45	2.6	2.8	8.3
Angola	1,180	19	48	2.9	3.1	2.8
Central African Republic	1,160	20	36	1.6	2.4	13.8
Kenya	1,010	14	35	2.1	3.0	14.0
Benin	980	13	39	2.6	3.0	2.5

(Continued)

Table 15.4 (Continued)

Country (ranked by column 1)	Income per capita 2000 ($PPP) (1)	Crude death rate – 2000 (per 1,000 population) (2)	Crude birth rate – 2000 (per 1,000 population) (3)	Rate of natural increase (%) (4)	Average annual population growth 1980–2000 (%) (5)	Percentage of adults with HIV/AIDS (late 1990s) (6)
Burkina Faso	970	19	44	2.5	2.4	6.4
Eritrea	960	13	39	2.6	2.7	2.9
Rwanda	930	22	40	1.8	2.5	11.2
Chad	870	16	45	2.9	2.7	2.7
Madagascar	820	12	40	2.8	2.8	0.2
Mozambique	800	20	40	2.0	1.9	13.2
Nigeria	800	16	40	2.4	2.9	5.1
Mali	780	20	46	2.6	2.5	2.0
Zambia	750	21	40	1.9	2.8	20.0
Niger	740	19	51	3.2	3.3	1.4
Ethiopia	660	20	44	2.4	2.7	10.6
Malawi	600	24	46	2.2	2.6	16.0
Burundi	580	20	40	2.0	2.5	11.3
Republic of the Congo	570	14	43	2.9	3.0	6.4
Tanzania	520	17	39	2.2	3.0	8.1
Sierra Leone	480	23	44	2.1	2.2	3.0
Democratic Republic of the Congo	–	17	46	2.9	3.2	5.1

Source: World Bank (2004); World Bank (2002a), pp. 48–50.

GDP decline. As countries like South Africa experience all these effects, democratic institutions are sure to feel the strain. In other parts of the continent, demographic collapse induced by HIV/AIDS could easily eliminate all chances of rising out of poverty and establishing decent, democratic governments (Lamptey *et al.*, 2002).

15.3 Agricultural Development

In terms of economic performance as well as demographic trends, the general picture in Sub-Saharan Africa is bleak, notwithstanding a scattering of positive signs here and there. The same pattern holds for agricultural development. Encouraging movement has happened in a few places. By and large, however, the results of recent decades have been disappointing. Moreover, agriculture's future outlook, like everything else in the region, is clouded by HIV/AIDS and other maladies.

As indicated at the beginning of the chapter, limited agricultural development south of the Sahara is partly the result of unpromising geography. Since much of the region is of ancient geological origin, fertile strata were removed long ago by the wind and driving rains, so that subsoil and even basement rock are all that remain. Also, much of the remaining soil contains aluminum in high enough concentrations that plant growth is stunted. The contrast is dramatic with the foothills of the Himalayas, which are geologically recent, and downstream river valleys in South Asia (Voortman *et al.*, 2000).

Just as good land is in short supply, precipitation tends to be deficient. Two-thirds of Sub-Saharan Africa is dry, with droughts occurring often in the Sahel (just south of the Sahara Desert) and the eastern and southern parts of the continent. At the same time, pests such as tsetse flies prey on livestock, just as labor productivity is sapped by malaria and other diseases. Moreover, actions by government – cheap-food policies administered by state marketing boards and currency over-valuation, to give a pair of examples – have diminished the incentives for agricultural production.[6]

Exceptions to the broad, Sub-Saharan pattern of agricultural stagnation have arisen because of favorable geography as well as public policies that, at the very least, do not discriminate against economic activity in the countryside. All too unusual in the region, this combination of circumstances has been present in South Africa, where farms established by the white minority and employing up-to-date technology (made available by agricultural research and extension) produce sugar, citrus and deciduous fruits, wine, and sunflower oil for international markets. Cut-flower enterprises and tobacco farms enjoyed similar success in Zimbabwe until the late 1990s,

---

[6] For an excellent review of agricultural issues in Sub-Saharan Africa, see Delgado (1998), who traces out the major post-independence policy initiatives in agriculture relating to the changing fashions of foreign-aid providers. Another useful contribution to the literature is Binswanger and Townsend (2000), in which colonial legacies and post-colonial realities bearing on agricultural development are underscored. Finally, Sahn *et al.* (1997) highlight ways that structural adjustment programs have had a positive impact on agricultural development, which is contrary to what many believe.

when commercial agriculture was largely destroyed by rural invasions instigated by the Mugabe dictatorship. Closer to the equator, cocoa producers in Ghana and the Ivory Coast are internationally competitive, as are farms yielding pineapples and other horticultural products in the same two nations and Kenya. Additional examples of commercial agriculture linked to global markets include coffee, tea, and oil palm estates in a number of countries.

Otherwise, Africa's farmers struggle. Facing difficult environmental conditions as well as neglect or worse by national governments, they generally avoid the use of purchased inputs, and thus are barely able to feed themselves and their families.

Factor Use

Conceivably, environmental limitations, such as infertile soils and inadequate rainfall, could be overcome in Sub-Saharan Africa by using fertilizer and other inputs in greater quantities and investing in infrastructure. However, such measures have been applied infrequently. For much of the time since independence, input use and investment have been discouraged by public policy. Governments have intervened repeatedly in commodity markets to keep food cheap, just as it has been common for the production of tradable commodities to be discouraged by export taxes and currency over-valuation (Collier and Gunning, 1999). Under these circumstances, producers have had little reason to spend money to raise output. But even if policy-induced distortions are avoided, investment is difficult to effect because rural areas are impoverished, sparsely populated (Table 15.5, column 1), or a combination of the two. Conditions such as these make local financing of roads and bridges (which are needed to move input to farmers and output to market) and dams, canals, and other irrigation works all but impossible. Even if a foreign donor or some other outside source covers capital expenses, operations and maintenance are often a severe financial challenge for local communities.

As emphasized in Chapter 5, trends in fertilizer use south of the Sahara contrast sharply with trends in the rest of the developing world. Instead of increasing in recent years, application rates have declined since 1980 in a number of countries (Table 15.5, columns 2 and 3). Rates have gone up in several places – dramatically so in relative terms in some parts of Central and West Africa. However, fertilizer applications throughout the continent are very low, averaging 8 kilograms per hectare instead of the developing-world norm of 107 kilograms per hectare (Dixon and Gulliver with Gibbon, 2001, p. 44). Henao and Baanante (1999) contend that application rates in Africa fall short of nutrient uptake by crops. Where farmers are, in effect, mining nutrients, land degradation is the typical result.

Where fertilizer use has declined, the blame is often assigned to diminished input subsidies, which in turn are linked to the structural adjustment programs that insolvent governments have had to implement. However, this analysis ignores other aspects of structural adjustment, including currency devaluation, which strengthen production incentives and increase the use of purchased inputs in the countryside. That fertilizer application rates are well below 30–40 kilograms per

Table 15.5 Agricultural inputs in Sub-Saharan African countries, for selected periods, 1979 through 2001

Country (ranked by income per capita in 2000)	Rural population density (people per sq. km 1999)	Fertilizer use (kg per ha)		Irrigated land (% of cropland)		Agricultural machinery (tractors per sq. km of arable land)	
		1979–1981	1999–2001	1979–1981	1997–1999	1979–1981	1997–1999
	(1)	(2)	(3)	(4)	(5)	(6)	(7)
Mauritius	691	255	332	15.0	18.2	0.3	0.4
South Africa	129	87	53	8.4	8.5	1.4	0.6
Botswana	233	3	12	0.5	0.3	0.5	1.8
Namibia	146	0	0	0.6	0.9	0.4	0.4
Gabon	73	2	1	2.4	3.0	0.4	0.5
Swaziland	448	105	33	34.0	38.3	1.7	1.7
Lesotho	450	15	17	—	—	0.5	0.6
Zimbabwe	252	61	55	3.1	3.5	0.7	0.7
Guinea	556	2	3	7.9	6.4	0.0	0.1
Ghana	325	10	5	0.2	0.2	0.2	0.1
Mauritania	230	6	1	22.8	9.8	0.1	0.1
Cameroon	127	6	7	0.2	0.5	0.0	0.0
Sudan	119	5	4	14.4	11.5	0.1	0.1
Ivory Coast	286	26	31	1.0	1.0	0.2	0.1
Senegal	222	10	12	2.6	3.1	0.0	0.0
Togo	134	1	8	0.3	0.3	0.0	0.0
Uganda	368	0	1	0.1	0.1	0.1	0.1
Angola	283	5	1	2.2	2.1	0.4	0.3
Central African Republic	112	1	0	—	—	0.0	0.0
Kenya	499	16	35	0.9	1.5	0.2	0.4

(Continued)

Table 15.5 (Continued)

Country (ranked by income per capita in 2000)	Rural population density (people per sq. km of arable land) 1999	Fertilizer use (kg per ha)		Irrigated land (% of cropland)		Agricultural machinery (tractors per sq. km of arable land)	
		1979–1981	1999–2001	1979–1981	1997–1999	1979–1981	1997–1999
	(1)	(2)	(3)	(4)	(5)	(6)	(7)
Benin	210	1	26	0.3	0.6	0.0	0.0
Burkina Faso	265	3	14	0.4	0.7	0.0	0.1
Eritrea	654	–	17	–	4.8	–	0.1
Rwanda	901	0	0	0.4	0.4	0.0	0.0
Chad	163	1	4	0.4	0.6	0.0	0.0
Madagascar	417	3	3	21.5	35.1	0.1	0.1
Mozambique	339	11	2	2.1	3.2	0.2	0.2
Nigeria	250	6	6	0.7	0.8	0.0	0.1
Mali	162	6	8	4.5	3.0	0.1	0.1
Zambia	105	15	9	0.4	0.9	0.1	0.1
Niger	168	1	0	0.7	1.3	0.0	0.0
Ethiopia	520	–	16	–	1.8	–	0.0
Malawi	458	20	27	1.1	1.4	0.1	0.1
Burundi	792	1	4	4.5	6.7	0.0	0.0
Republic of the Congo	642	3	27	0.6	0.5	0.5	0.4
Tanzania	640	11	8	3.1	3.3	0.4	0.2
Sierra Leone	653	6	2	4.1	5.4	0.1	0.0
Democratic Republic of the Congo	518	1	0	0.1	0.1	0.0	0.0

Source: World Bank (2004).

hectare in all but a handful of Sub-Saharan nations undoubtedly has a lot to do with a general lack of transportation infrastructure, which is a great barrier to specialization and trade of all sorts.

The causes of limited irrigation are similar. Aside from constraints on the local financing of infrastructure development associated with poverty, low population densities, or both, many parts of Africa lack groundwater reserves and good reservoir sites. The latter limitation is common where rainfall is sparse and the topography is generally flat, as opposed to a landscape featuring narrow gorges that can be dammed easily as well as rivers passing through those gorges that can fill up newly created reservoirs in short order. The challenge of agricultural water development in Sub-Saharan Africa is indicated by the fact that just five of the countries listed in Table 15.5 irrigated more than 10 percent of their cropland 25 years ago (column 4) and, because of a major increase in rain-fed farmland in the Sahelian nation of Mauritania, only four did so in the late 1990s (column 5). Two of the four are insular: Madagascar (a large island off Africa's southeast coast, where rice is grown in paddies) and Mauritius. Swaziland benefits from excellent conditions for the irrigated production of sugar and other crops. Irrigated farming of fertile land alongside the Nile River, which Egyptians have been doing for millennia (Chapter 13), also takes place upstream in Sudan, with cotton and groundnuts being major crops. Outside of these four nations and a couple of others, at least 19 out of every 20 cultivated hectares are rain fed.

As with fertilizer use, Sub-Saharan irrigation is paltry relative to what one finds in Asia. However, one must be cautious about this inter-regional comparison, avoiding in particular the idea that Green Revolution technology from places like India can be transferred without modification to Africa. Aside from accounting for differences in factors such as rural population density, which is much higher in agricultural settings in Asia, intensification will have to fit with local environmental realities. For example, new crop varieties suited to African soils and weather conditions must be developed, along with the right guidelines for fertilization and other kinds of land treatment (Voortman *et al.*, 2000). To carry out such tasks, agricultural research and extension networks will need considerable strengthening.

Finally, just as agricultural intensification has proven to be a great challenge, few Sub-Saharan nations have experienced significant mechanization. In Botswana, where GDP per capita is high by regional standards and the ratio of rural population to arable land is modest, tractor use has risen since the 1980s (Table 15.5, columns 6 and 7). Also, production of sugar and other commercial crops is somewhat mechanized in Swaziland. The ratio of tractors to arable land has declined in South Africa, mainly because the country has reduced subsidies for agriculture and farm inputs (captured entirely by white producers) in order to qualify for WTO membership. Elsewhere, farm machinery, which must be imported at great cost along with the fossil fuels these implements run on, is rare. Even animal traction, which has the side benefit of organic fertilization, is limited, both because many farmers are two poor to afford oxen and other draft livestock and because of tsetse flies and other pests.

Intensification, Extensification, and Trends in Output

With commercial fertilizer used sparingly and irrigation unheard of in many places, Sub-Saharan food supplies vary primarily because of changes in land use. For each of the crops listed in Table 15.6, annual growth in production from the 1970s through the 1990s (column 4) has resulted much more from increases in planted area (column 2) than from improved yields (column 3). Annual yield increases have equaled or exceeded 1.0 percent for just two commodity groups: roots and tubers and maize. In contrast, yields of millet and pulses (important sources of nutrition for the rural poor) have changed little over the years, while fruit production per hectare has not varied at all.

In line with modest trends for other crops, cereal output per hectare has grown slowly. Indeed, yields in more than a dozen of the countries listed in Table 15.7 were lower at the close of the twentieth century than what these had been two decades earlier (columns 1 and 2). In a number of other places, grain yields stagnated. Only in Benin, Burkina Faso, Ghana, and Mauritania, all in West Africa, and Cameroon, the Central African Republic, Mauritius, and Mozambique did output per hectare go up by more than 50 percent.

Aside from not growing quickly, cereal yields have been very low. Output per hectare exceeds 1.5 tons per hectare – a level long since surpassed in other developing regions – in just seven nations: Cameroon, Gabon, Madagascar (which as already

Table 15.6 Trends in crop area, yield, and output in Sub-Saharan African countries, 1970–2000

	2000 production (million tons) (1)	Average annual change, 1970–2000 (%)		
		Area (2)	Yield (3)	Production (4)
A. Crops				
Roots and tubers	154	1.7	1.0	2.8
Fruits	47	1.6	0.0	1.6
Maize	38	1.5	1.2	2.7
Vegetables	22	1.9	0.8	2.6
Sorghum	18	1.2	0.5	1.6
Millet	14	1.4	0.4	1.8
Rice	11	2.4	0.6	2.9
Pulses	7	1.6	0.2	1.9
Oil crops	6	0.9	0.7	1.6
B. Animal products				
Total milk	19	–	–	1.8
Total meat	8	–	–	2.0
Total eggs	1	–	–	3.7
Cattle hides	0.5	–	–	1.7

Source: Dixon and Gulliver with Gibbon (2001), pp. 44–45.

Table 15.7 Agricultural and food output in Sub-Saharan African countries, 1979 through 2000

Country (ranked by income per capita in 2000)	Cereal yield (kg per ha)		Arable and permanent cropland (1,000 ha)		Food production index (1989–1991 = 100)		Food production per capita (average annual % growth)			Agricultural productivity (value added per worker – 1995$)	
	1979–1981	1998–2000	1980	1999	1979–1981	1998–2000	1975–1984	1985–1989	1990–1999	1979–1981	1998–2000
	(1)	(2)	(3)	(4)	(5)	(6)	(7)	(8)	(9)	(10)	(11)
Mauritius	2,536	5,094	107	106	89.7	104.0	−1.7	1.0	−0.9	3,087	4,977
South Africa	2,105	2,332	13,572	15,712	92.6	103.4	−1.6	2.1	−1.4	2,899	3,866
Botswana	203	196	1,360	346	87.2	94.2	−2.8	−5.2	−2.7	630	688
Namibia	377	285	657	820	107.2	97.0	−5.2	2.5	−3.1	919	1,468
Gabon	1,718	1,662	452	495	79.0	114.0	−0.1	−0.1	−1.3	1,814	1,882
Swaziland	1,345	1,836	204	180	80.2	91.0	0.3	−1.8	−4.1	1,671	1,731
Lesotho	977	974	292	325	89.1	98.6	−1.5	0.7	−1.7	723	540
Zimbabwe	1,359	1,184	2,539	3,350	83.3	105.2	−4.6	0.8	−0.5	307	336
Guinea	958	1,312	1,570	1,485	96.3	143.9	−0.8	−4.3	1.3	–	292
Ghana	807	1,306	2,760	5,300	68.7	162.9	−4.0	0.9	2.9	670	558
Mauritania	384	916	195	500	86.5	105.7	0.3	0.7	−1.8	299	480
Cameroon	849	1,551	6,930	7,160	79.9	129.6	−2.1	−0.9	0.1	834	1,104
Sudan	645	514	12,417	16,900	105.1	158.4	−1.7	−1.2	3.5	–	–
Ivory Coast	865	1,136	3,880	7,350	70.8	130.5	0.2	0.3	0.8	1,074	1,136
Senegal	690	721	5,225	2,266	74.0	114.2	−6.3	5.5	−1.3	336	304
Togo	729	933	1,420	2,300	77.1	135.9	−1.5	0.6	1.4	345	538
Uganda	1,555	1,377	5,680	6,810	70.4	116.6	−4.5	1.5	−1.4	–	353
Angola	526	646	3,500	3,500	90.0	144.1	−3.3	−1.6	0.8	–	121
Central African Republic	529	1,084	1,945	2,020	79.7	132.3	−0.4	2.4	1.2	377	469

(Continued)

Table 15.7 (Continued)

Country (ranked by income per capita in 2000)	Cereal yield (kg per ha)		Arable and permanent cropland (1,000 ha)		Food production index (1989–1991 = 100)		Food production per capita (average annual % growth)			Agricultural productivity (value added per worker – 1995$)	
	1979–1981	1998–2000	1980	1999	1979–1981	1998–2000	1975–1984	1985–1989	1990–1999	1979–1981	1998–2000
	(1)	(2)	(3)	(4)	(5)	(6)	(7)	(8)	(9)	(10)	(11)
Kenya	1,364	1,434	2,275	4,520	67.5	105.3	−1.6	3.6	−1.9	262	225
Benin	698	1,056	1,795	1,850	63.1	151.3	0.3	0.6	2.5	311	586
Burkina Faso	575	868	2,563	3,450	62.7	135.5	−0.1	3.8	0.7	134	180
Eritrea	–	822	–	500	–	139.4	–	–	1.6	–	–
Rwanda	1,134	930	975	1,116	85.3	91.6	0.8	−3.5	−1.8	371	235
Chad	587	650	3,150	3,550	80.1	152.0	−1.3	1.6	2.1	155	227
Madagascar	1,664	1,891	3,000	3,108	83.8	109.4	−1.5	−1.7	−1.9	197	181
Mozambique	603	919	3,080	3,350	100.9	131.0	−4.1	0.3	0.8	–	134
Nigeria	1,265	1,206	30,385	30,738	57.2	152.3	−2.2	4.2	2.5	414	672
Mali	804	1,163	2,050	4,650	76.7	125.7	0.8	3.2	0.2	241	283
Zambia	1,676	1,391	5,108	5,279	72.9	100.8	−4.3	5.7	−2.3	196	214
Niger	440	379	3,350	5,000	97.9	141.7	−0.6	3.1	−0.2	222	214
Ethiopia	–	1,141	13,880	10,728	–	119.9	–	–	1.7	–	138
Malawi	1,161	1,514	2,320	2,000	93.2	152.7	−1.4	−4.2	3.3	109	140
Burundi	1,081	1,283	1,305	1,100	79.9	90.3	−2.1	0.5	−3.0	177	141
Republic of the Congo	838	687	669	220	82.3	117.1	−0.5	−1.1	−0.8	385	475
Tanzania	1,063	1,295	5,160	4,650	76.7	106.0	0.4	−0.9	−2.3	–	189
Sierra Leone	1,249	1,116	1,766	540	84.5	87.0	−1.2	−0.3	−2.7	367	341
Democratic Republic of the Congo	807	785	6,314	7,880	72.2	92.0	−0.8	0.1	−4.1	241	252

Source: FAO Production Yearbook (1983), pp. 45–47; Production Yearbook (2000), pp. 3–5; World Bank (2001a), p. 221; World Bank (2002a), pp. 142–144.

noted has experienced substantial irrigation development), Malawi, Mauritius, South Africa, and Swaziland. In many more countries, fewer than 1,000 kilograms of grain are harvested from each cultivated hectare.

With crop yields that are minimal, growing slowly, or both, the primary response to increases in food demand (resulting entirely from accelerated demographic expansion since living standards have stagnated for decades) has been to expand the geographic domain of farming, and rapidly so. Agricultural land use declined between 1980 and 1999 (Table 15.7, columns 3 and 4) in a dozen or so Sub-Saharan nations, including Mauritius and Swaziland (both of which have small national territories), Botswana and Senegal (where tens of thousands of arable hectares were lost to spreading deserts), Ethiopia (which lost arable land when Eritrea gained independence after a long and bloody civil war), and Sierra Leone (which was racked by brutal internal conflict). Farmed area increased by one-half or more during the final two decades of the twentieth century in five West African nations – Ghana, the Ivory Coast, Mali, Mauritania, and Togo – as well as Kenya. Of the fourteen countries listed in Table 15.7 where crop yields went down during this same period, agricultural land use declined only in tiny Lesotho, Sierra Leone, the Republic of Congo (the DRC's northwestern neighbor), and Botswana. In Sudan, Rwanda (where deforestation was already at an advanced cumulative stage many years ago), and eight other nations, agricultural extensification took place in the 1980s and 1990s.

Rapid expansion of farmed area and modest intensification have had a positive impact on food production (Table 15.7, columns 5 and 6). However, the increases, which averaged a little less than 25 percent during the 1990s, were only about half the output growth occurring in Asia that same decade (Chapter 11). Furthermore, human numbers were growing considerably faster in Africa, so that per-capita production actually declined during the final decade of the twentieth century in more than half the countries listed in Table 15.7 (column 9). There were several exceptions to this trend in West Africa, such as Benin, Ghana, the Ivory Coast, and Togo. Elsewhere, weak output growth has resulted from the full range of impediments to agricultural development reviewed in this section, from inhospitable environments to policy-induced market distortions to civil conflict.

Finally, the same impediments to development have held down the productivity of rural labor in practically the entire continent. Reflecting per-capita GDPs that are high by regional norms, agricultural value added (Table 15.7, columns 10 and 11) is above $1,000 per worker in Mauritius, South Africa, and a few of the latter's small, Anglophone neighbors. This threshold is also exceeded in the plantation economies of Cameroon and the Ivory Coast. Elsewhere, however, value added averages just a few hundred dollars for everyone who farms.

15.4 Consumption Trends and Food Security

Going along with declining per-capita food production in many Sub-Saharan nations has been a deterioration of human diets. Using Rask's (1991) methodology to analyze FAO data, as is done in each of the five preceding chapters, one finds that per-capita consumption of plant products, which is not very sensitive to changes in income, varied modestly in most of the region during the last two decades of the twentieth century (Table 15.8, column 1). Due to good weather and policy reform, substantial increases occurred in a number of West African nations. In Burkina Faso and Chad, average intake of milk, eggs, and meat (column 2) grew along with per-capita consumption of plant products. But in Benin, Ghana, and Mali, agricultural development and limited poverty alleviation allowed people to improve their diets not by eating more animal products, but instead by consuming plant products in greater quantities. In Mauritania and Nigeria, living standards did not improve and combined food intake per capita (column 3) went down, with plant products being substituted for animal products.

Since income elasticities of demand are higher for eggs, meat, and milk than for other foods and since most Africans are very poor, per-capita consumption of animal products went up significantly in just three of the countries listed in Table 15.8: Botswana, Chad, and Mauritius. In many more Sub-Saharan nations, however, this consumption went down, often without any compensating increase in the consumption of plant products. This was the trend during the 1980s and 1990s in Gabon, South Africa, and a few other places where the average person still consumes more than 500–600 cereal-equivalent kilograms per annum. But there are other places where per-capita consumption was at or below this threshold around 1980 and yet diets subsequently deteriorated. Zimbabwe is one such place. So is Uganda. Food consumption is particularly deficient in the poorest nations in Central, East, and Southern Africa. Armed struggle has taken a severe toll on human diets in Ethiopia and Eritrea, in Lusophone Angola and Mozambique, and the DRC, as well as Sierra Leone in West Africa. Acute resource scarcity and violent ethnic conflict have combined in Burundi and Rwanda to create an appreciable decline not in per-capita consumption of meat, eggs, and milk, which has always been negligible, but rather in average intake of plant products. Warfare has been avoided in Tanzania as well as Malawi and Zambia, which are the two poorest Anglophone nations in Southern Africa. The citizens of these lands have little food because of occasional droughts and also because development is not really happening.

Many of the poorest countries in the region (and the entire world) actually have relatively high indices of food self-sufficiency (Table 15.8, column 4). However, this is no mark of success. Instead, national food consumption is barely above domestic production because the financial wherewithal for imports is lacking – because not enough hard currency is being earned from exports, to be specific. By the same token, a falling ratio of domestic production over national consumption does not necessarily indicate failure. For example, South Africa had a high ratio 25 years ago, in part because of an apartheid-era emphasis on self-reliance. Insofar as the decline since then has resulted from increased specialization and trade, with the country importing more food as it exports more non-agricultural goods in which it holds a comparative advantage, this change is welcome. As in the Middle

Table 15.8 Food consumption and production trends in Sub-Saharan African countries, 1979 through 1999

Country (ranked by income per capita 2000)	Years	Plant product consumption per capita (in cereal-equivalent tons) (1)	Animal product consumption per capita (in cereal-equivalent tons) (2)	Total food consumption per capita (in cereal-equivalent tons) (3)	Total food self-sufficiency* (4)
Mauritius	1979–1981	0.28	0.44	0.72	1.22
	1989–1991	0.29	0.57	0.85	1.01
	1997–1999	0.30	0.60	0.90	0.88
South Africa	1979–1981	0.28	0.78	1.05	1.13
	1989–1991	0.29	0.73	1.02	1.00
	1997–1999	0.29	0.60	0.88	0.94
Botswana	1979–1981	0.21	0.55	0.75	1.75
	1989–1991	0.22	0.77	0.99	1.16
	1997–1999	0.22	0.67	0.89	1.02
Namibia	1979–1981	0.22	0.96	1.18	1.31
	1989–1991	0.22	0.63	0.85	1.41
	1997–1999	0.22	0.42	0.64	1.35
Gabon	1979–1981	0.24	0.92	1.16	0.55
	1989–1991	0.25	0.84	1.09	0.51
	1997–1999	0.25	0.72	0.97	0.58
Swaziland	1979–1981	0.25	0.83	1.07	1.31
	1989–1991	0.27	0.66	0.92	1.02
	1997–1999	0.26	0.68	0.93	0.95
Lesotho	1979–1981	0.24	0.37	0.61	0.78
	1989–1991	0.25	0.35	0.59	0.71
	1997–1999	0.25	0.28	0.53	0.66

(Continued)

Table 15.8 (Continued)

Country (ranked by income per capita 2000)	Years	Plant product consumption per capita (in cereal-equivalent tons) (1)	Animal product consumption per capita (in cereal-equivalent tons) (2)	Total food consumption per capita (in cereal-equivalent tons) (3)	Total food self-sufficiency* (4)
Zimbabwe	1979–1981	0.24	0.33	0.56	1.21
	1989–1991	0.22	0.29	0.51	1.10
	1997–1999	0.23	0.23	0.45	1.04
Guinea	1979–1981	0.25	0.13	0.37	0.87
	1989–1991	0.22	0.11	0.34	0.76
	1997–1999	0.25	0.14	0.39	0.80
Ghana	1979–1981	0.19	0.22	0.41	0.72
	1989–1991	0.22	0.24	0.46	0.71
	1997–1999	0.28	0.19	0.48	0.76
Mauritania	1979–1981	0.18	0.89	1.07	0.82
	1989–1991	0.24	0.82	1.06	0.80
	1997–1999	0.26	0.64	0.91	0.74
Cameroon	1979–1981	0.26	0.29	0.54	0.82
	1989–1991	0.24	0.33	0.56	0.78
	1997–1999	0.25	0.30	0.55	0.80
Sudan	1979–1981	0.21	0.62	0.83	0.99
	1989–1991	0.21	0.51	0.72	0.91
	1997–1999	0.22	0.62	0.85	0.87
Ivory Coast	1979–1981	0.31	0.34	0.65	0.77
	1989–1991	0.27	0.25	0.52	0.79
	1997–1999	0.29	0.17	0.46	0.83

Senegal	1979–1981	0.24	0.35	0.60	0.67
	1989–1991	0.24	0.40	0.64	0.67
	1997–1999	0.24	0.42	0.66	0.69
Togo	1979–1981	0.26	0.15	0.40	0.85
	1989–1991	0.26	0.20	0.45	0.82
	1997–1999	0.28	0.19	0.46	0.79
Uganda	1979–1981	0.23	0.30	0.53	0.98
	1989–1991	0.25	0.25	0.50	0.99
	1997–1999	0.24	0.22	0.45	0.97
Angola	1979–1981	0.22	0.37	0.59	0.71
	1989–1991	0.18	0.35	0.53	0.72
	1997–1999	0.20	0.28	0.48	0.78
Central African Republic	1979–1981	0.25	0.31	0.57	0.98
	1989–1991	0.20	0.44	0.64	0.94
	1997–1999	0.21	0.46	0.66	0.96
Kenya	1979–1981	0.23	0.42	0.65	0.94
	1989–1991	0.19	0.45	0.64	0.97
	1997–1999	0.20	0.38	0.58	0.90
Benin	1979–1981	0.23	0.22	0.45	0.97
	1989–1991	0.26	0.21	0.47	0.93
	1997–1999	0.28	0.21	0.49	0.95
Burkina Faso	1979–1981	0.18	0.18	0.36	0.92
	1989–1991	0.23	0.23	0.46	0.96
	1997–1999	0.25	0.24	0.49	0.95
Eritrea	1997–1999	0.19	0.22	0.41	0.80
Rwanda	1979–1981	0.26	0.11	0.37	0.73
	1989–1991	0.23	0.10	0.34	0.67
	1997–1999	0.23	0.11	0.34	0.62

(Continued)

Table 15.8 (Continued)

Country (ranked by income per capita 2000)	Years	Plant product consumption per capita (in cereal-equivalent tons) (1)	Animal product consumption per capita (in cereal-equivalent tons) (2)	Total food consumption per capita (in cereal-equivalent tons) (3)	Total food self-sufficiency* (4)
Chad	1979–1981	0.17	0.34	0.51	0.96
	1989–1991	0.18	0.40	0.58	0.96
	1997–1999	0.23	0.38	0.61	0.98
Madagascar	1979–1981	0.25	0.48	0.72	0.97
	1989–1991	0.22	0.42	0.64	0.98
	1997–1999	0.21	0.37	0.58	0.97
Mozambique	1979–1981	0.21	0.12	0.33	0.91
	1989–1991	0.20	0.11	0.31	0.84
	1997–1999	0.22	0.09	0.31	0.89
Nigeria	1979–1981	0.22	0.27	0.49	0.69
	1989–1991	0.27	0.19	0.46	0.66
	1997–1999	0.32	0.18	0.50	0.64
Mali	1979–1981	0.18	0.45	0.62	0.97
	1989–1991	0.24	0.43	0.68	0.97
	1997–1999	0.23	0.44	0.67	0.97
Zambia	1979–1981	0.25	0.26	0.50	0.88
	1989–1991	0.22	0.22	0.44	0.99
	1997–1999	0.21	0.18	0.40	0.84
Niger	1979–1981	0.23	0.41	0.63	0.73
	1989–1991	0.23	0.27	0.49	0.60
	1997–1999	0.22	0.26	0.48	0.63

Country	Period			
Ethiopia	1979–1981	0.20	0.30	0.50	1.00
	1989–1991	0.18	0.25	0.43	0.95
	1997–1999	0.20	0.23	0.42	0.98
Malawi	1979–1981	0.25	0.12	0.37	1.01
	1989–1991	0.22	0.10	0.32	0.93
	1997–1999	0.24	0.09	0.33	0.93
Burundi	1979–1981	0.23	0.12	0.35	0.91
	1989–1991	0.21	0.11	0.31	0.92
	1997–1999	0.18	0.08	0.27	0.91
Republic of the Congo	1979–1981	0.24	0.30	0.54	0.77
	1989–1991	0.23	0.34	0.57	0.71
	1997–1999	0.23	0.26	0.50	0.67
Tanzania	1979–1981	0.25	0.27	0.51	0.99
	1989–1991	0.23	0.29	0.52	1.00
	1997–1999	0.21	0.25	0.46	0.95
Sierra Leone	1979–1981	0.23	0.18	0.41	0.85
	1989–1991	0.22	0.12	0.34	0.85
	1997–1999	0.23	0.13	0.37	0.76
Democratic Republic of the Congo	1979–1981	0.23	0.12	0.36	0.90
	1989–1991	0.24	0.11	0.36	0.90
	1997–1999	0.19	0.09	0.28	0.90

Note: *Total food self-sufficiency equals total domestic food production divided by total domestic food consumption.
Source: Derived from FAOSTAT Agriculture Data, commodity balances and food supply data from 1979 to 1999 (www.fao.org).

Table 15.9 Recent food-insecurity indicators in Sub-Saharan African countries

Country (ranked by column 1)	Total food consumption per capita 1997–1999 (in cereal-equivalent tons)	Infant mortality rate – 2000 (per 1,000 live births)	Child malnutrition 2000 (% children <5 years)	Poverty rate late 1990s/early 2000s (% population below $1 per day)	Gini index (circa 1990s)
	(1)	(2)	(3)	(4)	(5)
Gabon	0.97	58	–	–	–
Swaziland	0.93	89	–	–	60.9
Mauritania	0.91	101	23	28.6	37.3
Mauritius	0.90	16	15	–	–
Botswana	0.89	58	17	33.3	–
South Africa	0.88	63	9	11.5	59.3
Sudan	0.85	81	34	–	–
Mali	0.67	120	27	72.8	50.5
Central African Republic	0.66	96	23	66.6	61.3
Senegal	0.66	60	13	26.3	41.3
Namibia	0.64	62	–	34.9	–
Chad	0.61	101	39	–	–
Kenya	0.58	78	22	26.5	44.9
Madagascar	0.58	88	40	49.1	38.1
Cameroon	0.55	76	22	33.4	47.7
Lesotho	0.53	91	16	43.1	56.0
Republic of the Congo	0.50	68	–	–	–

Nigeria	0.50	84	27	70.2	50.6
Benin	0.49	87	29	–	–
Burkina Faso	0.49	104	34	61.2	55.1
Angola	0.48	128	41	–	–
Ghana	0.48	58	25	44.8	40.7
Niger	0.48	114	40	61.4	50.5
Ivory Coast	0.46	111	24	12.3	36.7
Tanzania	0.46	93	29	19.9	38.2
Togo	0.46	75	25	–	–
Uganda	0.45	83	26	–	37.4
Zimbabwe	0.45	69	13	36.0	50.1
Ethiopia	0.42	98	47	31.3	40.0
Eritrea	0.41	60	44	–	–
Zambia	0.40	115	24	63.7	52.6
Guinea	0.39	95	23	–	40.3
Sierra Leone	0.37	154	–	57.0	62.9
Rwanda	0.34	123	27	35.7	28.9
Malawi	0.33	103	30	–	–
Mozambique	0.31	129	26	37.9	39.6
Democratic Republic of the Congo	0.28	85	34	–	–
Burundi	0.27	102	–	–	42.5

Source: Table 15.8 and World Bank (2004).

East and North Africa, indices of food self-sufficiency well under 1.00 in Angola, Gabon, and Nigeria reflect Dutch Disease, with petroleum exports driving up domestic currency values.

Since overall food consumption is low, stagnating, or both in so many countries south of the Sahara, standard indicators of food security are alarming. Outside the region, infant mortality rates above 100 per thousand live births are practically unknown. But in the world's poorest continent, 13 nations are in this category (Table 15.9, column 2), mainly because improved infant survival is positively tied to food consumption per capita. Except for Chad and Mali, each of these nations is characterized by average food intake of less than 600 kilograms per annum, as is to be expected where adequate diets are economically inaccessible for the poor.

With respect to another food-security indicator, Sub-Saharan Africa does not compare quite so poorly. Childhood malnutrition, as measured by the incidence of severely underweight children (Table 15.9, column 3), is common in the region, although percentages in South Asia are comparable if not higher (Chapter 11). On the other hand, no part of the world has a higher percentage of the population living on less than one dollar a day (Column 4), although estimates of this indicator are lacking for many places. Analysis of economic inequality is similarly spotty. There are some countries, such as Rwanda, where GDP per capita and Gini coefficients (column 5) are both very low. However, there are many more where extreme poverty as bad as anything encountered in South Asia coincides with income inequality of Latin American proportions. Burkina Faso is one such place. Sierra Leone is another. Gini coefficients for each of these nations and several others are about equal to that of South Africa, where earnings for the white minority continue to be an order of magnitude above incomes for the rest of the population.

15.5 Summary

Africa was the cradle of humankind. No matter where we or our immediate ancestors were born or moved from, each and every one of us has a family tree with roots south of the Sahara. Yet African environments, with some exceptions, are not especially hospitable to people. As a rule, soils are infertile and rainfall sparse. Where precipitation is heavy, disease-bearing organisms proliferate. Past the middle of the twentieth century, human settlement reflected these conditions and population densities tended to be low.

Problematical geography was bound to impede economic development in Sub-Saharan settings, which 50 years ago were no poorer than many other parts of the developing world – especially in Asia. However, the prolonged slide in per-capita GDP rankings that has happened in Africa since then has resulted as much from the mistakes of government as from the limitations of nature. Poorly prepared for independence, some nations embraced Marxist totalitarianism, with disastrous results. Elsewhere, warfare broke out, often with rival ethnic groups pitted against one another and sometimes with different sides receiving external assistance during the Cold War. Even where autocracy, central planning, and armed struggle were avoided, governments interfered regularly with market forces. Far more often than not, this interference was detrimental to agriculture, which has been the dominant economic sector in numerous countries.

Africa's poor economic record is reflected in a demographic transition that is not very far advanced. Other than in Anglophone nations at the southern tip of the continent, human fertility has not moved very far toward the replacement level. As a result, natural increase is elevated – indeed, far more rapid than in any other part of the world. During the years to come, about the only countries where birth rates could align with death rates may be those suffering a calamitous upswing in mortality because of HIV/AIDS.

The signs of economic underdevelopment are unmistakable in the Sub-Saharan countryside. Even where governments have refrained from pricing and other policies that harm rural producers, there has been little investment either in infrastructure for irrigation and transportation or in the science and technology base for agriculture. Most farmers engage in subsistence production. Using few if any purchased inputs, they depend entirely on what they are able to wrestle from the unforgiving ground with their own sweat. They are, in effect, mining environmental nutrients, at great cost to themselves and producing barely enough to nourish their families. Per-capita food consumption is miserably low in most of the continent, as are standard indicators of food security.

Except in Sudan, armed conflict is at a low ebb in Africa and the continent's worst experiments with autocracy and market suppression are fading into the past, in part because the Cold War and the proxy conflicts it engendered in the developing world are over. Since the middle 1990s, democracy has spread and GDP has grown faster than human numbers in a number of countries. There is a chance that the standard model (Chapter 8) can be applied, as will be needed if satisfactory steps toward solving problems such as poverty, hunger, and disease are ever to be taken.

Study Questions

1. In what ways do environmental realities add to the challenge of economic development in Sub-Saharan Africa?
2. In what ways does the legacy of colonialism impede development in the same region?
3. How have public policies and government action (or inaction) affected Sub-Saharan agriculture and its contributions to overall development?
4. Where are rates of HIV infection particularly high and what are the likely demographic impacts of AIDS in these places?

5. Explain why human fertility is higher in Sub-Saharan Africa than anywhere else in the world?
6. Why has Africa not experienced a Green Revolution similar to Asia's?
7. Why is there very little mechanization of agriculture south of the Sahara?
8. Why are ratios of food self-sufficiency not that low in a number of African nations with acute food insecurity?

16

The Global Food Economy in the Twenty-First Century

Population growth and food availability are age-old concerns. During the second half of the twentieth century, however, feeding the human race posed an unprecedented challenge. In absolute as well as relative terms, the rate of demographic expansion peaked, with human numbers climbing from slightly less than 2.5 billion in 1950 to a little more than 6.0 billion in 2000. Yet this growth was exceeded by increases in agricultural output. As a result, average food consumption rose substantially and the undernourished portion of the human race shrank (Johnson, 2000).

As Tweeten (1998) observes, humankind has dodged a Malthusian bullet. Moreover, this achievement has come about almost entirely because of yield growth tied to productivity gains in agriculture. Thanks to technological improvements that have allowed more and more crops to be raised without a corresponding increase in input use, per hectare production of cereals, which directly or indirectly account for more than half of what people eat, has gone from a little under 1.5 metric tons per hectare worldwide in the early 1960s to more than 3.0 tons at present. During the Green Revolution as well as more recently, cereal yields (not just output) consistently have risen faster than the world's population. Food consequently has become less scarce, as indicated by declining real commodity prices. Also, the share of the world's resources required to feed the human population has fallen, which has created a substantial improvement in living standards.

The Green Revolution had its greatest impacts where the threat of hunger was most severe. In the late 1960s and early 1970s, alarm over Asia's prospects reached a crescendo, with many convinced that mass starvation was unavoidable. But in these same years, the region's farmers were beginning to sow semi-dwarf varieties developed by scientists to yield more rice and wheat in response to irrigation and fertilization. Since then, no other part of the world has experienced a larger relative gain in living standards. Nowhere have greater strides been made toward food security.

Hundreds of millions of people remain poor and hungry in the world's most populous continent, especially in India and neighboring countries. However, food insecurity is now worse south of the Sahara, with rural areas suffering disproportionately. Various aspects of deprivation in the African countryside are readily identified: rampant disease, the neglect of education, deficient infrastructure, unimproved agricultural technology, and armed conflict. Either because national governments have lacked the means to address pressing needs, or all too often because of the havoc unleashed by local strongmen, rural people have had no choice other than to rely on their own muscle power and whatever natural fertility the environment provides. Emergency rations arrive from outside the region when mundane deprivation is punctuated by civil war or genocide. Otherwise, international efforts to raise living standards and reduce hunger have met with little success.

No one can regard the plight of Sub-Saharan Africa with indifference. As a practical matter, diseases that are uncontained in one setting inevitably spread worldwide because international travel, for commerce or pleasure, has become routine. Likewise, people made desperate because of war or famine do not stay put, but instead move to places where they can work or at least avoid starvation. Still relatively modest, migration through the Sahara and across the Mediterranean to Europe could swell if Africa's decline is not reversed. Furthermore, the same decline is putting enormous pressure on national states. As the case of Afghanistan made clear at the dawn of the twenty-first century, the hoodlums who seize control once states collapse do not just bully the impoverished locals: rather, they can strike anywhere. Thus, governmental deterioration south of the Sahara could cause the region to emerge as a base for international terrorism.

After evaluating hunger, disease, and other global problems, an expert panel calling itself the Copenhagen Consensus (2004) has recommended a limited number of initiatives to deal with critical challenges facing the human race. One of these is freer international trade. Since many of the panel's members are economists, including four Nobel laureates, the call for unencumbered commercial exchange, which has been repeated often since the days of Adam Smith and David Ricardo, is only to be expected. But more to the point, poverty cannot be alleviated in places like the African countryside if producers in these areas are denied full access to markets around the world.

The Copenhagen Consensus recognizes that free markets allow maximum living standards to be derived from available resources since rational specialization and trade are encouraged in such markets. Also stressed, however, is the enhancement of productive capacity in impoverished settings. Much of this enhancement has to do with human health. For example, anemia currently prevents hundreds of millions of people, mainly women and children, from leading active and productive lives. A good way to deal with this problem would be to spend $12 billion distributing iron tablets. Another $27 billion are proposed for the containment of the HIV and alleviation of the symptoms of AIDS. Controlling malaria, which kills nearly 3 million per annum and sickens 100 times that number every year, is another top priority (Copenhagen Consensus, 2004).

Aside from combating illnesses that kill many, sap the strength of or incapacitate countless others, and burden victims' families, the Copenhagen Consensus (2004) advocates supporting agricultural research and development in the tropics and subtropics, where practically all the world's food-insecure live. Among those who would agree is Gordon Conway, who led the Rockefeller Foundation from 1998 through 2004. However, Conway contends that simply repeating the Green Revolution, which his organization supported from its inception, is not viable. Inorganic fertilizers already pollute ground and surface waters in some parts of the world. At the same time, some insects and other organisms that harm crops are developing resistance to toxic pesticides. Thus, there are places where using more agricultural chemicals would take a toll on the environment and human health and would not necessarily have the desired effect on yields (Conway, 1997, pp. 89–90; 132).[1] Also, improvements in agricultural technology cannot be limited to places with fertile soils and adequate water resources, as tended to be the case during the Green Revolution. Instead, improvements are needed in a diverse array of environments – including less-favored areas where the world's poorest farmers are concentrated (Conway, 1997, pp. 134–135).

Conway calls for a "Doubly Green Revolution," which must have two main facets. One is the application of sound ecological principles in the management of land, water, and biological resources; the other facet relates to biotechnology. No one denies the importance of recognizing and containing the risks of genetic alteration of living organisms, such as the emergence of wild relatives of genetically modified crops which subsequently are able to out-compete other plants and are difficult to control (Conway, 1997, pp. 157–158). Nevertheless, biotechnology creates unique opportunities for environmentally sound improvements in agricultural yields. It also facilitates soil conservation and diminishes farmers' reliance on agricultural pesticides (Rauch, 2003).

Another environmental benefit will result if agricultural yields increase faster than demographic expansion, as has occurred in recent decades. That is, steady growth in production per hectare arrests the conversion of tropical forests and other natural habitats into cropland and pasture. As already indicated, cereal yields more than doubled between the 1960s and 1990s. This allowed farmed area to increase by only 11 percent during the same period, in the face of much faster growth in population and food demand (Chapter 3). If agricultural intensification does not continue, renewed extensification is bound to occur during the first half of the twenty-first century, when food demand will rise by at least 50 percent (Chapter 2).

Of course, agricultural intensification that is environmentally sound is also critical for achieving food security throughout the world. As Senauer and Sur's (2001) analysis suggests, economic expansion reduces the number of people whose earnings are

---

[1] These observations are not universally applicable. As pointed out in Chapter 15, inorganic fertilizer is severely under-utilized, not over-utilized, in Sub-Saharan Africa. So is irrigation. As a result, the Green Revolution has failed to take hold in the region, which also suffers from widespread land degradation.

so meager that they cannot afford a minimally adequate diet, particularly if the benefits of that expansion are widely shared. However, the greatest reductions in undernourishment occur if broad-based economic growth is coupled with lasting reductions in food prices.

Impressive though they are, past success in reducing global poverty and hunger are no reason for current or future complacency. In percentage terms, crop and livestock yield increments are slowing, and only the deceleration of population growth is averting a rise in real food prices. Since human numbers will continue increasing for a few more decades at least and since income growth is driving up food demand in many parts of the world, prices will surely increase if production per hectare stagnates. The only way to avoid a reversal of the trend in recent times toward diminished food scarcity is for demand growth to be exceeded by yield increases, which require agricultural research and development.

A notable lesson of this volume is that alleviating poverty and hunger requires a holistic, integrated approach. It is not enough to improve agricultural technology alone, or schools alone, sanitation alone, or infrastructure alone either. The standard model (Chapter 8) calls for all of these in time. But above all, sound overall economic policy, defined broadly to include the rule of law and the curtailment of corruption, is essential. Holistic and integrated, the standard model creates the economic growth that in turn allows for the simultaneous improvement of schools, roads, sanitation, nutrition, and the environment. For example, an economy that can sustain income growth is able to fund the development and transfer of agricultural technology, which in turn raises yields so that people can eat well and cropping does not need to extend onto environmentally fragile lands. Also, higher incomes and increased food supplies open up new opportunities for charitable groups to feed the needy.

16.1 Victims of Our Own Success?

No one would object to spending money to combat anemia, malaria, and HIV/AIDS. The sums required are tiny relative to global economic output. In addition, the benefits of disease control – consisting of GDP growth, not to mention diminished human suffering – far outweigh the costs. Likewise, the desirability of scientific advances aimed at the sustainable intensification of agriculture is also uncontested. As this happens, though, new issues emerge, including some that are entirely novel from the perspective of the broad sweep of human experience. The increased incidence of obesity is one of these issues.

When addressing the problem of obesity, one must never forget that the urge to consume excessive calories is etched in our genes and therefore very powerful. Our distant ancestors often faced food shortages, which meant that the ability to store calories as fat in flush times was a distinct advantage. People who easily converted calories into fat did not have to worry about being overweight because hard physical labor kept everyone trim. Since such people were in a better position to reproduce, thereby passing along their genetic traits, the urge we have today to over-eat is "somewhere between the propensity to breathe and the propensity to have sex," according to Stephen Bloom, the director of metabolic medicine at the University of London's Imperial College. Breaking the habit of over-eating, Bloom adds, is "much worse than stopping smoking" (Associated Press, 2004).

The innate drive for nourishment may be especially strong among people who suffered chronic hunger as infants or while in the womb. Take China, for example. Tens of millions starved to death there in the late 1950s and early 1960s; all of them the direct or indirect victims of the forced industrialization ordered by Mao Tse-tung (Chapter 7). Now in their forties, Chinese born during this period appear to be unusually prone to gain weight. Moreover, people who went hungry during the "Great Leap Forward" tend to be anxious about the nutritional status of their offspring and grandchildren, and so feed them too much if they have the means to do so. In Beijing, where incomes compare favorably with average earnings in the rest of the country, one-third of the population is overweight (Duncan, 2003).

Powerful though our urges to over-eat may be, the incidence of obesity has been driven up by changing technology and economic realities. Fewer jobs require great physical exertion, and many more can be accomplished while sitting comfortably at a desk. In affluent parts of the world, many people now shop online, rather than walking or even driving to a store. They send e-mails or use cellular telephones, which are always within reach, instead of dropping by their friends or associates for a visit. For a sizable part of the population in places like the United States, physical activity has declined, even as per capita caloric intake has risen (Rashad and Grossman, 2004).

In addition to making our lives more sedentary, technological change has promoted the production of foods that are cheap, tasty, and fast. As emphasized throughout this book, improvements in agricultural productivity have brought down the prices of edible goods. In a statistical analysis, Lakdawalla and Philipson (2002) find that this price decline is responsible for approximately two-fifths of the body mass index (BMI)[2] increases that the US population has experienced since 1980. Complementing the technologically induced trend toward diminished commodity scarcity have been innovations such as vacuum packing, microwave ovens, and improved food preservation, which "have enabled food manufacturers to cook food centrally and ship it to consumers for rapid consumption" (Cutler *et al.*, 2003, p. 94). This has greatly reduced the time that people spend preparing food, precisely as the opportunity cost of time has risen. To be more specific, changes in food-preparation technology have diminished the time spent in the kitchen by women, whose opportunity costs have gone up due to economic expansion and the receding of gender-related barriers to employment. As the "time cost" of food in general has declined, and particularly as the time cost of fattening products

---

[2] As explained in Chapter 10, a person's BMI is calculated by dividing his or her weight in kilograms by the square of his or her height in meters, or by multiplying weight in pounds by 704.5 and then dividing by the square of height in inches.

(i.e., those containing a lot of fat, oil, and sugar) has diminished, Americans and others have grown heavier.

Characterizing reductions in the time cost of edible products as "revolutionary," Cutler *et al.* (2003) refer us to the case of the potato.

> "Before World War II, Americans ate massive amounts of potatoes, largely baked, boiled, or mashed. They were generally consumed at home. French fries were rare, both at home and in restaurants, because the preparation of French fries requires significant peeling, cutting, and cooking. Without expensive machinery, these activities take a lot of time. In the postwar period, a number of innovations allowed the centralization of French fry production. French fries are now typically peeled, cut, and cooked in a few central locations using sophisticated new technologies. They are then frozen at −40 degrees and shipped to the point of consumption, where they are quickly reheated either in a deep fryer (in a fast-food restaurant), in an oven, or even a microwave (at home). Today, the French fry is the dominant form of potato and America's favorite vegetable.[3] This change shows up in consumption data. From 1977 to 1995, total potato consumption increased by about 30 percent, accounted for almost exclusively by increased consumption of potato chips and French fries (p. 94)."

Statistical analysis has been undertaken of the impacts of changes in food-preparation technology, of the sort that has caused consumption of French fries to skyrocket in recent decades. Among other things, Cutler *et al.* (2003) find that recent increases in per capita calorie intake are mainly a consequence of eating more snacks and meals, not ingesting more calories at a single sitting. This is consistent with the new directions that food-preparation technology has taken. Likewise, consumption of mass-produced food has increased more than consumption of other edible products, including what people cook from scratch in their own kitchens. The incidence of obesity has gone up more among women than among men, as is to be expected since members of the former sex tend to be more involved in food preparation and therefore are more directly affected by a reduction in this activity's time requirements. In addition, international differences in the incidence of obesity are related to access to new food technologies and to processed food.

This last result of empirical study highlights that reductions in the time cost of food have by no means been limited to the United States. Neither have they been confined to affluent places, such as OECD members. To the contrary, changes in food-preparation technology are also happening in the developing world. As mentioned in Chapter 12, supermarkets and other purveyors of convenience foods have proliferated throughout Latin America in recent years (Reardon and Berdegué, 2002). Likewise, changes of the same sort are occurring in other parts of the developing world with an urban middle class. In short, US-style eating habits are spreading, as are waistlines.

---

[3] Contrary to what many people suppose, the potato is a tuber, not a vegetable. Needless to say, deep-fried French fries, which should be avoided by anyone with a cholesterol problem, are no substitute for things like broccoli and peas.

That excess weight is, indeed, a global phenomenon is underscored by data gathered by the International Obesity Task Force (IOTF), which was created by the World Health Organization (WHO). True, the incidence of obesity is very high in wealthy nations. In the United States, for example, the portion of adult males with a BMI at or above 30 (i.e., the obesity cut-off) rose from 10.4 percent in the early 1960s to 19.9 percent in the early 1990s, as the rate for adult females went from 15.1 percent to 24.9 percent. However, the obese share of the population is alarmingly large outside the OECD – not so far in the poorest and most food-insecure places, but certainly among some of the leading nations of the developing world. Between 1975 and 1989, the portion of Brazilian men with a BMI of 30 or more increased from 3.1 percent to 5.9 percent and the incidence for women rose from 8.2 percent to 13.3 percent. As of 1989, 8 percent of black men and 44 percent of black women in South Africa's Cape Peninsula were above the same BMI threshold (WHO, 2000, pp. 21–22).

Obesity is unhealthy in various ways. People who are extremely overweight are more susceptible to type-2 diabetes, for example. They also run greater risks of cardiovascular disease and kidney failure. Drawing on recent research, the IOTF provides estimates of the expense of diagnosing and treating illnesses related to obesity, reporting for example that this expense in the United States was nearly $46 billion per annum, or 6.8 percent of total health-care costs, in the early 1990s (Wolf and Colditz, 1994, cited in WHO, 2000, p. 83). No such estimates are provided for developing countries. However, more recent investigations by the IOTF suggest that more than 115 million people in the developing world suffer from obesity-related ailments. For example, the economic cost of chronic diseases related to diet currently exceeds 2 percent of GDP in China, which is higher than the economic cost of undernourishment in that country (Burslem, 2004).

For the sake of limiting mortality and health-care costs, governments could apply various measures to reduce obesity. Education about the causes of weight gain gives people ideas about how to avoid the problem, and disseminating more information about health consequences should help motivate people to eat right and exercise. Fiscal instruments are also an option. Health insurance premiums could be raised for heavy people, for example. Alternatively, foods rich in fats, oil, and sugar could be taxed, just as is done with tobacco products. Fiscal approaches such as these could be justified if people who get sick because they are too heavy do not pay all their treatment costs.

However, applying fiscal measures to correct for the non-internalization of medical costs seems impolitic.[4] People place great value on their freedom to make their own eating decisions. Also, some of the cost and most of the inconvenience of obesity is shouldered by fat people themselves. Furthermore, taxation of fattening products, aside from modest assessments on soft drinks and snacks, may be inequitable if these items are favored by poorer segments of the population.

---

[4] One supposes that non-internalization of health-care costs contributes, perhaps substantially, to obesity. To say the least, insurance companies have limited ability to raise the premiums of customers as they gain weight. No internalization at all occurs for people whose medical bills are paid by the taxpayer.

Conceding that the obesity epidemic is unlikely to be combated mainly with taxes, Rashad and Grossman (2004) recommend subsidies for exercise or workout facilities as an alternative. However, this approach also seems unpromising, in light of the trouble and expense of monitoring people who have agreed to exercise in return for a financial reward. A better option might be to pay people who lose weight and do not gain it back, although here again monitoring costs could be appreciable.

No single policy measure seems adequate, then. Aside from educational initiatives and taxes, support could be provided for advances in food science, aimed both at lessening our craving for food and at yielding healthy alternatives to the fattening products we crave.

16.2 The New Food Economy

The study by Cutler *et al.* (2003) reveals much about how life is changing these days, for at least part of the human population. The desire people have to eat well is, as emphasized earlier, innate. But because of powerful currents in the modern economy, many of us have unprecedented access to fattening products. Choices about diet and exercise have yet to adjust appropriately.

Needless to say, weight gain is not the only consequence of current trends. Antle (1999) stresses that improvements in living standards, resulting during the past two centuries mainly from the accumulation of knowledge (Johnson, 2000), have raised the opportunity cost of the scarcest of all things – namely, our time.

> "The increasing opportunity cost of time goes hand in hand with an increasing specialization of labor, and together these factors drive the transition from the industrialized to the service- and information-based economy. The opportunity cost of time and the specialization of labor explain the growth in demand for an array of increasingly specialized goods and services, including the diversity of manufactured foods and food services that substitute for time spent preparing foods and raw ingredients in the home. The increasing opportunity cost of time also explains the explosive growth in the utilization of information technology that allows all kinds of repetitive tasks to be carried out by machines rather than people, further reinforcing the continued specialization of labor that leads to yet higher incomes. Just as significantly, the transition to a service- and information-based economy is associated with a rapidly declining cost of information that has a profound impact on the structure and efficiency of markets" (Antle, 1999, p. 994).

Although Antle (1999) does not address the issue, improved living standards and the concomitant increase in the scarcity value of time cannot be separated from female economic empowerment, which coincides with a decline in human fertility (Chapter 2) as well as fewer hours spent in the kitchen and other changes in the food economy mentioned in the quoted passage.

Surveying the economic literature on food and agriculture, Antle (1999) shows that every part of the food economy is being transformed as affluence spreads, technological innovation is making transmitting information virtually free, and services are becoming the dominant economic sector. For instance, the variety of products is exploding. Not so long ago, it was expensive by today's standards for suppliers to discern individual consumers' preferences about convenience attributes, nutritional content, and so forth. Likewise, the information management that suppliers required to respond to heterogeneous preferences by producing and delivering a diverse set of quality-differentiated goods was burdensome. Accordingly, a limited number of uniform commodities were available in the food economy, with buyers and sellers engaged mainly in haggling over price. No more! Advances in information technology now allow for easy communication between consumers and producers as well as fine-tuning of production processes. The result is an awesome variety of products, each of these featuring its own price and qualitative attributes.

In light of the returns created by transmitting and making use of information, the provision of services, which revolves largely around these activities, is becoming an increasingly dominant part of the food business, exactly as is happening in other parts of the economy. An unmistakable sign of this trend is that the value of unprocessed commodities is falling over time relative to total consumption expenditures, generally, and what people spend on food, specifically. In the United States, for example, the value of unprocessed ingredients has fallen from 41 percent of food expenditures in 1950 to 19 percent in 2000 (ERS-USDA, 2002). The long-term decline in real commodity prices is only part of the story. Just as important is that demand for preparation services, including though not limited to the services provided by restaurants, is rising at a fast pace.

The recent proliferation of fast-food chains in various parts of the world has been difficult to ignore. Nevertheless, this sort of business is probably constrained by Engel's Law. That is, income elasticities of demand for hamburgers, fried chicken, and other mass-produced fare sold at a low price are both a fraction of 1.0 and inversely related to GDP per capita, particularly if that fare is fattening or otherwise nutritionally undesirable. However, increases in household earnings may cause patronage of other kinds of establishments to rise proportionately or more, which suggests that Engel's Law (Chapter 2) may not apply. In the United States, for example, sales revenues of full-service restaurants, which offer entertainment of a sort along with meals cooked by trained chefs, are expected to grow in real terms by 18 percent between 2000 and 2020, which is comparable to increases in GDP per capita during the same period. The same 18 percent increase is well above the 6 percent real growth in revenues projected for fast-food eateries during the first two decades of the twenty-first century (Stewart *et al.*, 2004).

Demographic change has something to do with these trends. For example, the number of domiciles with either a single person or multiple adults but no children is increasing faster than the number of households with children. The former group tends to eat more meals and snacks away from home. Also, a growing segment of the population is elderly, and old people are not the main customers of fast-food chains. Nevertheless, the primary driver of increased food consumption outside the home is income growth and associated increases in the opportunity cost of time. For example, five-sixths of the aforementioned 18 percent rise in the sales revenues of full-service restaurants is expected to occur because of higher earnings (Stewart *et al.*, 2004).

Throughout the OECD as well in other parts of the world where an urban middle class exists, farming and ranching have ceased to dominate the food economy in all respects other than the utilization of land and water resources. Specialization and trade are very far advanced, with the value of inputs and post-harvest services provided by agribusinesses greatly exceeding the value that farmers add to food output. Already sizable, the value of preparation services provided by restaurants and other enterprises is growing. Looking to the future, expansion of the latter services is likely to be the main engine of expansion in the food economy.

16.3 The Changing Role of Government

Aside from becoming an ever more important part of the food economy, post-harvest services are characterized by substantial economies of scale. A typical consequence is that processing, preparation, and related lines of commerce are concentrated, with a limited number of large firms accounting for a sizable portion of total service output. The supermarket business, for instance, is currently dominated by a few large chains and is likely to become even more concentrated in the years to come, as some of these businesses merge and others cease operating (Duncan, 2003).

While retail prices are pulled down as economies of scale are captured, monopolistic behavior, which raises prices, is a possibility in markets served by a handful of firms. Antitrust and other policies can be applied to counteract monopolization. However, the economic case for this kind of corrective intervention by government is not all that compelling. Surveying various empirical studies, Antle (1999) concludes that long-term trends in wholesale and retail prices appear to be explained mainly by increased product differentiation and other factors, and not the exercise of monopoly power by a concentrated group of suppliers. Apparently, competition among firms that together account for a large share of sales in selected markets as each takes full advantage of scale economies is vigorous enough to keep prices close to efficient levels – not just the prices households pay for retail products, but the prices growers receive for output as well.

Tentative though the economic case for vigorous governmental action to combat monopolization of post-harvest services may be, it nevertheless is stronger than any economic rationale offered for commodity programs (e.g., price supports and deficiency payments) aimed at propping up agricultural incomes. As reported in Chapter 4, the average annual return on assets used in the commercial production of crops and livestock has held fairly steady at around 10 percent for many years (Hopkins and Morehart, 2002). This performance is what one expects of a competitive industry in long-run equilibrium. Also, agricultural production is increasingly concentrated in the hands of a limited number of large farmers. As a result, the beneficiaries of commodity programs are more affluent as a rule than their fellow citizens.

While it is next to impossible to make the case that commodity programs serve efficiency, equity, or environmental goals cost effectively, large farmers, who comprise an ever smaller portion of the total population, may find it easier to organize politically for "entitlements" from the non-agricultural majority of taxpayers. Still, these transfers are likely to diminish over time. Now that obesity is reaching epidemic proportions in many nations, making food artificially inexpensive with subsidies, which contributes to over-eating, is hard to justify. Moreover, cheap food is against the interest of poor farmers in developing countries whose cash-strapped governments cannot afford to provide countervailing subsidies.

Commodity programs of the sort examined in Chapter 4 are likely to be more constrained by international agreements in the future. As reported in Chapter 6, farm products are now a major focus of multilateral trade negotiations, with agricultural protectionism and subsidies in the OECD coming in for pointed criticism, especially by developing nations that are now active WTO members. At the same time, the European Union's (EU's) enlargement to include poorer nations and elevated US budget deficits are creating pressure for a more market-oriented agriculture in Europe and the United States, one that depends less on taxpayers and trade protectionism (Chapter 10).

While WTO negotiations will create pressure for freer markets throughout the global food economy, it is reasonable to expect governments in affluent nations, and even some countries with lower living standards, to find ways to support farmers. However, this support – which will often be presented as "decoupled" from commodity production, implying that the consequences for world markets and prices are negligible – is likely to come with strings attached. Some of these strings will be environmental. In every part of the world, not least in wealthy nations that subsidize farmers, agriculture is a major source of water pollution. Income subsidies to farmers are never fully decoupled and hence always increase cropping, chemical applications, soil erosion, and water pollution. Recognizing this, the non-farming majority and its political representatives will insist on sound management of natural resources in return for providing income assistance to food growers. Hence, environmental compliance is sure to be an ever more important feature of agricultural legislation.[5]

Support for agriculture will also hinge on food safety. Unnevehr (2004) points out the various reasons why demand for food safety is increasing. Scientific and technological advances, including improvements in our ability to detect harmful

---

[5]In some countries, farmers may earn payments from the government or (under a "cap-and-trade" market) from private firms facing a mandate to cut CO_2 emissions in return for sequestering carbon in soils. One purpose will be to build soil organic matter, thereby changing soil structure in ways that diminish erosion, retain moisture, and protect water quality. A second purpose will be to cut atmospheric concentrations of CO_2, which is the principal greenhouse gas contributing to global warming. However, the scope for economically sequestering carbon on cropland is limited; sequestration in forests offers greater promise (Tweeten and Tanner, 2005).

components, have had an effect. Consumption of fish and meat, which are subject to spoilage, has increased and the marketing channels connecting ranches and ports to consumers' tables have grown more complex, which can introduce or spread hazards. Even the rising importance of food preparation raises questions about safety, since the responsibility for, say, adequate cooking has thereby shifted from households to businesses.

Any doubts about the depth of interest in food safety are dispelled by consumers' reactions to diseases like "mad cow," or Bovine Spongiform Encephalopathy (BSE). For example, one animal was diagnosed with BSE in Canada in 2003. This led to an immediate cessation of beef exports to the United States, which formerly consumed half of total Canadian production. In December of the same year, the disease was detected in Washington state, in a single cow that had been imported from Canada. Japan and many other countries quickly terminated beef purchases from the United States. Although no other animals in Canada or the United States were found to be infected with BSE, these episodes cost the two countries' food economies billions of dollars.

While consumers are right to worry about BSE, concerns about food safety can be exploited for protectionist ends. This exploitation is supposed to be limited by the Codex Alimentarius, which the FAO and WHO established in 1963 to develop food standards that protect consumers' health while simultaneously ensuring "fair trade practices" (Codex Alimentarius Commission, 2004). Accepting the Codex approach, the WTO allows trade restrictions based on sound science. The conundrum is that countries do not always agree about what sound science is, not to mention safety standards for edible products. For example, the United States and the EU have disputed the latter's restrictions on genetically modified foods, with the former nation claiming that these restrictions have little objective basis and are instead political and protectionist. Even within the United States, however, worries about food safety, though not backed up by scientific evidence, sometimes impede technological improvements. For example, McDonalds, fearing consumer resistance, announced in 2000 that it would not use any genetically modified products. As Unnevehr (2004) points out, this has arrested the spread of Bt potatoes, which resist pests and have other beneficial properties.

Unnevehr (2004) commends harmonization of food-safety rules, as exemplified by the Codex Alimentarius, and other institutional arrangements at the global level. Uniform standards help preempt protectionism at the national or regional level that masquerades as something else. Provided of course that the rules have a demonstrable scientific basis, institutional harmonization also helps the world's food economy overcome the barriers to technological progress that arise if standards in some places permit new products (e.g., those resulting from the use of biotechnology) while those of other jurisdictions are restrictive. At the same time, uniform and reasonable rules governing the global food economy facilitate the specialization, trade, and productivity growth that, as has been emphasized throughout this book, have allowed a growing population to eat better during the

1800s and 1900s and that are essential if all people are to be well fed during the twenty-first century.

We close this volume by observing that available resources, technology, and economic knowledge are adequate to end poverty and hunger in every country around the world. Poverty and hunger are not destiny, but are instead the result of unfortunate policy choices. Responsibility for making better decisions rests largely with poor nations, although rich countries can be helpful. The single most important contribution of the latter may be education regarding sound economic policies. It would be difficult, for example, to overstate the importance of promoting multilateral free trade, among rich and poor nations alike. In addition, financial and technical assistance provided by rich countries to agricultural research, of the sort undertaken by the Consultative Group on International Agricultural Research (CGIAR), has a proven economic payoff and warrants expansion. Humanitarian assistance to help poor countries to avert famine and to combat diseases such as HIV/AIDS, malaria, and tuberculosis is also critical. The ideal, however, is to promote policy reform so that countries that are currently poor will have, in the not too distant future, sufficient income to address their own humanitarian problems, including food insecurity.

Study Questions

1 Summarize and explain the main recommendations of the Copenhagen Consensus.
2 Explain the Doubly Green Revolution and what Gordon Conway means by the term.
3 Analyze the causes of the obesity epidemic and evaluate the various proposals that have been made for dealing with it.
4 Relate the growing emphasis on processing and preparation services in the food economy to fundamental trends in the entire economy.
5 Does a strong rationale exist for applying antitrust laws in response to growing concentration of grocery stores and other agribusinesses?
6 Explain the purpose of the Codex Alimentarius.

Abbreviations and Acronyms

ABC	Uganda's program for the prevention of HIV–AIDS
AHDR	Arab Human Development Report
AIDS	Acquired Immune Deficiency Syndrome
BMI	body mass index
BSE	Bovine Spongiform Encephalopathy ("mad cow")
CAFO	concentrated animal feeding operation
CBR	crude birth rate
CDC	US Centers for Disease Control and Prevention
CDR	crude death rate
CGIAR	Consultative Group on International Agricultural Research
CIMMYT	International Maize and Wheat Improvement Center
CS	consumers' surplus
DRC	Democratic Republic of Congo, formerly Zaire
EBRD	European Bank for Reconstruction and Development
EKC	Environmental Kuznets Curve
EMBRAPA	Empresa Brasileira de Pesquisa Agrícola
EU	European Union
FAO	Food and Agriculture Organizations of the United Nations
FDI	foreign direct investment
FTAA	Free Trade Area of the Americas
GATT	General Agreement on Tariffs and Trade
GDP	gross domestic product
GLASOD	Global Land Assessment of Degradation
GMO	genetically modified organism
GNI	gross national income
HIV	Human Immunodeficiency Virus
HRS	Household Responsibility System (of post-Mao China)
IFPRI	International Food Policy Research Institute
IITA	International Institute of Tropical Agriculture
IMF	International Monetary Fund
IOTF	International Obesity Task Force of the World Health Organization
IRRI	International Rice Research Institute
ISI	Import-Substituting Industrialization

KGB	Committee for State Security of the Soviet Union
MC	marginal cost
MITI	Japan's Ministry of International Trade and Investment
MP	marginal product
MU	marginal utility
MV	marginal value
NAFTA	North American Free Trade Agreement
NATO	North Atlantic Treaty Organization
NEP	New Economic Policy (of the early Soviet Union)
NEV	net economic value
NGO	non-governmental organization
NPC	nominal protection coefficient
OECD	Organization for Economic Cooperation and Development
OPEC	Organization of Petroleum Exporting Countries
PPF	production possibilities frontier
PPP	purchasing power parity
PS	producers' surplus
PSE	producer subsidy equivalent
SOE	state-owned enterprise
TFP	total factor productivity
TFR	total fertility rate
UNDP	United Nations Development Program
UNPD	Population Division of the United Nations
USAID	US Agency for International Development
USDA	US Department of Agriculture
USLE	Universal Soil Loss Equation
WHO	World Health Organization
WTO	World Trade Organization
WTP	willingness-to-pay

Map Annex

Africa and Southwest Asia: Political

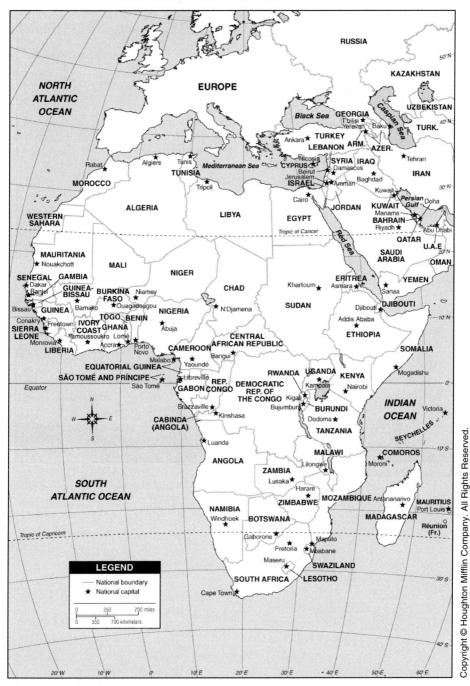

Asia and the South Pacific: Political

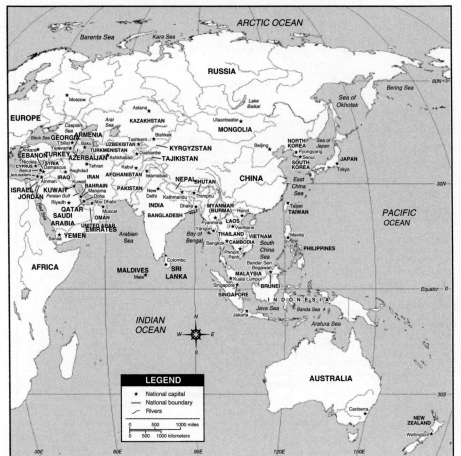

Education Place: http://www.eduplace.com

European Countries

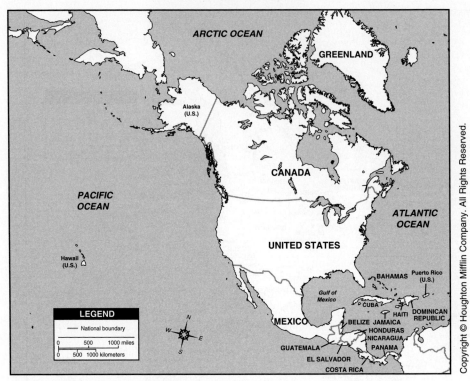

South America: Political

References

Ahmad, Sameena (2004) "Behind the Mask: A Survey of Business in China," *The Economist*, 20 March, 370: 1–20.

Ahuluwalia, Montek S. (2002) "Economic Reforms in India since 1991: Has Gradualism Worked?" *Journal of Economic Perspectives*, 16: 67–88.

Alexandratos, Nikos (ed.), (1995) *World Agriculture towards 2010*. Rome: Food and Agriculture Organization of the United Nations.

Alston, Julian M. and Philip G. Pardey (1996) *Making Science Pay: The Economics of Agricultural R&D Policy*, Washington, DC: AEI Press.

Ames, Bruce and Lois Gold (1989) "Pesticides, Risk, and Apple Sauce," *Science*, 244: 755–757.

Anderson, Jock R., Robert W. Herdt, Grant M. Scobie, Carl E. Pray and Hans E. Jahnke (1985) "*International Agricultural Research Centers: A Study of Achievements and Potential*" (Bulletin Number 32). Department of Agricultural Economics and Business Management, University of New England, Armidale.

Anderson, Terry and P.J. Hill (1975) "The Evolution of Property Rights," *Journal of Law and Economics*, 18: 163–179.

Anonymous (2004) "Asphalt and the Jungle," *The Economist*, 24 July, 372: 33–35.

Antle, John M. (1999) "The New Economics of Agriculture," *American Journal of Agricultural Economics*, 81: 993–1010.

Askari, Hossein and John Cummings (1976) *Agricultural Supply Response: A Survey of the Empirical Evidence*, New York: Praeger.

Aslund, Anders (2002) *Building Capitalism: The Transformation of the Former Soviet Union*, Cambridge: Cambridge University Press.

Associated Press. (2004) "Obesity Becoming Major Global Problem," *USA Today*, 8 May (http://www.usatoday.com/news/health/2004-05-08-global-obesity_x.htm?POE=click-refer).

Avery, Dennis (1996) "Why the Food Summit Failed," *Global Food Quarterly*, 18: 3–7.

Babinard, Julie and Per Pinstrup-Andersen (2001) *Shaping Globalization for Poverty Alleviation and Food Security: Nutrition* (Policy Brief). Washington, DC: International Food Policy Research Institute.

Ballard, Charles L., John B. Shoven and John Whalley (1985) "General Equilibrium and Computations of the Marginal Welfare Costs of Taxes in the United States," *American Economic Review*, 75: 128–138.

Barbier, Edward (2001) "The Economics of Tropical Deforestation and Land Use," *Land Economics*, 77: 155–171.

Barnett, Harold J. and Chandler Morse (1963) *Scarcity and Growth: The Economics of Natural Resource Availability*, Baltimore, MD: Johns Hopkins University Press.

Barraclough, Solon (1991) *An End to Hunger?* London: Zed Books.
Bates, Robert H. (2000) "Ethnicity and Development in Africa: A Reappraisal," *American Economic Review*, 90: 399–404.
Beaumont, Peter (1998) "Restructuring of Water Usage in the Tigris-Euphrates Basin: The Impact of Modern Water Management Policies," in Jeff Albert, Magnus Bernhardsson and Roger Kenna (eds.), *Transformations of Middle Eastern Natural Environments: Legacies and Lessons* (Bulletin Number 103) (www.yale.edu/environment/publication/bulletin/103pdfs/103beaumont.pdf). New Haven, CT: Yale School of Forestry and Environmental Studies.
Becker, Charles M. (2002) "Fertility Decline in Sub-Saharan Africa: Introduction," *Journal of African Policy Studies*, 7(2/3): 1–16.
Bennett, Adam (2003) "Failed Legacies," *Finance and Development*, 40: 22–25.
Bichel Committee (1999) *Organic Scenarios for Denmark: Report from the Interdisciplinary Group*. Miljøstyrelsen: Energiministeriet.
Binswanger, Hans and Robert Townsend (2000) "The Growth Performance of Agriculture in Sub-Saharan Africa," *American Journal of Agricultural Economics*, 82: 1075–1086.
Bishop, Joshua and Jennifer Allen (1989) "The On-Site Costs of Soil Erosion in Mali" (Environment Working Paper Number 21), World Bank, Washington, DC.
Bourguignon, Françoise and Christian Morrisson (2002) "Inequality among World Citizens, 1880–1992," *American Economic Review*, 92: 727–744.
Brainerd, Elizabeth and David M. Cutler (2005) "Autopsy on an Empire: Understanding Mortality in Russia and the Former Soviet Union," *Journal of Economic Perspectives*, 19: 107–130.
Broadman, Harry D. (2003) "Review of What's behind Foreign Investment in China?" *Finance and Development*, 40(4): 50.
Brown, Lester (1994) "How Could China Starve the World? Its Boom Is Consuming Global Food Supplies," *Washington Post*, 28 August, 708: C01.
Burslem, Chris (2004) "The Changing Face of Malnutrition," *IFPRI Forum*, October: 1, 9–12 (http://www.ifpri.org/org/pubs/newsletters/ifpriforum/if200410.htm).
Carr, Geoffrey (2003) "Climbing the Helical Staircase: A Survey of Biotechnology," *The Economist*, 29 March, 366: 1–24.
Carroll, Rory (2002) "Zambia Slams Door Shut on Genetically Engineered Food Aid," *The Guardian*, 30 October (http://www.guardian.co.uk/gmdebate/Story/0,2763,822141,00.html).
Chenery, Hollis B. and Moises Syrquin (1975) *Patterns of Development, 1950–1970*, London: Oxford University Press.
Cipolla, Carlo M. (1965) *The Economic History of World Population*, Baltimore, MD: Penguin Books.
Clark, Colin (1967) *Population Growth and Land Use*, London: Macmillan.
Coase, Ronald (1960) "The Problem of Social Cost," *Journal of Law and Economics*, 3: 1–44.
Cochrane, Willard (1965) *The City Man's Guide to the Farm Problem*, Minneapolis, MN: University of Minnesota Press.
Cochrane, Willard (1979) *The Development of U.S. Agriculture: A Historical Analysis*, Minneapolis, MN: University of Minnesota Press.
Codex Alimentarius Commission (2004) *FAO/WHO Food Standards: Codex Alimentarius* (http://www.codexalimentarius.net/web/index_en.jsp). Rome: U.N. Food and Agriculture Organization.
Collier, Paul and Jan Willem Gunning (1999) "Why has Africa Grown Slowly?" *Journal of Economic Perspectives*, 13: 23–40.

Conquest, Robert (1986) *The Harvest of Sorrow: Soviet Collectivization and the Terror-Famine*, Oxford: Oxford University Press.

Conway, Gordon (1997). *The Doubly Green Revolution: Food for All in the 21st Century*, London: Penguin Books.

Copenhagen Consensus (2004) *Copenhagen Consensus 2004: Today's Challenge – Tomorrow's Opportunity* (www.copenhagenconsensus.com).

Córdoba, José (2004) "Ravaged Colombia Sees Glint of Hope as Killings Fall Off," *Wall Street Journal*, 10 August, 616: A1 and A7.

Council for Economic Planning and Development (1999) *Taiwan Statistical Data Book*, Taipei, Republic of China: Council for Economic Planning and Development.

Craig, Barbara J., Klaus W. Deininger and Philip G. Pardey (1994) "An Embodiment Approach to Measuring Productivity Growth in U.S. Agriculture," Department of Agricultural and Applied Economics, University of Minnesota, St. Paul, MN.

Crook, Clive (2001) "A Survey of Globalization," *The Economist*, 29 September, 360: 1–32.

Cropper, Maureen and Charles Griffiths (1994) "The Interaction of Population Growth and Environmental Quality," *American Economic Review*, 84: 250–254.

Crosson, Pierre and Jock R. Anderson (1992) "Resources and Global Food Prospects: Supply and Demand for Cereals to 2030" (Technical Paper Number 184), World Bank, Washington, DC.

Cutler, David M., Edward L. Glaeser and Jesse M. Shapiro (2003) "Why Have Americans Become More Obese?" *Journal of Economic Perspectives*, 17: 93–118.

Dalrymple, Dana G. (1985) "The Development and Adoption of High-Yielding Varieties of Wheat and Rice in Developing Countries," *American Journal of Agricultural Economics*, 67: 1067–1073.

Dasgupta, Susmita, Benoit Laplante, Hua Wang and David Wheeler (2002) "Confronting the Environmental Kuznets Curve," *Journal of Economic Perspectives*, 16: 147–168.

Datt, Gaurav and Martin Ravallion (2002) "Is India's Economic Growth Leaving the Poor Behind?" *Journal of Economic Perspectives*, 16: 89–108.

Davis, Otto A. and Andrew B. Whinston (1965) "Piecemeal Policy in the Theory of Second Best," *Review of Economic Studies*, 34: 323–331.

Deevey, Edward S. (1960) "The Human Population," *Scientific American*, 203: 194–204.

Deininger, Klaus and Lyn Squire (1997) "Income Growth and Income Inequality," *Finance and Development*, 34(1): 38–41.

Delgado, Christopher L. (1998) "Africa's Changing Agricultural Development Strategies: Past and Present Paradigms as a Guide to the Future," *Brown Journal of World Affairs*, 5(1): 175–214.

Desai, Padma (2005) "Russian Retrospectives on Reforms from Yeltsin to Putin," *Journal of Economic Perspectives*, 19: 87–106.

de Soto, Hernando (2000) *The Mystery of Capital: Why Capitalism Triumphs in the West and Fails Everywhere Else*, New York: Basic Books.

de Vivo, G. (1987) "Ricardo, David," in John Eatwell, Murray Milgate and Peter Newman (eds.), *The New Palgrave: A Dictionary of Economics*. London: Macmillan Press.

Diamond, Jared (1997) *Guns, Germs, and Steel*, New York: W.W. Norton.

Diao, Xinshen, Agapi Somwaru and Terry Roe (2001) "A Global Analysis of Agricultural Trade Reform in WTO Member Countries," in *Background for Agricultural Trade Reform in the WTO: The Road Ahead*. Washington, DC: Economics Research Service, U.S. Department of Agriculture.

Dixon, John and Aidan Gulliver with David Gibbon (2001) *Farming Systems and Poverty: Improving Farmers' Livelihoods in a Changing World*. Rome: U.N. Food and Agriculture Organization.

Dixon, John and Kirk Hamilton (1996) "Expanding the Measure of Wealth," *Finance and Development*, 33(4): 15–17.

Drezner, Daniel (2004) "The Outsourcing Bogeyman," *Foreign Affairs*, 83: 22–34.

Duncan, Emma (2003) "Spoilt for Choice: A Survey of Food," *The Economist*, 13 December 369: 1–16.

Easterly, William (2001) *The Elusive Quest for Growth*, Cambridge: Massachusetts Institute of Technology Press.

Eberstadt, Nicholas (1995) "Population, Food, and Income: Global Trends in the Twentieth Century" in Ronald Bailey (ed.), *The True State of the Planet*. New York: The Free Press.

Eberstadt, Nicholas (2000) "World Depopulation," *Milken Institute Review*, 2: 37–48.

Economic Research Service of the U.S. Department of Agriculture (ERS-USDA) (2002) "Food Marketing and Price Spreads: USDA Marketing Bill." Washington,DC: ERS-USDA (http://www.ers.usda.gov/briefing/foodpricespreads/bill/index.htm).

ERS-USDA (2003) *International Food Consumption Patterns*. Washington, DC: ERS-USDA (http://www.ers.usda.gov/data/InternationalFoodDemand/).

Ehrlich, Paul R. (1968) *The Population Bomb*, San Francisco: Sierra Club.

Engerman, Stanley L. and Kenneth L. Sokoloff (1997) "Factor Endowments, Institutions, and Differential Paths of Growth among New World Economies: A View from Economic Historians of the United States" in Stephen Haber (ed.), *How Latin America Fell Behind: Essays on the Economic Histories of Brazil and Mexico*, Stanford: Stanford University Press.

Epplin, Francis and Joseph Musah (1987) "A Representative Farm Planning Model for Liberia," in *Proceedings of Liberian Agricultural Policy Seminar 1985* (Report B-23). Stillwater: Oklahoma State University Department of Agricultural Economics, Agricultural Policy Analysis Project.

European Bank for Reconstruction and Development (EBRD) (2002) *Transition Report: Agriculture and Rural Transition*, London: EBRD.

Fischer, Günther and Gerhard K. Heilig (1997) "Population Momentum and the Demand on Land and Water Resources," *Philosophical Transactions of the Royal Society, Series B*, 352: 869–889.

Fischer, Stanley (2001) "Exchange Rate Regimes: Is the Polar View Correct?" *Journal of Economic Perspectives*, 15: 3–24.

Food and Agriculture Organization of the United Nations (FAO) (1976, 1981, 1982, 1983, 1993, 2000, 2002A and 2003A) *FAO Production Yearbooks*. Rome: FAO.

FAO (1992) *Water for Sustainable Food Production and Rural Development – The UNSED Agenda: Targets and Cost Estimates*. Rome: FAO.

FAO (1996) *Data of Food and Agriculture 1996*. Rome: FAO.

FAO (1997) *Food Security: Some Macroeconomic Dimensions*. Rome: FAO.

FAO (2002B) *The State of Food Security in the World 2001*. Rome: FAO.

FAO (2003B) *The State of Food Insecurity in the World 2003*. Rome: FAO (ftp://ftp.fao.org/docrep/fao/006/j0083e/j0083e00.pdf).

FAO (2003C) *The State of the World's Forests 2003*. Rome: FAO (http://www.fao.org/DOCREP/005/Y7581E00.HTM).

FAO. (2004) *FAOSTAT Data*. Rome: FAO (http://faostat.fao.org).

Foster, Phillips and Howard D. Leathers (1999) *The World Food Problem*, Second Edition, Boulder: Lynne Rienner Publishers.

Fraga, Arminio (2004) "Latin America since the 1990s: Rising from the Sickbed?" *Journal of Economic Perspectives*, 18: 89–106.

Freedom House (2002) *Freedom in the World Country Rankings* (http://www.freedomhouse.org/ratings/index/).

Gardner, Bruce L. (2002) *American Agriculture in the Twentieth Century: How it Flourished and What it Cost*, Cambridge: Harvard University Press.

Gardner, Edward (2003) "Wanted: More Jobs," *Finance and Development*, 40(1): 18–21.

Gibson, Paul, John Wainio, Daniel Whitley and Mary Bohman (2001) "Profiles of Tariffs in Foreign Agricultural Markets" (Agricultural Economic Report Number 796), U.S. Department of Agriculture, Economic Research Service, Washington, DC.

Green, Edward C. (2005) "AIDS in Africa – A Betrayal," *The Weekly Standard*, 10(19): 27–29.

Griliches, Zvi (1958) "Research Costs and Social Returns: Hybrid Corn and Related Innovations," *Journal of Political Economy*, 66: 419–431.

Guriev, Sergei and Andrei Rachinsky (2005) "The Role of Oligarchs in Russian Capitalism," *Journal of Economic Perspectives*, 19: 131–150.

Hausler, Gerd (2002) "The Globalization of Finance," *Finance and Development*, 39(1): 10–12.

Hayami, Yujiro and Vernon W. Ruttan (1970) "Factor Prices and Technical Change in Agricultural Development: The United States and Japan, 1880–1960," *Journal of Political Economy*, 78: 1115–1141.

Hayami, Yujiro and Vernon W. Ruttan (1985) *Agricultural Development*, Revised Edition, Baltimore, MD: Johns Hopkins University Press.

Heath, John and Hans Binswanger (1996) "Natural Resource Degradation Effects of Poverty and Population Growth Are Largely Policy Induced: The Case of Colombia," *Environment and Development Economics*, 1: 65–84.

Henao, Julio and Carlos Baanante (1999) "Estimating Rates of Nutrient Depletion in Soils of Agricultural Lands of Africa," International Fertilizer Development Center, Muscle Shoals, AL.

Henneberry, Shida Rastegari and Luther Tweeten (1991) "A Review of International Agricultural Supply Response," *Journal of International Food and Agribusiness Marketing*, 2: 49–68.

Herbst, Jeffrey (2000) *States and Power in Africa: Comparative Lessons in Authority and Control*, Princeton, NJ: Princeton University Press.

Heritage Foundation (2002) *Index of Economic Freedom* (http://www.heritage.org/research/features/index/).

Hertel, Thomas, Bernard Hoekman and Will Martin (2002) "Developing Countries and a New Round of WTO Negotiations," *World Bank Research Observer*, 17: 113–140.

Holmquist, Frank (2003) "Kenya's Post-Election Euphoria and Reality," *Current History*, 102: 200–205.

Hopkins, Jeffrey and Mitchell Morehart (2002) "An Empirical Analysis of the Farm Problem: Comparability of Rates of Return," in Luther Tweeten and Stanley R. Thompson (eds.), *Agricultural Policy for the 21st Century*. Ames, IA: Iowa State University Press.

Huang, Jikun, Scout Rozelle and Mark W. Rosegrant (2000) "China's Food Economy to the Twenty-First Century: Supply, Demand, and Trade," *Economic Development and Cultural Change*, 47: 737–766.

Huffman, Wallace E. and Robert E. Evenson (1989) "Supply and Demand Functions for Multiproduct U.S. Cash Grain Farms: Biases Caused by Research and Other Policies," *American Journal of Agricultural Economics*, 71: 761–773.

Johnson, D. Gale (2000) "Population, Food, and Knowledge," *American Economic Review*, 90: 1–14.

Johnson, Glenn and C. Leroy Quance (1972) *The Overproduction Trap in U.S. Agriculture*, Baltimore, MD: Johns Hopkins University Press.

Johnston, Bruce F. (1966) "Agriculture and Economic Development," *Food Research Institute Studies*, 6: 251–312.

Johnston, Bruce F. and John W. Mellor (1961) "The Role of Agriculture in Economic Development," *American Economic Review*, 51: 566–593.

Kanwar, Ramesh S. (2003) "Water Quality and Agricultural Chemicals," in Rattan Lal, David Hansen, Norman Uphoff and Steven Slack (eds.), *Food Security and Environmental Quality in the Developing World*, Boca Raton, FL: Lewis Publishers.

Kerr, Richard A. (1998) "Acid Rain Control: Success on the Cheap," *Science*, 282: 1024–1027.

Kotkin, Stephen (2001) *Armageddon Averted: The Soviet Collapse, 1970–2000*, Oxford: Oxford University Press.

Kremer, Michael (2002) "Pharmaceuticals and the Developing World," *Journal of Economic Perspectives*, 16: 67–90.

Kuran, Timur (2004) "Why the Middle East is Economically Underdeveloped: Historical Mechanisms of Institutional Stagnation," *Journal of Economic Perspectives*, 18: 71–90.

Kurlantzick, Joshua (2002) "Asia Minor: Is China's Economic Boom a Myth?" *New Republic*, 16 December, 227: 20–25.

Kuznets, Simon (1955) Economic Growth and Income Inequality, *American Economic Review*, 45: 1–28.

Kuznets, Simon (1965) *Economic Growth and Structure*, New York: W.W. Norton.

Kynge, James (2002) "World Report – China," *The Financial Times*, 12 December, 598: I–IV.

Lamptey, Peter, Merywen Wigley, Dara Carr and Yvette Collymore (2002) "Facing the HIV/AIDS Pandemic," *Population Bulletin*, 57: 1–40.

Lakdawalla, Darius and Tomas Philipson (2002) "The Growth of Obesity and Technological Change: A Theoretical and Empirical Examination" (Working Paper W8946), National Bureau of Economic Research, Chicago, IL.

Landes, David S. (2000) "Culture Makes almost all the Difference," in Samuel Huntington and Lawrence Harrision (eds.), *Culture Matters: How Values Shape Human Progress*, New York: Basic Books.

Landes, David S. (1998) *The Wealth and Poverty of Nations*, New York: W.W. Norton.

Lee, Ronald (2003) "The Demographic Transition: Three Centuries of Fundamental Change," *Journal of Economic Perspectives*, 17: 167–190.

Lewis, Bernard (2002) *What Went Wrong? Western Impact and Middle Eastern Response*, New York: Oxford University Press.

Lloyd, Robert B. (2002) "Zimbabwe's Autocratic Democracy," *Current History*, 101: 219–224.

Lomborg, Bjørn (2001) *The Skeptical Environmentalist: Measuring the True State of the World*, Cambridge: Cambridge University Press.

Long, Simon (2004) "India's Shining Hopes: A Survey of India," *The Economist*, 21 February 370: 1–20.

Lora, Eduardo (2001) "Structural Reform in Latin America: What Has Been Reformed and How to Measure It" (Working Paper Number 466), Research Department, Inter-American Development Bank, Washington, DC.

Maddison, Angus (1995) *Monitoring the World Economy, 1820–1992*, Paris: Organization for Economic Cooperation and Development.

Maddison, Angus (2001) *The World Economy: A Millennial Perspective*, Paris: Organization of Economic Cooperation and Development.

Mahar, Dennis (1989) *Government Policies and Deforestation in Brazil's Amazon Region*, Washington, DC: World Bank.

Malthus, Thomas (1963) *An Essay on the Principle of Population*, Homewood, IL: R.D. Irwin.

Mankiw, N. Gregory (2001) *Principles of Economics*, Second Edition, Fort Worth, TX: Harcourt College Publishers.

Matras, Judah (1977) *Introduction to Population: A Sociological Approach*, Englewood Cliffs, NJ: Prentice-Hall, Incorporated.

McInerney, John (1976) "The Simple Analytics of Natural Resource Economics," *Journal of Agricultural Economics*, 27: 31–52.

McKinnon, Ronald I. (1991) *The Order of Market Liberalization: Financial Control in the Transition to a Market Economy*, Baltimore, MD: Johns Hopkins University Press.

Meadows, Donella H., Dennis L. Meadows, Jørgen Randers and William W. Behrens III (1972) *The Limits to Growth*, New York: Universe Books.

Mellor, Johns W. (ed.) (1995) *Agriculture on the Road to Industrialization*, Baltimore, MD: Johns Hopkins University Press.

Mitchell, Donald O., Merlinda D. Ingco and Ronald C. Duncan (1997) *The World Food Outlook*, Cambridge: Cambridge University Press.

Morgan, Dan (1979) *Merchants of Grain*, New York: Viking Press.

Naughton, Barry (1995) *Growing out of the Plan: Chinese Economic Reform, 1978–1993*, Cambridge: Cambridge University Press.

Ndulu, Benno and Stephen A. O'Connell (1999) "Governance and Growth in Sub-Saharan Africa," *Journal of Economic Perspectives*, 13: 41–66.

Neely, Christopher (1999) "How Big is Japan's Debt," *International Economic Trends* (http://research.stlouisfed.org/publications/iet/19990201/cover.pdf).

Nord, Mark, Margaret Andrews and Steven Carlson (2003), "Household Food Security in the United States, 2002" (FANRR-35), Economic Research Service, U.S. Department of Agriculture, Washington, DC.

Nordhaus, William and Joseph Boyer (2000) *Warming the World: Economic Models of Global Warming*, Cambridge: MIT Press.

North, Douglass (1973) *Institutions, Institutional Change, and Economic Performance*, Cambridge: Cambridge University Press.

Ocampo, José Antonio (2004) "Latin America's Growth and Equity Frustrations during Structural Reforms," *Journal of Economic Perspectives*, 18: 67–88.

Oldeman, L.R., R.T.A. Hakkeling and W.G. Sombroek (1991) *World Map of the Status of Human-Induced Soil Degradation: An Explanatory Note*, Second Revised Edition, Wageningen, NL: International Soil and Reference Information Centre.

Organization for Economic Cooperation and Development (OECD). *PSE/CSE Database* Paris: OECD (http://www.oecd.org).

O'Rourke, P.J. (1994) *All the Trouble in the World*, New York: Atlantic Monthly Press.

Pardey, Philip G. and Nienke M. Beintema (2001) *Slow Magic: Agricultural R&D a Century after Mendel*. Washington, DC: International Food Policy Research Institute.

Perkins, Dwight H., Steven Radelet, Donald R. Snodgrass, Malcolm Gillis and Michael Roemer (2001) *Economic Development*, Fifth Edition, New York: W.W. Norton.

Persaud, Suresh and Luther Tweeten (2002) "Impact of Agribusiness Market Power on Farmers," in Luther Tweeten and Stanley R. Thompson (eds.), *Agricultural Policy for the 21st Century*, Ames, IA: Iowa State University Press.

Pingali, Prabhu L., Cynthia B. Marquez, Agnes C. Rola and Florencia G. Palis (1995) "The Impact of Long-Term Pesticide Exposure on Farmer Health: A Medical and Economic Analysis in the Philippines," in Prabhu L Pingali and Pierre A. Roger (eds.), *Impact of Pesticides on Farmer Health and the Rice Environment*, Boston: Kluwer Academic Publishers.

Pinstrup-Andersen, Per (2002) *More Research and Better Policies Are Essential for Achieving the World Food Summit Goal*. Washington, DC: International Food Policy Research Institute.

Pollan, Michael (2002) "The Unnatural Idea of Animal Rights," *New York Times Magazine*, 10 November, 92: 58–64, 100–102.

Powell, David E. (2002) "Death as a Way of Life: Russia's Demographic Decline," *Current History*, 101: 344–348.

Pullen, J.M. (1987) "Malthus, Thomas Robert," in John Eatwell, Murray Milgate and Peter Newman (eds.), *The New Palgrave: A Dictionary of Economics*, London: Macmillan Press.

Rabalais, Nancy N., R. Eugene Turner, Dubravko Justić, Quay Dortsch and William J. Wiseman (1999) *Characterization of Hypoxia: Topic I Report for the Integrated Assessment on Hypoxia in the Gulf of Mexico*, Silver Spring, MD: National Oceanic and Atmospheric Administration.

Rashad, Inas and Michael Grossman (2004) "The Economics of Obesity," *The Public Interest*, 156 (Summer): 104–112.

Rask, Kolleen and Norman Rask (2004) "Reaching Turning Points in Economic Transition: Adjustments to Distortions in Resource-Based Consumption of Food," *Comparative Economic Studies*, 46: 542–569.

Rask, Norman (1991) "Dynamics of Self-Sufficiency and Income Growth," in Fred J. Ruppel and Earl D. Kellogg (eds.), *National and Regional Self-Sufficiency Goals: Implications for International Agriculture*, Boulder, CO: Lynne Rienner Publishers.

Rauch, Jonathan (2003) "Will Frankenfood Save the Planet?" *Atlantic Monthly*, October, 292: 103–108.

Reardon, Thomas and Julio A. Berdegué (2002) "The Rapid Rise of Supermarkets in Latin America: Challenges and Opportunities for Development," *Development Policy Review*, 20: 371–388.

Reed, Thomas C. (2004) *At the Abyss: An Insider's History of the Cold War*, New York: Ballantine Books.

Ricardo, David (1965) *The Principles of Political Economy and Taxation*, London: Dent and Sons, Limited.

Richards, R. Peter, David B. Baker and Donald J. Eckert (2002) "Trends in Agriculture in the LEASEQ Watersheds, 1975–1995," *Journal of Environmental Quality*, 31: 17–24.

Richardson, Stanley Dennis (1990) *Forests and Forestry in China: Changing Patterns of Resource Development*, Washington, DC: Island Press.

Righter, Rosemary (2002) "The Rot behind China's Bank Scandals," *The Times of London*, 12 February, 67372: 31.

Rosegrant, Mark, Mercedita Agcaoili-Sombilla and Nicostrada D. Perez (1995) "Global Food Projections to 2020: Implications for Investment" (Discussion Paper 5), International Policy Research Institute, Washington, DC.

Rosegrant, Mark, Ximing Cai and Sarah Cline (2002) *Global Water Outlook to 2025*, Washington, DC: International Food Policy Research Institute.

Roy, Amit H. (2003) "Fertilizer Needs to Enhance Production – Challenges Facing India," in Rattan Lal, David Hansen, Norman Uphoff and Steven Slack (eds.), *Food Security and Environmental Quality in the Developing World*, Boca Raton, FL: Lewis Publishers.

Rudel, Thomas (2001) "Did a Green Revolution Restore the Forests of the American South?" in Arild Angelsen and David Kaimowitz (eds.), *Agricultural Technologies and Tropical Deforestation*, Wallingford, CT: CABI Publishing.

Runge, C. Ford, Benjamin Senauer, Philip Pardey and Mark Rosegrant (2003) *Ending Hunger in our Lifetime: Food Security and Globalization*, Baltimore, MD: Johns Hopkins University Press.

Ruttan, Vernon W. (2002) "Productivity Growth in World Agriculture: Sources and Constraints," *Journal of Economic Perspectives*, 16: 161–184.

Sachs, Jeffrey (1997) "The Limits of Convergence," *The Economist*, 14–20 July, 344: 19–22.
Safavian, Mehnaz S., Douglas H. Graham and Claudio Gonzalez-Vega (2001) "Corruption and Microenterprises in Russia," *World Development*, 29: 1215–1224.
Sahn, David, Paul Dorosh and Stephen D. Younger (1997) *Structural Adjustment Reconsidered: Economic Policy and Poverty in Africa*, Cambridge: Cambridge University Press.
Salih, Mohamed Abdul Rahim M. (2001) *African Democracies and African Politics*, London: Pluto Press.
Samuelson, Paul A. (1969) "The Way of an Economist," in Paul A. Samuelson (ed.), *International Economic Relations: Proceedings of the Third Congress of the International Economic Association*, London: Macmillan.
Sanchez, Pedro A. and Terry J. Logan (1992) "Myths and Science about the Chemistry and Fertility of Soils in the Tropics," in Rattan Lal and Pedro A. Sanchez (eds.), *Myths and Science of Soils of the Tropics*, Madison, WI: Soil Science Society of America.
Schepf, Randall, Erik Dohlman and Christine Bolling (2001) "Agriculture in Brazil and Argentina: Developments and Prospects for Major Field Crops" (Agriculture and Trade Report WRS-01-3), Economic Research Service, U.S. Department of Agriculture, Washington, DC.
Scherr, Sara J. and Satya Yadav (1997) "Land Degradation in the Developing World: Issues and Policy Options for 2020" (2020 Brief Number 44), International Food Policy Research Institute, Washington, DC.
Schneider, Robert (1992) "Brazil: An Analysis of Environmental Problems in the Amazon" (Report 9104-Br), World Bank, Washington, DC.
Schneider, Robert (1995) "Government and the Economy on the Amazon Frontier" (Environment Paper 11), World Bank, Washington, DC.
Schultz, Theodore W. (1960) "Value of U.S. Farm Surpluses to Underdeveloped Countries," *Journal of Farm Economics*, 42: 1019–1030.
Schultz, Theodore W. (1961) "Investment in Human Capital," *American Economic Review*, 51: 1–17.
Schultz, Theodore W. (1964) *Transforming Traditional Agriculture*, New Haven, CT: Yale University Press.
Schweizer, Peter (1994) *Victory*, New York: Atlantic Monthly Press.
Sen, Amartya (1981) *Poverty and Famines: An Essay on Entitlement and Deprivation*, Oxford: Clarendon Press.
Sen, Amartya (2000) "Culture and Development" (mimeo), *World Bank Meeting*, Tokyo.
Senauer, Benjamin and Mona Sur (2001) "Ending Global Hunger in the 21st Century: Projections of the Number of Food Insecure People," *Review of Agricultural Economics*, 25: 68–81.
Short, Philip (1999) *Mao: A Life*, New York: Henry Holt.
Simpson, David, Roger Sedjo and John Reid (1996) "Valuing Biodiversity: An Application to Genetic Prospecting," *Journal of Political Economy*, 104: 163–185.
Singer, Max (2003) "Saudi Arabia's Overrated Oil Weapon," *The Weekly Standard*, 18 August, 8: 22–25.
Skidelsky, Robert (1996) *The Road from Serfdom: The Economic and Political Consequences of the End of Communism*, New York: Penguin Books.
Smith, Adam (1964) *The Wealth of Nations*, London: J.M. Dent and Sons.
Smith, Lisa and Lawrence Haddad (2000) "Overcoming Child Nutrition in Developing Countries" (Discussion Paper 30), International Food Policy Research Institute, Washington, DC.

Sohngen, Brent and Robert Mendelsohn (2003) "An Optimal Control Model of Forest Carbon Sequestration," *American Journal of Agricultural Economics*, 85: 448–457.

Solow, Robert (1957) "Technical Change and the Aggregate Production Function," *Review of Economics and Statistics*, 39: 312–320.

Song, Sora (2004) "We're Fat. Now What?" *Time*, 22 March, 163: 73.

Southgate, Douglas (1994) "The Causes of Tropical Deforestation and Agricultural Development in Latin America," in Katrina Brown and David W. Pearce (eds.), *The Causes of Tropical Deforestation: The Economic and Statistical Analysis of Factors Giving Rise to the Loss of Tropical Forests*, London: University College London Press.

Southgate, Douglas and Morris Whitaker (1994) *Economic Progress and the Environment: One Developing Country's Policy Crisis*, New York: Oxford University Press.

Sowell, Thomas (1998) *Conquests and Cultures*, New York: Basic Books.

Stelzer, Irwin M. (2005) "The Axis of Oil," *Weekly Standard*, 10(20): 25–28.

Stewart, Bobby, Rattan Lal and Samir El-Swaify (1991) "Sustaining the Resource Base of an Expanding World Agriculture," in Rattan Lal and Frans Pierce (eds.), *Soil Management for Sustainability*, Ankeny, IA: Soil and Water Conservation Society.

Stewart, Hayden, Noel Blisard, Sanjib Bhuyan and Rodolfo M. Nayga (2004) "The Demand for Food away from Home: Full Service or Fast Food?" (Agricultural Economic Report Number 829), Economic Research Service, U.S. Department of Agriculture, Washington, DC.

Tanzi, V. and L. Schuknecht (1995) "The Growth of Government and the Reform of the State in Industrial Countries" (Working Paper), International Monetary Fund, Washington, DC.

Tawney, R.H. (1966) "Religion and the Rise of Capitalism," in Stanley Cohen and Forrest Hill (eds.), *American Economic History*. New York: Lippincott.

Thurow, Lester (1992) *Head to Head: The Coming Economic Battle among Japan, Europe, and America*, New York: William Morrow.

Tietenberg, Thomas (2003) *Environmental and Natural Resource Economics*, Sixth Edition, Chicago, IL: Addison Wesley Publishing.

Timmer, C. Peter, Walter P. Falcon and Scott R. Pearson (1983) *Food Policy Analysis*, Baltimore, MD: Johns Hopkins University Press.

Tolstoy, Leo (1992) *Anna Karenina*, New York: Barnes and Noble.

Truth about Trade and Technology (2004) *The Truth about Organic Food*, Des Moines, IA (http://www.truthabouttrade.org/print.asp?id=2384).

Tweeten, Luther (1998) "Dodging a Malthusian Bullet in the 21st Century," *Agribusiness*, 14: 15–32.

Tweeten, Luther (1999) "The Economics of Global Food Security," *Review of Agricultural Economics*, 21: 473–488.

Tweeten, Luther and George Brinkman (1976) *Micropolitan Development*, Ames, IA: Iowa State University Press.

Tweeten, Luther and Jeffrey Hopkins (2003) "Economies of Size, Government Payments and Land Costs" in Charles Moss and Andrew Schmitz (eds), *Government Policy and Farmland Markets*, Ames, IA: Iowa State Press.

Tweeten, Luther and Donald McClelland (eds.). (1997) *Promoting Third-World Agricultural Development and Food Security*, Westport, CT: Praeger Publishers.

Tweeten, Luther and Mariah Tanner (2005) "Technological Choice and the Changing Structure of Agriculture: Farming to Sequester Carbon and Reduce Global Warming," in Rattan Lal (ed.), *Climate Change and Global Food Security*, Boca Raton, FL: CRC Press.

Tweeten, Luther and Stanley R. Thompson (eds.) (2002) *Agricultural Policy for the 21st Century*, Ames, IA: Iowa State University Press.

United Nations Development Programme (UNDP) (2002A) *Arab Human Development Report 2002: Creating Opportunities for Future Generations*. New York: United Nations Development Programme, Regional Bureau for Arab States.

UNDP (2002B) *Human Development Report 2002*. New York: Oxford University Press.

United Nations Population Division (UNPD) (1993) *World Population Prospects: The 1992 Revision*. New York: UNPD.

UNPD (2001) *World Population Prospects: The 2000 Revision, Volume 1 (Comprehensive Tables)*, New York: UNPD.

UNPD (2003) *World Population Prospects: The 2002 Revisions*, New York: UNPD (www.un.org/esa/population/unpop.htm).

U.S. Department of Agriculture (USDA) (2001) *Agricultural Outlook*, June/July: 58.

USDA (2002) *Agricultural Productivity in the United States*, Washington, DC: USDA Economic Research Service (http://www.ers.usda.gov/data/agproductivity).

U.S. Environmental Protection Agency (USEPA) (2000) *National Water Quality Inventory, 2000 Report*. Washington, DC: USEPA.

Unnevehr, Laurian J. (2004) "Mad Cows and Bt Potatoes: Global Public Goods in the Food System," *American Journal of Agricultural Economics*, 86: 1159–1166.

Van Lynden G.W.J. and L.R. Oldeman (1997) *The Assessment of the Status of Human-Induced Soil Degradation in South and Southeast Asia*, Wageningen, NL: ISRIC.

von Braun, Joachim (1989) "Commentary: Commercialization of Smallholder Agriculture – Policy Requirements for Capturing Gains for the Malnourished Poor," *International Food Policy Research Institute Report*, 11: 4.

Voortman, R.L., B.G.J.S. Sonneveld and M.A. Keyzer (2000) "African Land Ecology: Opportunities and Constraints for Agricultural Development" (Working Paper Number 37), Center for International Development, Harvard University, Cambridge, MA.

Watkins, Kevin and Joachim von Braun (2003) "Time to Stop Dumping on the World's Poor," International Food Policy Research Institute, Washington, DC.

Weber, Max (1930) *The Protestant Ethic and the Spirit of Capitalism*, London: Allen and Unwin.

Williams, Jeffrey C. and Brian D. Wright (1991) *Storage and Commodity Markets*, New York: Cambridge University Press.

Williamson, John (1990) "What Washington Means by Policy Reform," in John Williamson (ed.), *Latin American Adjustment: How Much Has Happened*? Washington, DC: Institute for International Economics.

Williamson, John (2000) "What Should the World Bank Think about the Washington Consensus?" *World Bank Research Observer*, 15: 251–264.

Wischmeier, Walter (1976) "Use and Misuse of the Universal Soil Loss Equation," *Journal of Soil and Water Conservation*, 31: 5–9.

Wolf, A.M. and G.A. Colditz (1994) "The Costs of Obesity: The U.S. Perspective," *PharmacoEconomics*, 5: 34–37.

Wood, Stanley, Kate Sebastian and Sara J. Scherr (2001) *Pilot Analysis of Global Ecosystems: Agroecosystems*. Washington, DC: International Food Policy Research Institute.

World Bank (2001A) *African Development Indicators 2001*, Washington, DC: World Bank.

World Bank (2000A) *Can Africa Claim the 21st Century?* Washington, DC: World Bank.

World Bank (2003A) *Global Economic Prospects: Realizing the Development Promise of the Doha Agenda*, Washington, DC: World Bank.

World Bank (1991) *Social Indicators of Development 1990*, Baltimore, MD: Johns Hopkins University Press.

World Bank (2001B, 2002A, 2003B) *World Development Indicators*, Washington, DC: World Bank.

World Bank (2004, 2005) *World Development Indicators – Online Version*. Washington, DC: World Bank (http://devdata.worldbank.org/dataonline/)*.

World Bank (1986, 1987, 1992, 2000B, 2000/2001, 2001C, 2002B, 2003C) *World Development Reports*. New York: Oxford University Press.

World Bank and U.N. Development Program (UNDP). (1990) *A Proposal for an Internationally Supported Programme to Enhance Research in Irrigation and Drainage Technology in Developing Countries*, Washington, DC: UNDP.

World Food Program (WFP) (2003) *INTERFAIS (International Food Aid Information System)*. Rome: WFP.

World Health Organization (WHO) (2000) *Obesity: Preventing and Managing the Global Epidemic* (Technical Report Series 894). Geneva: WHO.

WHO (n.d.) *Uganda Reverses the Tide of HIV/AIDS* (http://www.who.int/inf-new/aids2.htm).

World Resources Institute (WRI) (1992) *World Resources 1992–93*. Washington, DC: WRI.

Wright, Brian D. and Bruce L. Gardner (1995) *Reforming Agricultural Commodity Policy*, Washington, DC: AEI Press.

Yergin, Daniel (1991) *The Prize: The Epic Quest for Oil, Money, and Power*, New York: Simon and Schuster.

Yergin, Daniel and Joseph Stanislaw (1998) *The Commanding Heights: The Battle between Government and the Marketplace that Is Remaking the Modern World*, New York: Simon and Schuster.

Young, Alwyn (1995) "The Tyranny of Numbers: Confronting the Statistical Realities of the East Asian Growth Experience," *Quarterly Journal of Economics*, 110: 641–680.

Yousef, Tarik M. (2004) "Development, Growth, and Policy Reform in the Middle East and North Africa since 1950," *Journal of Economic Perspectives*, 18: 91–116.

---

*This is a proprietary website. Data reported in this source are also contained in the annual print version of *World Development Indicators*.

Index

ABC program
  in Uganda, 334
affluent nations, 29, 64, 139, 141, 206, 230, 361
  dietary change and consumption trends, 220–225
    obesity problem, 222–225
  food economy, 212–220
    agricultural subsidies and protectionism, 214–217
    production technology and output trends, 217–220
  population dynamics, 210–212
  standards of living, 207–210
    in Japan, 207, 209
    in Spain, 209
    in United States, 209–210
Africa, 21, 26, 46, 55, 56, 142, 146, 164, 175, 176, 203, 324, 355
  economic progress, foreign assistance in, 183
  farmers, 65–66, 80, 111, 113, 181
  fertilizer application rates, 111–112
  shifting cultivation, 46
  see also specific regions
agribusiness, 3, 47, 48–49, 365, 366
agricultural development
  and the environment, 119–123
  agribusiness, 3, 46–47, 48, 135–142, 214, 259
    gains, from freer trade, 139–142
    international trade negotiations, 137–139
  in affluent nations, 212–214, 217–220
  in Asia
    extensification and output growth, 242
    intensification and output growth, 239–242
    intensified production, 237–239
  in Eastern Europe and Former Soviet Union, 306–313
    factor use, 309–311
    yields and land use, 311–313
  for economic growth and diversification, 161–164
    Outward-Looking Strategy, 161, 162
    productivity growth, 163
  in Latin America and Caribbean, 258–265
    factor use, 261–262
    intensification and extensification 263–265
  in Middle East and North Africa, 281–288
    factor use, 282–284
    intensification and extensification, 284–288
  in Sub-Saharan Africa, 337–345
    factor use, 338–341
    intensification and extensification, 342–345
  subsidies and protectionism, 214–217
  in United States, 50–51, 52
agriculture
  and economic development, 148–165
  and environment, 119–123, 217
    deterioration, 109–119
    market failure, 102–109
    trade-offs, 99–102
  governmental intervention, in developing nations, 160–161
  and globalization, 124–146

growth and economic structure, 148–155
  standard of living and income distribution, 153–155
  structural transformation, diversity of, 151–153
and international trade negotiations, 137–139
nature of, 42–49
  fertilization, 43
  soils and climate, 43–45
  specialization and diversity, 45–49
product scarcity
  historical trends, 83–86
response, to demand growth
  input mix, changes in, 198–201
  production trends, 201–202
supply increments, 49
  extensification, 52–56
  intensification, 56–66
  TFP growth, 49, 50
yields and land use and output trends, 311–313
Albania, 297, 305
Alfisols, 44
Algeria, 21, 22, 25, 27, 149, 152, 154, 277, 278, 282, 284, 287
Anderson, Kym, 142
Angola, 169, 324, 328, 346
apparatchiks, 299
Argentina, 143–146, 250, 251, 253, 255, 257, 261, 266
Armenia, 300, 309, 313, 316
Asia, 21, 26, 46, 56, 142, 164, 176, 228
  agricultural development
    extensification and output growth, 242
    intensification and output growth, 239–242
    intensified production, 237–239
  consumption trends, 242–148
  dietary change, 242–248
  economic progress, foreign assistance in, 183
  energy requirement, minimal, 173n
  farmers, 80
  fertilizer application rates, 111
  food security, 242–248
  forest area, 117–118

GDP per capita trends, 228–231
  in China, 231–232
  in India, 232–233
Green Revolution, 59
land degradation, 111
mechanization, 238–239
Outward-Looking Strategy, 161–162
population dynamics, 233–237
  human fertility reduction, 234–235
  natural increase, 236–237
rice production, 45
water scarcity, 238
see also specific regions
Australia, 46, 51, 209, 210, 212, 214, 217, 219, 222, 226, 261
  comparative advantage, in agriculture, 214–215, 219
Azerbaijan, 299, 309

Bacillus thuringiensis (Bt) cotton, 120
Bahrain, 280
Bangladesh, 21, 22, 24, 27, 60, 149, 152, 154, 228, 238, 243
Belarus, 299
Benin, 324, 346
Bhopal disaster, 132
Biotechnology, agricultural, 3, 4, 64, 65, 67, 87, 181, 227, 359
  environmental benefits of, 64, 120, 359
  see also genetically modified organisms
birth control, 15, 22–24, 257, 333
Bloom, Stephen, 361
body mass index (BMI), 222, 361
Bolivia, 149, 152, 255, 259, 322
Borlaug, Norman, 59
Botswana, 193, 325, 328, 333, 341, 345
Bovine Spongiform Encephalopathy (BSE), 368
Brazil, 61, 115, 118, 250, 251, 253, 254, 255, 262, 265
Bread for the World, 166
Brown, Lester, 66n
Bulgaria, 303
Burkina Faso, 333, 342, 346, 354
Burundi, 118, 322, 346
bush-fallow farming, *see* shifting cultivation

Cambodia, 238, 242
Cameroon, 118, 342
Canada, 51, 215, 217, 219, 253, 261, 368
　comparative advantage, in agriculture, 215, 219
cap and trade, 108
capital formation, 50, 185, 207–209
Caribbean, 189, 190, 203, 250
　agricultural development, 258–265
　　factor use, 261–262
　　intensification and extensification, 263–265
　consumption trends, 265–271
　dietary change, 265–271
　food security, 265–271
　GDP per capita trends, 251–254
　population dynamics, 254–258
　　human fertility reduction, 255–257
　　natural increase, 257–258
Casey, William, 277n
Central African Republic, 342
Chad, 328, 346, 354
chemical fertilizers, 43
Chenery, Hollis, 153
Chile, 24, 32, 150, 151–152, 251, 254, 255, 259, 261, 262
China, 117–118, 169n, 228, 234, 237, 242, 243, 302, 361, 363
　GDP percapita trends, 231–232
　Household Responsibility System (HRS), 159, 231, 238–239, 246
Chávez, Hugo, 254
Clean Air Act, 108
Club of Rome, 1, 13–14, 85
Coase, Ronald, 106
Codex Alimentarius, 368
collectivization, 159, 231, 303, 307, 309
Colombia, 116, 230, 253, 254, 255, 261, 265
commodity programs, 366–367
　market impacts, 78–83
communism, 75, 158–159, 190, 295–297, 300–301
　and agriculture, 158–160
communitarian values
　in standard model, 183–185
　　economic equity, 184–185
comparative advantage
　illustration, 143–146

　net benefits of trade, distribution of, 126–128
　theory, 8, 125–128, 130n
competitive equilibrium, 76–78, 94
concentrated animal feeding operation (CAFO), 48
Congo, Republic of, 324, 345
Consultative Group on International Agricultural Research (CGIAR), 59, 369
consumers' surplus (CS), 76–77, 96–97, 127, 128, 129–130, 172
consumption and production, aligning, 74, 90–91
　agricultural product scarcity
　　historical trends, 83–86
　commodity programs, market impact, 78–83
　competitive equilibrium, 76–78
　decentralized decision-making, 89–97
　　demand and supply, shifts, 91–93
　　NEV and maximization, 94–97
　outlook, for twenty-first century, 86–89
Conway, Gordon, 359
Copenhagen Consensus, 358–359
Costa Rica, 262, 263, 266, 270
Croatia, 297
crude birth rate (CBR), 17–18, 20, 21, 22, 196, 210, 236, 257, 280, 329, 334
crude death rate (CDR), 17–18, 20, 21, 22, 196, 236, 257, 280, 305, 329
currency over-valuation, 80, 130–132, 156, 161, 180, 338
Czech Republic, 297, 316

decentralized decision-making, 89–97
　demand and supply, shifts, 91–93
　NEV and maximization, 94–97
　production and consumption alignment, 90–91
decoupling, of farm payments, 82
Deflation, 179
deforestation, 56
　in Brazilian Amazon
　　intervention failure, 115
　　Environmental Kuznets Curve, 121–122
　global warming, 116

in Latin America, 119, 199
in Sub-Saharan Africa, 118
demand for food, 10
 demographic transition, 16–25
  in developing world, 19–22
  human fertility revolution, 22–25
  human population trends, 25–28
 elasticity, 37, 38, 79, 80–81, 92, 93, 96, 104, 108
 food consumption and income, 28–31
 fundamental economics, 34–40
  changes, in demand, 37–40
  own-price elasticity, 36–37
 Malthusianism, 11–16
  classical view, 11–13
  criticism, 15–16
  neo-Malthusianism, 13–15
 shifts, 91–93
 trends and projections, 31–33
Democratic Republic of the Congo (DRC), 112, 118, 322–323, 324–325, 346
demographic transition, 16–25
 crude birth rate (CBR), 17–18, 20, 21, 22, 196, 210
 crude death rate (CDR), 17–18, 20, 21, 22, 196, 236, 257
 in developing world
  modern transition, 19–22
 in Eastern Europe and Former Soviet Union, 302–306
  human fertility, 303–305
  natural increase, 305–306
 human fertility revolution, 22–25
 human population trends, 25–28
 in Sub-Saharan Africa
  human fertility, 329–333
  natural increase, 334–337
 total fertility rate (TFR), 18, 22, 24, 25, 27
 see also population dynamics
Denmark, 114n, 221
developed world, *see* affluent nations
developing world
 agricultural extensification, 54, 55, 284–288
 agricultural intensification, 56, 58, 61, 284–288
 agricultural protectionism, 139
 deforestation, 118–119
 demographic transition, 19–22
 fertilizer application rates, 111–112, 338
 food insecurity, 167–170, 203
 foreign exchange earnings, 156
 foreign exchange rates, 180
 governmental intervention, 160–161
 modern transition, 19–22
 population growth, 198
 rural population, 199
 taxation, 180
Dhaka Declaration, 138
dietary change
 and consumption trends, in affluent nations, 220–225
  obesity problem, 222–225
  Rask's methodology, 221
 in Asia, 242
 in Eastern Europe and Former Soviet Union, 313
 in Latin America and Caribbean, 265
 in Middle East and North Africa, 288
dietary energy requirements, 173n
diminishing marginal utility, 36, 77
Doha round of trade negotiations, 138, 141
Dominican Republic, 263
Doubly Green Revolution, 359
Dust Bowl, 113n
Dutch disease, 189, 282, 293, 354
dynamic benefits of liberalized trade, 140–141

East Asia, 111, 199
Eastern Europe, 155, 173, 189–190, 191–192, 194, 195, 203, 295
 agricultural sector, 306–313
  factor use, 309–311
  yields and land use, 311–313
 consumption trends, 313–316
 demographic trends, 302–306
  human fertility, 303–305
  natural increase, 305–306
 dietary change, 313
 economic growth patterns, 296–302
 food security, 316–318
Eberstadt, Nicholas, 19
economic convergence, 209
 hypothesis, 190–191

economic diversification, *see* structural change
economic growth, 3, 13, 75, 121, 155–156, 157, 161–164, 183, 184, 228, 246, 251, 360
  and food prices, 173–175
  income distribution, 187–191
    differences and economic convergence, 190–191
    GDP per capita, regional trends, 189–190
  patterns
    in Eastern Europe and Former Soviet Union, 296–302
  and structure, 148–155
    standard of living and income distribution, 153–155
    structural transformation, diversity of, 151–153
Ecuador, 21, 22, 27, 98, 149, 152, 180, 262
efficiency, 76–78, 94, 96–97
  impeded by monopoly, 78, 95
  impeded by market failure, 78, 103–105
Egypt, 61, 280, 282, 284, 285, 291
Ehrlich, Paul, 1, 15n, 228
El Salvador, 180, 257, 262, 270
elasticity
  demand 37, 38, 79, 80–81, 92, 93, 96, 104, 108
  income, 29–31, 32, 39–40, 121, 148, 243, 346, 365
  own-price, 29–30, 36–37, 39
  supply, 70, 72, 79, 92, 93, 96
Engel's Law, 30, 121, 148, 150, 365
England, 20
Entisols, 44
environment and economy
  linkages and trade-offs, 99–102
environmental deterioration, 120
  in absence of agricultural intensification, 109
    agricultural extensification, 114–119
    land degradation, 110–114
Environmental Kuznets Curve (EKC), 121–122, 153
epidemiological troubles, in Sub-Saharan Africa, 329
Eritrea, 325, 346

Estonia, 299, 307
Ethiopia, 61–62, 323, 324–325, 328, 333, 345, 346
European Bank for Reconstruction and Development (EBRD), 307
European Union (EU), 129, 138–139, 141–142, 215, 217, 226, 367, 368
export cropping, 133–134
extensification, 41, 60, 61, 62, 199, 238, 259, 359
  agricultural supply, 52–56
  in Asia, 239–242
  environmental deterioration, 114–119
  in Latin America and Caribbean, 263–265
  in Middle East and North Africa, 284–288
  in Sub-Saharan Africa, 342–345
extension, agricultural, 49

factory farming, 48
fair trade, 134–135, 368
fertilization, 43, 53, 110–111, 113, 120, 140, 201, 240, 282, 284
fertilizer, 59, 66, 72, 162–163, 198, 199, 287, 311, 338–341
  application rates, 60, 61, 62, 111, 199–201, 217, 237, 239, 261, 262, 284
  chemical fertilizers, 43
  commercial fertilizer, 45, 52, 80, 112, 114, 120, 342
  subsidies, 111, 287
food aid, 170, 171–173, 181, 202, 246
Food and Agriculture Organization (FAO) of the United Nations, 1, 221, 288, 346, 368
  declaration on food security, 166
food consumption, 6, 10, 74, 87, 166, 213, 357, 365–366
  and income, 28–31
  Rask's methodology, 221
  trends
    in affluent nations, 220–222
    in Asia, 243–248
    in Eastern Europe and Former Soviet Union, 313–318
    in Latin America and Caribbean, 265–271

in Middle East and North Africa,
    288–293
in Sub-Saharan Africa, 345–354
food crisis, of 1970
    lessons, 84–85
food scarcity, 4, 6, 41, 67, 98, 163, 166, 360
    historical trends, 66–67, 83–86
    impacts, 5
food security, 4, 8, 57, 134, 166, 357, 359
    achievement, 170–175
        economic growth and prices,
            173–175
        food aid, 171–173
    in Asia, 242–248
    in Eastern Europe and Former Soviet
        Union, 313–318
    FAO's declaration, 166
    insecurity, 167–170
    in Latin America and Caribbean,
        265–271
    in Middle East and North Africa, 168,
        169, 288–293
    in Sub-Saharan Africa, 345–354
    synthesis and economic development
        seven-step logical framework,
            176–177
        standard model, 178–183
food self-sufficiency, 176n, 244–245
    and export cropping, 133–134
    in OECD, 222
Ford Foundation, 59
foreign aid, 156, 182, 183, 323
Former Soviet Union, 159, 190, 201, 203,
    231, 325
    agricultural sector, 306–313
        factor use, 309–311
        yields and land use, 311–313
    consumption trends, 313–318
    demographic trends, 302–306
        human fertility, 303–305
        natural increase, 305–306
    dietary change, 313–318
    economic growth patterns, 296–302
        Russian case study, 300–302
    food security, 316–318
France, 209, 280
free trade, 179, 215, 217, 226, 251
    gains, 139–142

versus fair trade, 134–135
see also liberalization
fundamental concepts
    demand, 34–40
        demand elasticity, 37–40
        own-price elasticity, 36–37
    market equilibrium and efficiency,
        76–78, 89–97
    supply, 49, 68–73
        agricultural supply, 72–73
        changes, 71–72
        elasticity, 70

Gabon, 330, 342
Gaidar, Yegor, 300
Gambia, 134
GDP per capita trends
    in affluent nations, 207–210
    in Asia, 228–233
        China, 231–232
        India, 232–233
    economic progress, foreign assistance in,
        183
    farmers, 65–66, 80
    in Latin America and Caribbean,
        251–254
    in Middle East and North Africa,
        274–278
    regional trends, 189–190
        Dutch disease, 189
    in Sub-Saharan Africa, 324–328
    see also standards of living
General Agreement on Tariffs and Trade
    (GATT), 136, 137, 138, 139
genetically modified organisms (GMOs)
    environmental benefits of, 64
    public acceptance of, 6, 67, 213, 213n
    see also biotechnology, agricultural
Georgia, 299, 305, 309, 313
Germany, 209, 210, 215, 280
Ghana, 112, 323, 329, 338, 339, 342, 345, 346
Gini coefficient, 154, 190–191
    in affluent countries, 190, 209
    in Asia, 190, 246–248
    in Latin America and the Caribbean,
        190, 270
    in the Middle East and North Africa,
        190, 291

Gini coefficient (*cont'd*)
  in Eastern Europe and Former Soviet Union, 190, 316–318
  in Sub-Saharan Africa, 190, 354
  see also income distribution and inequality
global food economy, in twenty-first century, 357
  freer international trade, 358
  government, changing role of, 366–369
  new food economy, 364–366
Global Land Assessment of Degradation (GLASOD) methodology, 110, 120
globalization and agriculture, 124
  agricultural trade, 135–142
    freer trade, gains, 139–142
    international trade negotiations, 137–139
  comparative advantage, 125–128, 130n, 135, 143–146
    net benefits of trade, distribution of, 126–128
  export cropping, 133–134
  food self-sufficiency, 133–134, 176n, 222, 246, 270
  free trade versus fair trade, 134–135
  multinational firms, 132–133
  trade distortions, net costs of, 128–132
government's role
  for global food economy, 366–369
    commodity programs, 366–367
  in public goods, 180–181
Great Depression, 135, 137, 179n, 209, 250
Green Revolution, 58–59, 62, 64, 65, 85, 111, 151, 232, 248, 357
  Doubly Green Revolution, 359
Group of Twenty, 125
Guatemala, 128, 134, 255, 257, 259, 270
Guinea, 324
Gulf of Mexico, 102–103

Haiti, 257, 262, 266, 270
Hatch Act of 1887, 49–50
health-care cost, 225
  government's role, 363
  obesity, 363
health cost, 167, 168

Hitler, Adolf, 26, 299n
HIV/AIDS, 333, 334
  in Central Africa, 392
  in Nigeria, 334
  in Southern Africa, 330
  in Sub-Saharan Africa, 198, 303, 329, 333, 334, 335–336
Honduras, 255, 270
Hong Kong, 157, 162, 206, 230, 234, 243
Household Responsibility System (HRS), 159, 238–239
  in China, 231
human capital
  formation, 3, 50, 151, 164, 185, 194, 202, 274
  regional indicators, 195
human fertility
  affluent nations, 210–212
  in Asia, 234–235
  in Caribbean, 255–257
  demographic transition, 22–25
  in Eastern Europe, 303–305
  female empowerment, 24, 25, 210, 280
  in Former Soviet Union, 303–305
  in Latin America, 255–257
  in Middle East, 278–280
  in North Africa, 278–280
  regional trends, 193–196
  standards of living, 24, 32, 210, 234, 255, 278–280
  in Sub-Saharan Africa, 329–333
human longevity increase and population dynamics, 16, 19, 191–193
Hungary, 297, 307, 311, 316
hunger, *see* food security
Hussein, Saddam, 275, 277

Import-Substituting Industrialization (ISI), 160, 161, 162, 250
Inceptisols, 44
income distribution and inequality, 153–155, 175, 190–191
  in affluent countries, 190, 209–210
  in Asia, 190, 246–248
  in Latin America and the Caribbean, 190, 270
  in the Middle East and North Africa, 190, 291–293

in Eastern Europe and Former Soviet
   Union, 190, 316–318
in Sub-Saharan Africa, 190, 354
see also Gini coefficient
income elasticity, 32, 121, 148, 243, 346, 365
   of food demand, 29–31, 39–40
India, 44, 60, 114, 120n, 132, 191, 228, 230,
   232–233, 246n
   Bhopal disaster, 132
   fertilizer subsidies, 111
   GDP per capita trends, 232–233
      Green Revolution, 232
   human longevity, 191
   income inequality, 155
   intervention failure, 120n
   liberalization, 233
Indonesia, 117, 228–230, 242, 243, 270
infant mortality, 16, 18, 24, 25, 191
   in Asia, 246
   in Latin America and the Caribbean,
      270
   in the Middle East and North Africa, 291
   in Eastern Europe and Former Soviet
      Union, 316
   in Sub-Saharan Africa, 354
inflation, 131, 179, 207, 209, 230, 251, 253,
   318
input mix, changes, 198–201
intensification, 52, 66
   in affluent nations, 217–220
   in Asia, 239–242
   environmental impacts of, 62–64,
      109–119, 120
   impacts on agricultural supply, 56–62,
      62–66
   in Latin America and Caribbean,
      263–265
   in Middle East and North Africa,
      284–288
   in Sub-Saharan Africa, 342–345
International Food Policy Research
   Institute (IFPRI), 134, 166
International Maize and Wheat
   Improvement Center (CIMMYT), 59
International Monetary Fund (IMF), 136n,
   178, 182, 251, 325
International Obesity Task Force (IOTF),
   363

International Rice Research Institute
   (IRRI), 59
intervention failure, 99, 111, 115, 120n
Iran, 61, 277, 280, 284, 285, 291
Iraq, 277, 282, 284, 285n, 287, 291
irrigation, 43, 44, 63, 64, 98, 110–111, 120,
   238, 239, 284, 285, 309, 341
Islamic regions, 273, 280
   see also individual entries
Israel, 189, 274, 275, 284, 287
Italy, 209, 210, 219, 221
Ivory Coast, 112, 329, 338, 339, 345

Jamaica, 253, 262
Japan, 51, 53, 58, 129–130, 177, 207, 209,
   215, 217, 219, 221, 368
   financial crisis, 207
   social mobility, 209
John Paul II, 297
Johnson, D. Gale, 15
Jordan, 277, 287, 291

Kazakhstan, 300
Kenya, 134, 329, 338, 345
Kuwait, 277, 281, 282, 287
Kuznets Curve, 121–122, 153–154
Kyrgyz Republic, 305, 313

land degradation, 109–110, 110–114, 119,
   120, 338
Laos, 21, 24, 149, 150, 151–152, 242
Latin America, 21, 25, 26, 46, 55, 56, 66,
   137, 142, 164, 176, 168, 250
   agricultural development, 258–265
      factor use, 261–262
      intensification and extensification,
         263–265
   GDP per capita trends, 251–254
   population dynamics, 254–258
      human fertility reduction, 255–257
      natural increase, 257–258
Latvia, 299, 305
Lebanon, 275, 277, 287, 291
Lenin, Vladimir, 158, 160
Lesotho, 333, 345
Levant, 284
liberalization
   in China, 231

liberalization (cont'd)
  in Eastern Europe, 307
  in Former Soviet Union, 307
  in India, 233
  in OECD countries, benefits, 140–141, 217
  trade policy
    standard model, 180
  see also free trade
Liberia, 323, 325
Libya, 112, 277, 282, 284, 287, 291
Lithuania, 299, 316
livestock, 29n, 43, 114
  water pollution, 43
Lomborg, Bjorn, 66n

Macedonia, 297, 299, 305
Macondo, 130–131
macroeconomic policies, 179–180, 189
  in affluent countries, 207–209
  in Asia, 230
  in Latin America and the Caribbean, 251–254
  in the Middle East and North Africa, 274
  in Eastern Europe and Former Soviet Union, 297–301
  in Sub-Saharan Africa, 324–325
Madagascar, 324, 341, 342
Mahgreb, 278, 280, 287
Malawi, 345, 346
Malaysia, 162, 234, 236, 238, 239, 240, 242, 243
Mali, 113, 345, 346, 354
malnutrition, see food security
Malthusianism, 11–16
  erroneous predictions of, 15–16
  modern variants of, 13–15
  principle of population, 11–13
  see also Club of Rome, Paul Ehrlich
Mandela, Nelson, 323
Mankiw, Gregory, 125, 150
market economy
  institutional framework, 178–179
  tendency toward competitive equilibrium, 76, 89–91
market failure, 7, 102, 181
  correction, 105–109
  non-internalization and inefficiency, 103–105

Marxism, 61–62, 75, 182, 302–303, 318, 319, 321, 325, 355
Mauritania, 341, 342, 345, 346
Mauritius, 112, 325, 330, 341, 342, 345, 346
mechanization, 199
  in Asia, 238–239
  Sub-Saharan Africa, 341
Mendel's genetics, 52
Mexico, 59, 250, 253, 254, 262, 266
Middle East
  agricultural development, 281–288
    factor use, 282–284
    intensification and extensification, 284–288
  consumption trends, 288–293
  dietary change, 288–293
  food security, 288–293
  GDP per capita trends, 274–278
  population dynamics, 278
    human fertility, 278–280
    natural increase, 280–281
minifundios, 259, 262
Mobutu Sese Seko, 322–323
Moldova, 299, 316
Mollisols, 44
Mongolia, 230, 238
monopoly, 78, 95, 179, 366
Morocco, 277, 278, 284, 288
Morrill Act, 49, 50
mortality, 13, 20, 22, 25, 44, 93, 94
  infant, 16, 246, 270
Mozambique, 143–146, 169, 324, 325, 333, 342, 346
Mugabe, Robert, 328
multinational firms, 132–133
Myanmar, 117, 236, 243

Namibia, 325, 329
Nepal, 230, 242, 243
net economic value (NEV), 76, 78, 104–105
  maximization, 94–97
  and trade, 126–128
The Netherlands, 219, 221, 280
New Economic Policy, of the early Soviet Union, 158–159, 160, 160n
new food economy, 364–366
New Zealand, 51, 207, 210, 215, 217–219, 222, 226, 261

Nicaragua, 255, 270
Niger, 329–330
Nigeria, 61, 101, 112, 322, 323–324, 334, 346
nomenklatura, 296, 299, 301
nominal protection coefficient (NPC), 214–217
North Africa, 54, 273
  agricultural development, 281–288
    extensification, 54, 284–288
    factor use, 282–284
    intensification, 56, 61, 237–238, 284–288
  consumption trends, 288–293
  dietary change, 288–293
  food security, 288–293
  GDP per capita trends, 274–278
  population dynamics, 278
    human fertility, 278–280
    natural increase, 280–281
North America, 212, 226
North American Free Trade Agreement (NAFTA), 253
North, Douglass, 106
Norway, 215

obesity, 360–364
  International Obesity Task Force (IOTF), 363
  overeating, 9, 202, 225
  reduction measures, 363
  repercussions, 363
  in rich countries, 222–225
Oman, 277, 280, 287
opportunity cost, 29, 100, 102, 126, 146, 364
Organization for Economic cooperation and Development (OECD), 139, 207, 208, 215
  food self-sufficiency, 222
  liberalization, 140–141, 217
Organization of Petroleum Exporting Countries (OPEC), 274, 277, 278
Ottoman, 273, 275
Outward-Looking Strategy, 161, 162
own-price elasticity, 29–30, 36–37, 38, 70
Oxisols, 44
O'Rourke, P.J., 24

Pakistan, 44, 60, 228, 238, 242, 243, 246
Panama, 180, 262, 263
Paraguay, 255
per-capita production trends, in agriculture, 66–67
Peru, 251, 254, 259, 262
Philippines, 62, 134, 234, 236
Poland, 150, 215, 297, 303, 311
polluter pays principle, 106–107
pollution, 105, 106–107, 122, 132
  agriculture, 102–103
  impacts, 109
  industrial and municipal, 63
  tax, 107–108
  water, 43, 120, 121, 367
population dynamics
  affluent nations
    human fertility, 210–212
  in Asia, 233–237
    human fertility reduction, 234–235
    natural increase, 236–237
  in Latin America and Caribbean, 254–258
    human fertility reduction, 255–257
    natural increase, 257–258
  in Middle East and North Africa
    human fertility, 278–280
    natural increase of, 280–281
  regional trends, in global food economy
    human fertility reduction, 193–196
    human longevity increase, 191–193
    natural increase, 196–198
  see also demographic transition
positive-sum game, 126, 145
poverty, 151, 176, 194
  in Asia, 246–248
  in Latin America and the Caribbean, 270
  in the Middle East and North Africa, 291
  in Eastern Europe and Former Soviet Union, 316
  in Sub-Saharan Africa, 354
poverty trap, 182
precautionary principle, 213
price control, 80
price distortions, 309
price signals, 75, 98
price stability, 79, 179, 180
price support, 81–82, 130

principle of population, 11–13
  human behavior, 11
producer subsidy equivalent (PSE), 214–215, 307
producers' surplus (PS), 76, 78, 96–97, 127–128, 129–130, 172
production possibilities frontier (PPF), 144
productivity growth, in agriculture, 163
protectionism, 125, 128–130, 136, 139, 214–217, 226
public goods, 47–48, 64–65, 156, 180–181
  TFP growth, in US agriculture, 49
public policy, 79
purchasing power parity (PPP), 24n, 154
Putin, Vladimir, 300, 301–302

Qatar, 280, 287

Rask's methodology for food consumption analysis, 221
Reagan, Ronald, 207, 277n
regional trends, synopsis
  agriculture's response, to demand growth
    input mix, changes, 198–201
    production trends, 201–202
  economic growth and income distribution, 187–191
    GDP per capita, 189–190
    income distribution differences, 190–191
  population dynamics
    human fertility reduction, 193–196
    human longevity increase, 191–193
    natural increase, 196–198
research and development, agricultural, 2, 121, 359
  economic returns of, 51–52
  funding of, 65, 181, 359
  in Japan, 58
  in United States, 49–51
  role of government, 3, 47, 64–65
  role of private sector, 3, 64
  see also biotechnology, Green Revolution
resource scarcity, 106, 164
Ricardo, David, 11, 82, 125, 130n, 358
Rockefeller Foundation, 59, 359
Romania, 309

Rosen, Joseph D., 213
Russia, 26, 193, 231, 296, 300–302, 303, 309, 316, 318, 319
  flat income tax, 301
  see also Former Soviet Union
Rwanda, 113, 118, 134, 322, 328, 345, 346

Samuelson, Paul, 125
Saudi Arabia, 277, 278, 280, 281, 287, 288, 291, 300n
Scandinavia, 20
Schroeder, Gerhard, 302
Schultz, Theodore W., 50
Sen, Amartya, 167
Senegal, 345
shifting cultivation, 46, 119
Sierra Leone, 90, 169, 323, 325, 345, 346, 354
Singapore, 157, 162, 206, 228–230, 230, 231, 234
slash-and-burn farming, see shifting cultivation
Slovenia, 297, 313, 316
Smith, Adam, 124, 358
Smith–Lever Act of 1914, 50
social mobility, 209
Solow, Robert, 49
Somalia, 323, 324
South Africa, 112, 323, 325, 329, 334, 345, 346
South America, 55, 118, 146, 259, 322
South Asia, 54, 60, 139, 173n, 189, 190, 198, 234, 354
South Korea, 29, 162, 164, 230, 231, 234, 239, 242, 243
Southern Mexico, 259
Soviet Union, see Former Soviet Union
Spain, 209, 210–211, 219, 221
specialization and diversity, in agriculture, 45–49
  economic forces and public policies, 45
  environmental variations, 45, 46
Sri Lanka, 238, 243, 246
state-owned enterprises (SOEs), 177, 231–232, 233, 251
Stalin, J., 159, 231, 303, 307
standard model
  communitarian values and economic equity, 183–185

liberal trade policy, 180
macroeconomic policies, 179–180
market economy, institutional
  framework, 178–179
public goods, government's role,
  180–181
standards of living, 31–32, 74, 230,
  253–254, 277, 295, 333, 367
  and economic structure, 149
  and environmental quality, 121
  and income distribution, 153–155
  in Japan
    financial crisis, 207
    social mobility, 209
  in Spain, 209
  in United States
    economic inequality, 209–210
    GDP per capita growth, 209
    social mobility, 209
    see also GDP per capita trends
static resource allocation gains of
  liberalized trade, 140
structural adjustment, 251, 338
structural change, 148–153, 157
  see also Import-Substituting
    Industrialization, Outward-Looking
    Strategy
Sub-Saharan Africa, 169, 321
  agricultural development, 337–345
    factor use, 338–341
    intensification and extensification,
      342–345
  consumption trends and food security,
    345–354
  demographic trends, 329
    human fertility, 329–333
    natural increase, 334–337
  GDP per capita trends, 324–328
  mechanization, 341
subsidies, 115, 131–132, 313, 367
  agricultural subsidies, 80, 293
    and protectionism, 214–217, 367
  communist-era subsidies, 309
  exports, 130, 138, 142, 143, 171
  farm, 103, 121, 137, 139, 217, 226, 282
  fertilizers, 111
  irrigation, 99
    environmental impacts, adverse, 288

production, 319
PSE, 214
removal of, 311
underwriting, 182, 287
Sudan, 345, 355
supply of food, 41
  agriculture
    increase, 49, 201–202
    nature, 42–49
    per-capita production trends, 66–67
  fundamental economics, 68–73
    changes, 71–72
    elasticity, 70, 72, 79, 92, 93, 96
    real and imagined characteristics,
      72–73
  management, 80–81
  shifts, 91–93
Swaziland, 325, 341, 345
Sweden, 18, 30, 149, 150, 151, 153, 155, 184,
  219, 222, 225
Switzerland, 215
Syria, 277, 284, 285n, 287

Taiwan, 162, 230, 231, 239
Tajikistan, 299, 305, 313–316
Tanzania, 21, 22, 30, 45, 149, 333
tariff, 129, 214
  GATT, 136, 217
technology development, see research and
  development, agricultural
technology transfer, see extension,
  agricultural
temperate zones, 44, 206n
Thailand, 21, 24, 149, 152, 228–230, 243
Thatcher, Margaret, 207
Togo, 345
total factor productivity (TFP), 2, 49, 50,
  83, 163, 265
  inflation-adjusted value
    of US agricultural output, 50–51
total fertility rate (TFR), 18, 193–194, 210,
  234, 255, 278, 303, 329
totalitarianism, 75, 93, 295, 302–303
trade barriers, 139, 141, 142, 160
trade distortions, net costs of
  currency over-valuation, 130–132
  protectionism, 128–130
Tunisia, 278, 284, 287

Turkey, 215, 278, 284, 285, 291, 297, 300
Turkmenistan, 313–316
Tyers, Rodney, 142

Uganda, 193, 324, 346
　ABC program, 334
Ukraine, 193, 299, 303, 316
Ulam, Stanislaw, 125
Ultisols, 44
undernourishment, 167, 168, 170, 272, 363
　causes, 176
UN Development Program (UNDP), 63
UN Population Division (UNPD), 27, 28, 33, 34, 55, 87
United Arab Emirates, 277, 280, 287, 291
United Kingdom, 207, 217
United States, 210, 215, 217, 219, 368
　comparative advantage, in agriculture, 215
　economic inequality, 209–210
　GDP per capita growth, 209
　social mobility, 209
Universal Soil Loss Equation (USLE), 113
urbanization
　human fertility, 22–24, 278
Uruguay, 136, 254, 255, 257, 261
US Agency for International Development (USAID), 167
USSR, *see* Former Soviet Union
Uzbekistan, 313, 316

Venezuela, 254, 255, 266
Vietnam, 60, 230, 236, 238, 242

Wales, 20
Washington Consensus, 178, 183
water scarcity, 238
Weinberger, Caspar, 277n
women empowerment, 24–25, 196, 234
World Bank, 27, 63, 136n, 167, 173, 175, 178, 182, 183, 230, 307, 325
World Food Summit, 170
World Health Organization (WHO), 363
World Trade Organization (WTO), 125, 138, 367
　free trade, 139
　liberalization, 141

Xaioping, Deng, 231

Yeltsin, Boris, 300, 301
Yemen, 277, 280, 288
Yukos, 302
Yushchenko, Victor, 299n, 302

Zaire, *see* Democratic Republic of the Congo
Zambia, 346
Zimbabwe, 193, 328, 328n, 346